KNOWLEDGE, TEXT A[ND]
ANCIENT TECHNI[QUE]

The relationship between theory and practice, between norms indicated in a text and their extra-textual application, is one of the most fascinating issues in the history and theory of science. Yet this aspect has often been taken for granted and never explored in depth. The essays contained in this volume provide a multi-layered and nuanced discussion of this relationship as it emerges in ancient Greek and Roman culture in a number of fields, such as agriculture, architecture, the art of love, astronomy, ethics, mechanics, medicine and pharmacology. The main focus is on the textuality of processes of transmission of knowledge and its application in various fields. Given that a text always contains complex and destabilizing aspects that cannot be reduced to the specific subject matter it discusses, to what extent can and do ancient texts support extra-textual applicability?

MARCO FORMISANO is Professor of Latin Literature at Ghent University. He has published extensively on ancient technical and scientific writing. His first monograph, *Tecnica e scrittura* (2001), was dedicated to late Latin scientific texts. He has studied the ancient art of war as a literary genre and its tradition (*Vegezio, Arte della guerra romana* (2003) and *War in Words: Transformations of War from Antiquity to Clausewitz*, co-edited with H. Böhme (2012)), as well as Vitruvius (*Vitruvius in the Round*, co-edited with S. Cuomo, special issue of *Arethusa*, 2016). He is the editor of a series devoted to late antique literature entitled 'The Library of the Other Antiquity'.

PHILIP VAN DER EIJK is Alexander von Humboldt Professor of Classics and History of Science at the Humboldt-Universität zu Berlin. He has published on ancient medicine, philosophy and science, comparative literature and patristics. He is the author of *Aristoteles: De insomniis, De divinatione per somnum* (1994); *Diocles of Carystus* (2000–1); *Philoponus on Aristotle on the Soul I* (2005–6); *Medicine and Philosophy in Classical Antiquity* (2005); and, with R. W. Sharples, of *Nemesius: On the Nature of Man* (2008). He has edited *Ancient Histories of Medicine* (1999) and *Hippocrates in Context* (2005), and co-edited *Ancient Medicine in Its Socio-Cultural Context* (1995).

KNOWLEDGE, TEXT AND PRACTICE IN ANCIENT TECHNICAL WRITING

EDITED BY

MARCO FORMISANO

Ghent University

PHILIP VAN DER EIJK

Humboldt-Universität zu Berlin

CAMBRIDGE
UNIVERSITY PRESS

University Printing House, Cambridge CB2 8BS, United Kingdom

One Liberty Plaza, 20th Floor, New York, NY 10006, USA

477 Williamstown Road, Port Melbourne, VIC 3207, Australia

314-321, 3rd Floor, Plot 3, Splendor Forum, Jasola District Centre, New Delhi - 110025, India

103 Penang Road, #05-06/07, Visioncrest Commercial, Singapore 238467

Cambridge University Press is part of the University of Cambridge.

It furthers the University's mission by disseminating knowledge in the pursuit of education, learning and research at the highest international levels of excellence.

www.cambridge.org
Information on this title: www.cambridge.org/9781316620625
DOI: 10.1017/9781316718575

© Cambridge University Press 2017

This publication is in copyright. Subject to statutory exception and to the provisions of relevant collective licensing agreements, no reproduction of any part may take place without the written permission of Cambridge University Press.

First published 2017
First paperback edition 2021

A catalogue record for this publication is available from the British Library

Library of Congress Cataloging in Publication data
Names: Formisano, Marco, editor. | Eijk, Ph. J. van der (Philip J.), editor.
Title: Knowledge, text and practice in ancient technical writing / edited by Marco Formisano, Philip van der Eijk.
Description: Cambridge : Cambridge University Press, 2017. | Includes index.
Identifiers: LCCN 2016035987 | ISBN 9781107169432
Subjects: LCSH: Technical writing – History and criticism. | Classical literature – History and criticism. | Technology – Greece – History. | Technology – Rome – History.
Classification: LCC T11 .K635 2017 | DDC 609.38 – dc23
LC record available at https://lccn.loc.gov/2016035987

ISBN 978-1-107-16943-2 Hardback
ISBN 978-1-316-62062-5 Paperback

Cambridge University Press has no responsibility for the persistence or accuracy of URLs for external or third-party internet websites referred to in this publication, and does not guarantee that any content on such websites is, or will remain, accurate or appropriate.

Contents

List of Illustrations		*page* vii
List of Contributors		viii
Acknowledgements		xii
List of Abbreviations		xiii

1 Introduction: From Words to Acts? 1
   Philip van der Eijk and Marco Formisano

2 Introduction: The Poetics of Knowledge 12
   Marco Formisano

3 Machines on Paper: From Words to Acts in Ancient
   Mechanics 27
   Markus Asper

4 *Si qui voluerit*: Vitruvius on Architecture as 'the Art of
   the Possible' 53
   Elisa Romano

5 Caesar's Rhine Bridge and Its Feasibility in Giovanni
   Giocondo's *Expositio pontis* (1513) 68
   Ronny Kaiser

6 From Words to Acts? On the Applicability of
   Hippocratic Therapy 93
   Pilar Pérez Cañizares

7 *Naso magister erat – sed cui bono?* On Not Taking the Poet's
   Teaching Seriously 112
   Alison Sharrock

8	From *technē* to *kakotechnia*: Use and Abuse of Ancient Cosmetic Texts Laurence Totelin	138
9	From Discourses to Handbook: The *Encheiridion* of Epictetus as a Practical Guide to Life Gerard Boter	163
10	The Problem of Practical Applicability in Ptolemy's *Geography* Klaus Geus	186
11	Living According to the Seasons: The Power of *parapēgmata* Gerd Graßhoff	200
12	*Auctoritas* in the Garden: Columella's Poetic Strategy in *De re rustica* 10 Christiane Reitz	217
13	The Generous Text: Animal Intuition, Human Knowledge and Written Transmission in Pliny's Books on Medicine Brooke Holmes	231
14	From Descriptions to Acts: The Paradoxical Animals of the Ancients from a Cognitive Perspective Pietro Li Causi	252

Index Locorum — 269
General Index — 279

Illustrations

3.1	Diagram from W. Schmidt, 'Herons von Alexandria Druckwerke und Automatentheater' in Heron Alexandrinus, *Opera*, vol. 1 (Leipzig 1899), 176 (= S.)	*page* 42
3.2	Diagram © Joyce van Leeuwen	46
10.1	*The Palk Strait*: https://de.wikipedia.org/wiki/Palkstra%C3%9Fe	189
10.2	Ptolemy's first reduction	190
10.3	Ptolemy's triangle	190
10.4	Ptolemy's reduction, Kory	191
11.1	Milet *parapēgma*, Antikensammlung Berlin	203
11.2	Ptolemy's selection of 30 bright fixed stars using the equatorial coordinate system	209
11.3	Weather events as a function of the Alexandrian calendar (in days) and divided between the cited authorities	210
11.4	Distribution of named, directional winds	211
11.5	Number and type of weather protasis cited by all the authorities	213
11.6	Flora (×) and fauna (dots) events	213

Contributors

MARKUS ASPER is Professor of Classics at the Humboldt-Universität zu Berlin. He has research interests in Greek literature, especially Hellenistic poetry, and in ancient science and its literary forms. He is the author of *Onomata allotria: Zur Genese, Struktur und Funktion poetologischer Metaphorik bei Kallimachos* (1997), *Kallimachos von Kyrene: Werke* (2005) and *Griechische Wissenschaftstexte* (2007), and is the editor of *Writing Science: Mathematical and Medical Authorship in Ancient Greece* (2013).

GERARD BOTER is Professor of Greek at the Vrije Universiteit Amsterdam. His main research interests are Plato, Epictetus and Philostratus. He has published critical editions of Epictetus' *Encheiridion* (2007) and its later adaptations (1999), and he is currently preparing a critical edition of Philostratus' *Life of Apollonius of Tyana*. Recent publications include 'Evaluating Others and Evaluating Oneself in Epictetus' *Discourses*' in I. Sluiter and R. M. Rosen (eds.), *Valuing Others in Classical Antiquity* (2010), and 'Studies in the Textual Tradition of Philostratus' *Life of Apollonius of Tyana*', *Revue d'Histoire des Textes*, n.s. 9: 1–49.

MARCO FORMISANO is Professor of Latin Literature at Ghent University. He has published extensively on ancient technical and scientific writing. His first monograph, *Tecnica e scrittura* (2001), was in particular dedicated to Latin scientific texts in late antiquity. He has studied the ancient art of war as a literary genre and its tradition (*Vegezio, Arte della guerra romana* (2003) and *War in Words: Transformations of War from Antiquity to Clausewitz*, co-edited with H. Böhme (2012)), and has worked on Vitruvius (*Vitruvius in the Round*, co-edited with S. Cuomo, special issue of *Arethusa* (October 2016)). He is the editor of a series entirely devoted to late antique literature entitled 'The Library of the Other Antiquity', published by Universitätsverlag Winter, Heidelberg. He was Frances Yates Fellow at the Warburg Institute in London and Research Associate at the Italian Academy, Columbia University.

List of Contributors

KLAUS GEUS is an ancient historian, philologist and geographer. He earned his PhD at Bamberg in 1991 and worked afterwards in Mannheim, Jena, Bayreuth and Tübingen. In 2009 he was appointed Full Professor of Historical Geography of the Ancient World at the Freie Universität Berlin. Geus works in the areas of ancient geography and astronomy. Up to now he has published 20 books and 250 papers and articles. Geus' most recent books are on ancient mapping (*Vermessung der Oikumene* (2013)), mental modelling (*Common Sense Geography* (2014)) and Ptolemy (*Travelling along the Silk Road* (2014)).

GERD GRAßHOFF is Professor for History and Philosophy of Science at the Humboldt-Universität zu Berlin. Since 2010 he has been Director of the Excellence Cluster TOPOI. His research fields cover the history of ancient science from Babylonian astronomy to modern times, methods of scientific discovery and philosophical models of causal reasoning. Among his publications are *The History of Ptolemy's Star Catalogue* (1990), *Naturgesetz und Naturrechtsdenken im 17. Jahrhundert: Kepler – Bernegger – Descartes – Cumberland* (2002), *Kausalität und kausales Schliessen: Eine Einführung mit interaktiven Übungen* (2004) and (with Alfred Stückelberger) *Klaudios Ptolemaios: Handbuch der Geographie* (2006).

BROOKE HOLMES is Professor of Classics and Director of the Interdisciplinary PhD Program in Humanistic Studies at Princeton University. She works at the intersection of the history of medicine and the life sciences, the history of philosophy and ancient literature. Her publications include *The Symptom and the Subject: The Emergence of the Physical Body in Ancient Greece* (2010) and *Gender: Antiquity and Its Legacy* (2012), as well as three co-edited volumes: *Aelius Aristides between Greece, Rome, and the Gods* (2008); *Dynamic Reading: Studies in the Reception of Epicureanism* (2012); and *The Frontiers of Ancient Science: Studies in Honor of Heinrich von Staden* (2015). She is currently at work on a book project on sympathy in antiquity.

RONNY KAISER works as research assistant at the Collaborative Research Center 644 'Transformations of Antiquity' (Humboldt-Universität zu Berlin). His main research interests are the strategies and functions of the appropriation of biographical, historiographical and ethnographical texts from classical antiquity in German humanism. He has published several articles on humanist commentary on the *Germania* of Tacitus, as well as on humanist historiography. Selected publications include:

'Caesar in Pommern – Transformation der Antike in Bugenhagens *Pomerania*' in Marcus Born (ed.), *Retrospektivität und Retroaktivität: Erzählen – Geschichte – Wahrheit* (2009), 47–68; and 'Understanding National Antiquity: Transformations of Tacitus's *Germania* in Beatus Rhenanus's *Commentariolus* (1519)' in Karl Enenkel (ed.), *Transformations of the Classics via Early Modern Commentaries* (2014), 261–77.

PIETRO LI CAUSI teaches classics in Palermo. His research focuses on the anthropology of the ancient world, especially Greek and Latin zoology and zooanthropology, as well as gift economy in late Republican and early Imperial Rome. In recent years, he has investigated the ancient debate on the moral status of animals (*L'anima degli animali* (2015)), theories of hybridization in Greek and Roman culture (*Generare in comune* (2008)), and the dynamics of gratitude and memory in Seneca's *On benefits* (*Il riconoscimento e il ricordo* (2012)).

PILAR PÉREZ CAÑIZARES holds a PhD in classics from the Complutense University in Madrid. She has published various papers on the Hippocratic writings and is preparing a critical edition of the treatise *Affections*, which is the outcome of a Wellcome Trust Research Fellowship she held at Newcastle University. She currently works as a lecturer at the Institute for Romance Languages of the Vienna University of Economics and Business.

CHRISTIANE REITZ studied classics and comparative linguistics at the Universities of Bonn and Heidelberg. After working as assistant professor and lecturer at the University of Mannheim and as deputy chair at the Universities of Gießen and Heidelberg, she was appointed to the Chair of Classical Latin Literature and Language at the University of Rostock in 1999. Her main research interests lie in Roman and Greek epic poetry, and the reception and transmission history of ancient literature, as well as authors of technical literature. She is one of the editors of the series *Litora Classica* and *Hypomnemata*. Her current research project, 'Structural Elements of Epic Poetry', is funded by the German Research Foundation.

ELISA ROMANO is Professor of Classical Philology at the University of Pavia. She works mostly on the cultural history of the late Roman Republic and of the Augustan age, and on scientific and technical Latin texts. She is the author of *La capanna e il tempio: Vitruvio o Dell'architettura* (1987), *Medici e filosofi: Letteratura medica e società altoimperiale* (1991), and of an Italian translation and commentary on Vitruvius' *De architectura*

(1997). She is currently working on a book about the idea of antiquity in the second and first centuries BCE.

ALISON SHARROCK is Professor of Classics at the University of Manchester. She has published widely on Latin poetry from Plautus to Ovid, with a significant strand in didactic poetry, including her first book, *Seduction and Repetition in Ovid's Ars Amatoria 2* (1994). Other works include *Reading Roman Comedy: Poetics and Playfulness in Plautus and Terence* (Cambridge University Press 2009), and the edited volumes *Intratextuality: Greek and Roman Textual Relations* (2000, with Helen Morales), *The Art of Love: Bimillennial Essays on Ovid's Ars Amatoria and Remedia Amoris* (2006, with Roy Gibson and Steve Green) and *Lucretius: Poetry, Philosophy, Science* (2013, with Andrew Morrison and Daryn Lehoux).

LAURENCE TOTELIN is Senior Lecturer in Ancient History at Cardiff University. Her research focuses on the history of Greek and Roman botany and pharmacology. She has a particular interest in ancient gynaecological and cosmetic recipes. Her publications include *Hippocratic Recipes: Oral and Written Transmission of Pharmacological Knowledge in Fifth- and Fourth-Century Greece* (2009) and, with Gavin Hardy, *Ancient Botany* (2016).

PHILIP VAN DER EIJK is Alexander von Humboldt Professor of Classics and History of Science at the Humboldt-Universität zu Berlin. He has published widely on ancient medicine, philosophy and science, comparative literature and patristics. He is the author of: *Aristoteles: De insomniis, De divinatione per somnum* (1994); *Diocles of Carystus* (2 vols., 2000–1); *Medicine and Philosophy in Classical Antiquity* (Cambridge University Press 2005); *Philoponus on Aristotle on the Soul 1* (2 vols., 2005–6); and, with R. W. Sharples, *Nemesius: On the Nature of Man* (2008). He has edited *Ancient Histories of Medicine* (1999) and *Hippocrates in Context* (2005), and co-edited *Ancient Medicine in Its Socio-Cultural Context* (2 vols., 1995). He is General Editor of the Cambridge Galen Translations.

Acknowledgements

This volume has arisen from an international conference held at the Institute for Cultural Inquiry in Berlin in May 2011. The conference was a collaborative activity of the projects *Transformationen der Antike* (Sonderforschungsbereich 644) and *Medicine of the Mind, Philosophy of the Body: Discourses of Health and Well-Being in the Ancient World* (Alexander von Humboldt-Professur). We are grateful to the Deutsche Forschungsgemeinschaft, the Alexander von Humboldt Foundation, the Humboldt-Universität zu Berlin and the Institute for Cultural Inquiry for their financial and institutional support. For practical assistance with the organization of the conference and the preparation of this volume, we are grateful to Ricarda Gäbel, Stefanie Jahnke, Friderike Senkbeil, Annette Schmidt, Dorothea Keller and Evangelia Nikoloudakis.

THE EDITORS

Abbreviations

Chapter 7

OCD   *Oxford Classical Dictionary*

Chapter 8

CML   *Corpus Medicorum Latinorum*
CMG   *Corpus Medicorum Graecorum*
RE   *Pauly's Realenzyklopädie der klassischen Altertumswissenschaft*
NP   *New Pauly*
PGM   *Papyri Graecae Magicae*

Chapter 9

LSJ   Liddell Scott Jones, *A Greek–English Lexicon*

Chapter 11

CDF   computable document format

Chapter 13

CAG   *Commentaria in Aristotelem Graeca*
SVF   *Stoicorum veterum fragmenta*, ed. H. von Arnim (Leipzig 1903–24)

CHAPTER I

Introduction
From Words to Acts?

Philip van der Eijk and Marco Formisano

The Scope of This Volume

In the current boom of studies devoted to ancient technical and scientific writing,[1] the notion of practical applicability has generally been taken for granted. Technical texts, or, more broadly, texts dealing with technical topics,[2] are believed to be the way they are because they are deemed applicable and capable of being used 'out there', outside the texts themselves (and, in the case of texts with didactic intentions, outside the lecture room). Indeed, the very concept of technical texts or *Fachtexte* is an expression of the presupposition that they are useful, or at least intended to be useful, in extra-textual reality, i.e. in the execution of a specific discipline or field of expertise (*technē*, *ars*). In short, usefulness and practicability (*to chrēsimon*, *utilitas*) have come to be regarded as prominent features of this kind of text.

Yet, upon closer inspection, both the principle of practical applicability and the category of technical texts itself are distinctly more complex and ultimately problematic when we consider them in the context of ancient textual culture more generally, and of so-called 'cognitive' or 'epistemic' texts more specifically.

The chapters included in this volume discuss by means of examples, though certainly not exhaustively, the issue of practical applicability across a wide spectrum of ancient texts (and their reception) belonging to different periods, disciplines, genres and intellectual traditions: medicine, mechanics, architecture, agriculture, cosmetics, pharmacology, philosophy, astronomy, geography, zoology, warfare, architecture and the art of love.

---

[1] For a discussion of some of the more salient trends in recent scholarship see Chapter 2.
[2] We have added this extension here in order to include texts that *prima facie* would not be regarded as technical texts, such as didactic poetry, dialogues or letters, which, at least on a formal level, appear to belong to the category of 'literary' texts, i.e. texts with aesthetic pretentions. Yet this distinction between literary and non-literary texts is profoundly problematic; for a discussion of this and related issues, see Chapter 2.

Despite the great variety of themes and forms, from different angles they all converge on the same point and ask one single question: to what extent does the principle of practical applicability determine the construction of the specific nature of each text?

It is worth emphasizing that this question is different from that of *real* applicability outside the textual sphere. In other words, the chapters in this volume are not so much concerned with the extent to which a given description of a procedure was or is *really* applicable, but rather in observing how the pretention of applicability itself is relevant or even vital to the construction of a specific textual discourse. Within this discourse, particular attention is given to the self-fashioning of the author, the positioning of the text in relation to the extra-textual world and the text's perceived role in modes and processes of teaching and instruction.

Throughout the chapters, two aspects can be discerned that receive attention from all contributors, though with varying degrees of emphasis. First, there is the aspect of textuality, the particular voices of the texts themselves in all their complexity and contradictory qualities. Secondly, there is the contextual dimension, the question of how a text positions itself within the historical and intellectual context in which it was produced (including tradition and the work of predecessors) and how it addresses the relationship between theory and practice. In Chapter 2, Marco Formisano analyses this aspect of textuality and the various issues it raises, and he illustrates this by reference to an ancient discipline in which practical applicability appears vital, i.e. the art of war; he also discusses the wider scholarly context in which this volume aims to situate itself. But, first, some introductory remarks (by Philip van der Eijk) on the context provided by the ancient discourse of science and practice may be appropriate.

Ancient Views on Text and Context, Theory and Practice

The relationship between theory and practice, universal knowledge and application in particular cases, is a recurring theme in ancient philosophical and scientific discourse. Treatises on specific *technai* or *artes* are the obvious place for discussion of this relationship, as a *technē*, or an *ars*, is by definition concerned with practical application or even production in the world of everyday life outside the text. A *technē* that is without practical use is a contradiction in terms, or so it would seem; indeed, one could argue that the proof of a *technē* lies in its successful application; and the proof of successful teaching of a *technē* becomes manifest in the extent to which it manages to equip its students with the ability to apply this in actual practice outside the classroom, and to develop this ability into genuine experience and skill.

Yet matters are not as straightforward as they seem. As the chapters in this volume show, there is great variety in the ways in which the relationship between theory and practice is discussed, handled, problematized and rhetorically exploited in a number of different disciplines, such as architecture, mechanics, measurement, agriculture, geography, weather prediction and medicine. This variety is not just a consequence of the different subject matter of each discipline and the requirements this raises. It is also, to a considerable extent, caused by non-technical issues, such as the literary ambitions of the author, the specific formal and rhetorical features of the text, and factors to do with the context or situation in which the text is produced, such as the audience it addresses and the strategic or opportunistic aims the author, by writing the text, wishes to achieve.

One recurring issue in this connection is the bridging of the gap between instruction by the teacher (or by the book), on the one hand, and independent application by the apprentice or reader in everyday life, on the other. How is that gap perceived, how is it discussed, problematized, negotiated, how is it resolved? And how is the extra-textual dimension reflected in the texts themselves? The contributors to this volume show that authors of technical texts dealt with these questions in a number of different ways, and they examine the various reasons for these differences, whether they lie in the discipline itself, in the author's specific take on it (e.g. the specific views an author may hold on controversial, subject-related issues), or rather in literary and contextual factors.

For some ancient authors, the relationship between theory and practice presented an interesting intellectual problem in its own right. Aristotle, in the much-discussed first chapter of his *Metaphysics*, distinguishes different levels of cognition and skill, and lists, in an ascending scale of scientific rigour and universality, sensation (*aisthēsis*), imagination (*phantasia*), memory (*mnēmē*), experience (*empeiria*), skill/art/expertise/competence (*technē*), and scientific knowledge (*epistēmē*). He uses the example of medicine to illustrate these different levels of practical and theoretical knowledge. He distinguishes between doctors who rely on experience only (the *cheirotechnai*), and doctors who build their medical practice on theoretical, scientific knowledge of the body (the *architektones*). As he puts it, the former only know that (*hoti*) a particular drug is effective in the treatment of particular diseases, while the latter also know why (*dioti*) it is effective precisely in these cases (981a5–b17). Yet while the latter kind of knowledge is superior from a scientific point of view, the former, Aristotle concedes, may be more successful and effective in actual therapeutic practice: and that, it seems, is ultimately what counts, for a patient needs to be treated by a competent, hands-on doctor rather than by someone who may have a profound philosophical understanding of medicine but who lacks practical experience.

That insight immediately raises questions about the right method of instruction, and about the extent to which texts, *qua* texts, are useful instruments in the process of teaching the practical arts. In this volume, our example is Ptolemy, who explicitly addresses the question of the practical usefulness (*to euchrēston*) of written instructions and graphic representations in the domain of geography, cartography and weather prediction (Chapters 10 and 11). For the problem of applicability is not confined to texts written on papyrus or parchment, but also applies to inscriptions, *parapēgmata* (almanacs inscribed on stone) and maps.

Oral and Written Instruction

Within the ancient discourse about the (in)adequacy of texts for purposes of practical application, a further issue raised by ancient authors is the question of oral and written instruction. The pros and cons of written versus oral knowledge were explicitly discussed, not only in such well-known passages as Plato's *Phaedrus* (275A ff.), but also in more technical contexts, for instance in the medical writings attributed to Hippocrates, in Diocles of Carystus and in Galen.[3] Throughout ancient medicine, we see a recurring tension between the two. On the one hand, there was a realization of the need for written instruction, enabling greater spread, dissemination and accessibility of knowledge, greater ease of reference and retrieval. On the other hand, there was awareness of the continuing need for oral instruction, allowing for interaction, question and answer between teacher and student, and immediate correction of misunderstanding or mistakes. Oral instruction also offered the scope, for the teacher, to show how to do things and the opportunity, for the student, to imitate this under the teacher's supervision, to repeat and repeat, and thus by trial and error to gain mastery of the skill and to develop experience and dexterity.

The difficulties arising here are easy to imagine when one thinks of a text about how to lay on a specific bandage, or a text on how to take the pulse and distinguish different kinds of pulse by means of the sense of touch.

---

[3] See my discussion in 'Towards a Rhetoric of Scientific Discourse: Some Formal Characteristics of Greek Medical and Philosophical Texts (Hippocratic Corpus, Aristotle)' in E. J. Bakker (ed.), *Grammar as Interpretation: Greek Literature in Its Linguistic Contexts* (Leiden 1997), 77–129, especially 93–9.

Ancient technical writers clearly tried to present their instructions in ways that would make them easily applicable, self-evident and self-explanatory even to the lay person, the *idiōtes*, who had no ambition to become an expert in the art in question but who would nevertheless benefit from the ability to apply them in practical cases. This is how the *encheirēsis*, the practical procedure or manual, was developed as a distinct genre. In the present volume, an example of this can be found in the Hippocratic work *On Affections* (see Chapter 6), which is explicitly intended for lay people to help themselves in cases of disease and injury when no expert physician is around, or to prevent disease by leading a healthy lifestyle.

Yet at the same time, it is clear, even from a work such as *On Affections*, that authors realized that written texts have their limitations. The inadequacy of the written word in medicine is emphasized more than once by Galen, who stresses the importance of the *phōnē zōousa*, the *viva vox*, which allows direct contact between the teacher and the student, interactive learning by dialogue, and which presents an indispensable dimension of the teaching of the arts.[4] Galen himself was a prolific and versatile writer, who realized that for the communication of his ideas the written word was essential. Yet he also complains about the misunderstandings and the abuse that other malicious or stupid doctors make of his written instructions.

One way of addressing the problem of inadequacy was the addition of supplementary material, such as anatomical illustrations and diagrams. The earliest examples are Apollonius' commentary on the Hippocratic surgical work *On Joints* and mechanical writings. Yet even here, the use of such materials was anything but straightforward, and scope for confusion and misunderstanding remained. Texts served at best as reminders or aide-mémoires, but could never entirely replace the oral, practical teaching situation.

The Art of Living and the Art of Loving

Medicine, and the other disciplines mentioned above, are all examples of what Aristotle would call practical or productive arts, and a number of these are represented in this volume. Aristotle often uses these arts to illustrate the relationship between theory and practice in another practical

---

[4] Galen, *On the Powers of Foodstuffs* 1.1.47 (6.480 K.); *On the Mixtures and Powers of Simple Drugs* 6, proem (11.971 K.); *On the Composition of Drugs according to Places* 6.1 (12.894 K.). Cf. also Diocles of Carystus' reply to someone who claimed to have purchased a medical book (*iatrikon biblion*) and therefore no longer to be in need of instruction: 'Books are reminders for those who have received teaching, but they are gravestones to the uneducated' (Diocles, fr. 6, in P. J. van der Eijk, *Diocles of Carystus: A Collection of the Fragments with Translation and Commentary*, 2 vols. (Leiden: Brill, 2000–1)).

field, which one may call the art of living, the *ars vivendi*: for the question of applicability of general principles arises also in the area of ethics and etiquette. How moral principles and courtesy rules translate into specific actions is by no means straightforward or easy. Again, Aristotle reflects on these matters on more than one occasion in his *Nicomachean Ethics*, e.g. in the final chapter (X.9), where, apart from medicine, he also uses the analogy of sailing: one cannot become a competent and skilful steersman through 'learning by the book', one has to be able to apply one's knowledge in actual practice, and to develop the ability to do so.[5] Likewise, in the field of ethics, Aristotle famously argues, one becomes a good person not by studying a philosophical treatise on ethics, but by doing good things (*Eth. Nic.* II.1–2). In the present volume, the example of Epictetus' *Encheiridion* illustrates some of the issues arising here (Chapter 9).

From ethics, it is only a small step to a further practical and productive art that is considered here, the *ars amandi*, the art of making love, an area where the gap between theory and practice is particularly sensitive, tangible, and potentially and dramatically delicate, as illustrated by Ovid's didactic poem *ars amatoria* (see Chapter 8). With Ovid (as with Galen), we also enter the domain of the literary exploitation of the problematic relationship between theory and practice. Thus some authors claim to aim for usefulness in one respect but have an additional side-agenda as well, like the writers (or compilers) of cosmetic recipe texts with pornographic sub-texts considered in Chapter 7. Other authors implicitly underplay the potential applicability of their material, such as (again) Ptolemy in his work on weather predictions, in which *parapēgmata* serve as vehicles of 'hybrid practical knowledge', communicating knowledge about signs occurring regularly but omitting other events linked to the weather in chains of signs, suggesting that information about such signs was either handed down orally or recorded in separate compendia (Chapter 11).

Making Sense of Technical Texts

Finally, the problem of the relationship between theory and practice, between the textual and extra-textual world presents itself to us, too, as

---

[5] *Nicomachean Ethics* 1181b2–6 (comparing legislation with medicine):

> Neither do men appear to become expert physicians on the basis of medical books. Yet they try to discuss not only general means of treatment, but also how one might cure and how one should treat each individual patient, dividing them according to their various habits of body; these [discussions] appear to be of value for men who have had practical experience, but they are useless for those who have no knowledge about the subject. (Cf. *Politics* 1287a35.)

students of the ancient world, as readers of and commentators on Greek and Latin technical texts. We often wonder, and try to envisage, how a technical text has functioned or worked. In doing so, we sometimes take recourse to such notions as the 'Sitz im Leben' of a text, or we develop hypotheses about how the text is supposed to have functioned in a wider didactic setting. We sometimes postulate the presence of diagrams or other audiovisual teaching aids which we assume must have accompanied the teaching. Yet that raises the question of how we determine that setting, other than through inferences based on features of the text. In the absence of independent, contextual information, a certain degree of circularity cannot be avoided here.

However, as said above, the chapters in this volume are not so much concerned with the question of how successfully or unsuccessfully these texts were actually applied outside the textual sphere, but rather with the extent to which the principle of applicability has shaped the text into the form it has acquired. That aspect has not received the attention it deserves, and this volume aims to address this gap in the rapidly expanding scholarly literature on the subject.

The Contents and Arrangement of This Volume

The contributions to this volume have been arranged by subject matter. We begin with the field of mechanics, and the challenges posed by the transition from 'machines on paper', as described in construction manuals, to actually working physical mechanisms. Markus Asper, in his account of Hellenistic-Roman mechanical writings (Chapter 3), points out that, as artifacts, machines result from an 'act', namely the process of construction. At the same time, machines are usually the result of 'words', that is, the product of explicit knowledge that regulates how to build such machines and what to do with them. Therefore, one assumes, writings on such bodies of knowledge may seem to be perfect candidates for the precarious process that turns words into acts. The chapter considers ancient Graeco-Roman mechanical writing, such authors as Ps.-Aristotle's *Mechanics*, Athenaeus mechanicus, Philo of Byzantium, Biton, Cato maior, Vitruvius and Hero. Throughout these texts, there seem to be several ways to understand 'act', i.e. what the author wants his readers to do. Four such 'acts' emerge: (i) practical construction; (ii) comprehensive knowledge about construction; (iii) decision-making with regard to the construction of machines; and (iv) scientific explanation of machines. The chapter illustrates these four 'acts' and their peculiar means of communication, with some examples.

From here, we move on to architecture and bridge-building. In her discussion of Vitruvius' work on architecture (Chapter 4), Elisa Romano argues that the text of the *De architectura* offers specific clues that one of Vitruvius' aims is the potential application of his instructions. The chapter examines the shape that the idea of applicability takes in the text and the means that Vitruvius offers his addressee to put into practice the rules he sets out. She argues for a higher level of translatability that coincides with the performance of all the potentials contained in the architect's *ars*. Vitruvius invites the architect to judge what is possible and to adapt himself to necessity: it is in that adaptability that he proves both his practical experience and his technical skill and intellectual qualities. In other words, in the adaptability to 'what is possible', the blending of theoretical aptitude and practical ability, of *ratiocinatio* and *fabrica*, is achieved, and so is one of the theoretical requirements of the architect's *scientia*.

Kaiser's discussion of Caesar's construction of a bridge over the Rhine, and Giocondo's reading of this (Chapter 5), shows not only a 'literary' reception but, more relevantly for this volume, an application of Caesar's text. The chapter considers how Caesar's well-known account of the Rhine bridge, which probably aimed at documenting Caesar's military virtues as well as the cultural superiority of the Romans, was read in early modern Europe. On the basis of the *Expositio pontis* (1513), a treatise by the Renaissance architect and humanist scholar Giovanni Giocondo, Kaiser examines especially the visual and textual strategies through which Caesar's chapter was removed from its ancient context and read within a humanist discourse of theory and practice. With his main focus on the technical issues of Caesar's text, Giocondo adapts it to his own contemporary circumstances and thus inverts the original relationship between acts and words.

The practical art of medicine is next (Chapter 6). Pilar Pérez Cañizares considers the transition from words to acts in the domain of Hippocratic therapeutics, with the aim of establishing whether therapeutic instructions (as conveyed in medical texts) could be interpreted and easily followed by their readers. In particular, she examines evidence that hints at specialized knowledge shared by writers and targeted readers, such as references to cauterization with red-hot irons or with fungi, incision, application of cupping vessels, administration of drugs and surgical interventions. The evidence shows that these texts rely on shared knowledge, i.e. empirical knowledge in the case of everyday measures, such as following a diet or administering drugs, and also knowledge obtained from more formal medical training, as in the case of surgical interventions, which could not have been carried out without prior understanding of anatomy and training in the use of medical instruments.

We then move to the world of human relationships with the art of love and beauty. In her discussion of Ovid's *Ars Amatoria* (Chapter 7), Allison Sharrock argues that whereas the topos of mixing *utile dulci* is perhaps the defining mark of didactic literature, the major difficulty for Ovid in making claims about the utility of his *Ars Amatoria* was that what he is teaching was illegal when he wrote it. One way of addressing, if not obviating, that problem is to claim that poetry is not useful because it is not actually teaching anything – with all the obvious ironies for a didactic poem. The chapter explores Ovid's rhetoric of utility in his erotodidaxis, including his use of strategies from the more technical end of the didactic spectrum. The claim for utility in Ovid's advice naturally brings it into conflict with the Augustan moral legislation, with ancient and modern debates about the extent to which character and behaviour can be taught, and finally with the extent of the responsibility that literature may hold for the abuse of its lessons by unsophisticated readers.

A related field is ancient cosmetics. As Laurence Totelin points out (Chapter 8), Greek and Roman authors often negatively associated cosmetics with frivolous women. Cosmetics was not a noble art, a *technē*, but a corrupted one, a *kakotechnia*. Totelin deconstructs this moralistic discourse and argues that, despite claims to the contrary, men too regularly used cosmetics in the ancient world. In addition, men may also have employed cosmetic treatises for gratification purposes: they may have read these texts as pornography. Indeed, there was much that pertained to sexuality in ancient cosmetic treatises, which are mostly known to us through fragments preserved in ancient medical texts. These texts included recipes to treat sexual diseases, as well as ingredients that were sexually connoted. Totelin argues that there was much overlap between ancient cosmetology, sexology and gynaecology, and that all three topics may have been covered within the same treatises. These texts were often attributed to women, who would have been seen as experts in these fields.

We then move to the art of living and practical philosophy, exemplified by Epictetus' ethical handbook (*Encheiridion*), a summary composed by Arrian based on the Stoic philosopher's lectures. As Gerard Boter shows (Chapter 9), the purpose of this short work is to provide the reader with practical advice on how to lead his life in accordance with the Stoic ideal of paying attention exclusively to the things within our control: that is, to the way in which we react to everything that happens to us. Discussion of the theoretical framework is confined to the first two chapters of the work. The *Encheiridion* deals with all aspects of everyday life, stressing the importance of constant philosophical training (*askēsis*), which leads to moral improvement (*prokopē*) and thus to perfect freedom and happiness.

Practical issues in geography, cartography and weather prediction are discussed in Chapters 10 and 11, both in relation to Ptolemy. Klaus Geus studies Ptolemy's use of the words *euchrēstia* and *to euchrēston* in the preface to his famous *Introduction to Geography* (c. AD 150) and shows that Ptolemy attaches great importance to the principle of practical applicability: his experience that maps deteriorate rapidly through the process of copying made him very circumspect in this regard. His work contains practical advice to future map-makers about how to draft maps on the basis of Ptolemy's instructions and tables of co-ordinates without the need to have access to a graphical master copy. In Chapter 11, Gerd Graßhoff shows how weather prognostication, being the oldest systematic inquiry into the regularities of nature and at the same time one with the most obviously practical purposes, laid the foundation of systematic observations of the changing rising and setting events of the sun and brightest stars. He analyses the systematic observational weather programme recorded in the Babylonian Astronomical Diaries and Ptolemy's comprehensive collection of historical weather rules in his *Phaseis*, and the extent to which a concern with practical application is present in the arrangement and presentation of the information.

This is followed by botany and the art of gardening, an area *par excellence* for issues of applicability of technical knowledge. As Christiane Reitz points out (Chapter 12), Columella's work on agriculture is clearly concerned with the practical reality of managing a large estate. To appeal to his readership, the author employs various rhetorical strategies and persuasive techniques in his argumentation. Quotations from poetic authorities, mainly Virgil, convey a certain *ornatus*. This ambitious display of education and learning reaches its peak in the garden poem, in Book 10. By assuming different roles and by means of sophisticated allusions, Columella presents himself as an accomplished poet. The garden poem subtly shifts between different poetic discourses of epic, elegy and hymn. The main aim – as had been modestly argued in the prose preface to Book 10 – is not so much an attempt at a humble follow-up to Virgil's *Georgics*. The author presents himself to his audience as an authority who masters a whole range of poetic codes and traditions that function as part of the overall didactic concept.

Finally, we get to zoology and the challenges the living natural world poses to human understanding and practical management of the environment. What, Brooke Holmes asks in Chapter 13, is the use value of Pliny's encyclopedia? Though the knowledge contained therein is clearly meant to be applied to the world, Pliny seems relatively unconcerned about the means of translating text into practice. Holmes approaches the text's commitment to teaching its readers the 'means of life' in the books on medicine by interrogating why Pliny thinks human beings require instruction in the means of life at all. She focuses on the role played by non-human animals both in acquainting people with sources of benefit and harm in the world and in providing a model of 'untaught' nature. She considers the problems posed by the human need for instruction to Pliny's idea of providential Nature and argues that it is only by understanding its role in mimicking the generosity of Nature that we can appreciate Pliny's conceptualization of its utility.

In the last Chapter (14), Pietro Li Causi addresses the cognitive challenges of descriptions of fantastic animals in writings on natural history more generally. Even though the zoological sections of ancient *naturales historiae* present no problems to readers translating what is read into practical action, there is always a process of negotiation taking place between the written text and extra-linguistic reality. In this respect, the story of the cultural representation of the Indian Unicorn might be read as paradigmatic. Ctesias is the first to describe this animal as a ferocious creature in the shape of a horse with a single polychrome horn on its forehead. The traits of this seemingly fantastic animal will undergo several changes, and its Cognitive Type will be 'applied' to several extra-linguistic referents in the course of the centuries.

CHAPTER 2

Introduction
The Poetics of Knowledge
Marco Formisano

This volume is dedicated to technical and scientific texts of Greek and Roman antiquity, a group of texts that over the past three decades has enjoyed a great deal of attention. This fascinating field has been cultivated by classical scholars in continental Europe, especially in France, Germany, Spain and Italy, and is now also becoming a fashionable topic in Anglo-American scholarship. While the texts discussed here used to be considered marginal to the ancient literary canons, this volume challenges the very distinction between literary and non-literary texts. Moreover, and perhaps more controversially, it critically reviews recent attempts in the study of technical texts to distinguish between their "literary" and their "non-literary" aspects. This is not to say that previous scholarship has not moved in this direction, but the discussion has been conducted almost exclusively from one, limited perspective. On the one hand, in fact, much work has been already done in order to emphasize that ancient technical and scientific texts share formal characteristics with texts belonging to other genres, i.e. that the language and style adopted by their authors are subject to certain formal peculiarities which are well integrated in ancient rhetoric and literary culture. Philip van der Eijk, in a study that has rightly become a standard in the field, argued for a "grammar of scientific discourse, a system of rules and conditions pertaining to the possibilities that are available to the users of scientific language in order to present knowledge in a certain way, with a certain purpose, and for a certain audience."[1] Since then, scholarship has moved on, following the path indicated by van der Eijk. But, on the other hand, precisely this emphasis on formal and rhetorical aspects enforced two implicit assumptions, i.e. that the literary quality of a technical or scientific text is appreciable only on the basis of certain formal, that is rhetorical, characteristics (style and strategies of argumentation etc.),

---

[1] Van der Eijk 1997, 82.

and, at the same time, that their literary nature is less concerned with the specific contents of the texts than with their form.

It is common to read in scholarship that the text of a given ancient author, e.g. a historian, a philosopher, a legal, medical or technical author, contains "literary polish" or that it is embellished with elements of "literary" style or peppered with citations from and allusions to "literary" texts. This approach implies two interrelated assumptions: that some ancient texts are *in themselves* not "literary," and that "literature" or "literary style" is something that can adorn or embellish a text of a lower status. On this view, the "literariness" of such texts is not an inherent element but something that can be *added*, mostly by stylistic or rhetorical means. But there is something of a paradox here, insofar as classical scholars frequently emphasize that these texts deserve attention *precisely* as literature (whatever that means) and not (only) as sources for the history of knowledge, science and technology.

In this part of the introduction I propose another approach, one that emphasizes the complexity of the relationship between knowledge and literature. Not only does a certain field of knowledge give sense to the text, but also, and more importantly for us, it is the text, *qua* text, that gives a literary substance to knowledge. Knowledge exists outside the text, but once it enters the textual universe, i.e. once it becomes text itself, its meanings multiply, so that even what we perceive as the most technical information can in fact bear a sophisticated literary meaning, not only by means of external factors such as formal or rhetorical polish, but also in and of themselves. This point, which we might call the textualization of knowledge, has been profitably discussed by a number of literary theorists. For instance, Bakhtin describes the interaction between primary speech genres (those of direct, everyday communication) and secondary speech genres, including the traditional genres of literary texts: when a text belonging to a secondary speech genre (for example, a medical treatise) contains utterances reproducing characteristics of a primary genre (in our case, a medical prescription), the latter can no longer be interpreted independently from the literary discourse – it now unavoidably belongs to the secondary speech genre in question.[2] Following these lines we can argue that as soon as a scientific discussion, even the most technical, enters the textual universe, its meaning and nature are irreversibly modified and the text thus opens itself to a broad spectrum of hermeneutical approaches.

---

[2] Bakhtin 1986.

The essays contained in this volume are almost entirely devoted to ancient technical and scientific texts, and they seek to discuss, scrutinize, nuance and perhaps challenge this cluster of assumptions by directly focusing on, or even unabashedly attacking, what is generally considered to be the primary characteristic of texts of this kind: their claim to be instruments applicable in extra-textual reality. It is, in fact, precisely the discourse of applicability that constitutes, as already mentioned above, what scholars usually tend to conceive of as the aspect most remote from the literariness within this kind of text. That claim is in this volume not taken for granted: it becomes the object of our scrutiny. The contributions to this volume aim at exploring the literariness of ancient technical and scientific texts not only by means of observing their rhetorical aspects, authorial strategies or intertextual connections to texts considered more literary (above all poetry), but more importantly by exploring precisely how literary their 'non-literary' core is, i.e. how the question of practical applicability is at work in these texts and affects the treatment of its subject matter.

There is a significant difference between the simple fact that a given text treats a given topic – for instance architecture or medicine – and the text itself. In other words, a text might well speak about architecture or medicine, but *as a text* it cannot be reduced to its topic, while the topic itself is unavoidably shaped by the textual form: medicine in Celsus and architecture in Vitruvius are not the same as medicine and architecture actually practiced "out there." Furthermore, readers of an ancient "technical" text will find a variety of meanings in the text that have little or nothing to do with the subject matter indicated by the title or by explicit authorial statements; here the gap between that which we call ancient "technical literature" or "manuals" on the one hand, and those present-day texts to which we give the same labels on the other, is especially noticeable. For instance, readers of the gromatic treatise by the late Roman author Agennius Urbicus will encounter a sophisticated theory of language,[3] and, as we shall see shortly, a leading motif of the military treatise by the early imperial Greek writer Onasander is contemplation (*theōria*). In turn, as any reader of ancient literatures knows, elements of technical discourses can regularly be found in poetical or philosophical texts.

As I noted above, classical scholarship has traditionally been aware of certain literary qualities of technical texts, but in order to integrate this

---

[3] Formisano 2001, 109–16.

kind of literariness within its own disciplinary discourse, it has drawn attention only to certain aspects and has developed strategies that have had the paradoxical effect of ignoring or sanitizing the textual nature of these texts as a whole. The literariness of ancient technical texts is generally described in terms of rhetoric and style; in terms of intertextual relationships with other, *more literary* texts (especially poetry); in relation to their social, political and ethical values; or by focusing on the figure of the author, particularly on his intentions and strategies of self-fashioning. All these methods of interpretation are of course both legitimate and productive, but they still do not do full justice to the literary quality of ancient technical writings as such. No classicist these days would affirm that agriculture is *the* (only) subject of Virgil's *Georgics* or love of Ovid's *Ars Amatoria*, yet many imply or openly state that precisely this is true of agriculture in Columella or of architecture in Vitruvius. As far as authorial declarations of intention are concerned, while today's scholarly readings of the *Georgics* do not take Virgil's declared intention to give instructions to the farmers seriously, many take for granted that Columella's text should be read only or mostly as a treatise on agriculture, because the author explicitly declares that this is his topic. Some scholars, aiming to shed light on the formal or literary aspects of ancient technical writing, programmatically reject readings oriented to the reconstruction of historical, social, political or economic contexts by means of these texts.[4] And yet I would argue that this "historicizing" method of reading can actually be very useful for the treatment of the literary features of those texts for one methodological reason above all: historians are accustomed to go under the surface of the text in search of traces confirming or negating their hypothesis, and those traces are not necessarily dependent either on the subject matter or on the intentions of the authors as they are formulated in the texts.

In what follows, I shall refer to some recent contributions to the scholarship that illustrate this tendency towards rendering opaque precisely that literary nature of technical texts which scholars have sought to describe. A common thread is this: no matter how experienced they are with poetic texts characterized by sophisticated language, complex meanings and destabilizing qualities, no matter how skeptical they are towards the explicitly declared intentions of a poet, when these interpreters turn to "technical" texts, they usually take literally what the texts and their authors say. Recent

---

[4] See for instance Doody and Taub 2009, 7, Fögen 2009, 3.

scholarship on what we call "literature of knowledge"[5] draws attention to these texts' rhetorical qualities, and indeed the analysis of rhetorical strategies and argumentative styles has played a central role in the renewal of interest in this body of textuality.[6] Markus Asper incisively formulates the perspective that informs this approach: "Science writers ancient and modern seek to persuade their audiences and will employ whatever device comes handy to reach the goal."[7] German scholars in particular have recently placed emphasis on *Wissensvermittlung*, the transfer of knowledge; Asper's monograph on Greek *Wissenschaftstexte* is exemplary in this sense since it describes at length generic, formal and structural characteristics of Greek technical and scientific texts from the Pre-Socratics to late antiquity. Although his book is limited to Greek texts, Asper's approach, arguments and conclusions can be in principle extended to Latin texts, precisely because he offers a systematic description and classification. Asper's project is comparative in nature, since it brings together texts on different ancient disciplines with the aim of overbridging the boundaries normally placed by modern scholars, and of concentrating on the commonalities of the rhetorical strategies and methods of systematization of knowledge followed by different ancient technical and scientific authors. All of these strategies, it goes without saying, go in one direction and have only one goal: the description of *Wissensvermittlung*, i.e. of the rhetorical and argumentative strategies adopted by Greek authors in order to transfer a set of knowledge related to a particular disciplinary field, such as medicine, mathematics or philosophy.[8] This very brief reference to Asper's impressive book must serve here not to enter into a detailed discussion, but only as an indicating factor of the relevance of *Wissensvermittlung* for a prominent strand of scholarship.

This more general focus has led to various more specific directions which are usefully summarized by Asper in the introduction to a collective volume entitled *Writing Science*:

> First, authorial self-representation and the construction of authority, which is, naturally, somehow connected to the way social authority in general is constructed in the society under investigation. Second, the workings of

---

[5] The term *Fachtext* applied by German scholars covers a slightly different meaning, on the one hand because it can be used virtually for any text in prose whose topic is a certain field or discipline, on the other because the word *Fach* indicates precise disciplinary boundaries and compartimentalization of knowledge. I prefer to use here "literature of knowledge," which seems to be a much looser definition, but it also emphasize "literature" rather than "technology."
[6] See for instance Kullmann and Althoff 1998, Meißner 1999, Formisano 2001, Horster and Reitz 2003 and 2005, Paniagua 2006, König and Whitmarsh 2007, Fögen 2009.
[7] Asper 2012, 2.   [8] Asper 2007, in part. 11–56.

Introduction: The Poetics of Knowledge 17

narrative as it impacts both on the organization of empirical data and on its representation in writing. A third potentially fruitful area is the impact of metaphor on terminology and, more generally, the ideology of terminology.[9]

The first point especially has received considerable attention, becoming the focal point of some recent work.[10] Aude Doody and Liba Taub in the introduction to the collection of papers *Authorial Voices in Greco-Roman Technical Writing* strongly insist on explicit authorial statements of intention when they point out that:

> In antiquity, as today, authors of writing on scientific and medical subjects had a range of *choices* available to them as they sought to convey their ideas and information... The volume addresses as a theme the *strategies adopted by authors* to create an impression of authority for their works... We are concerned with the *choices made by authors* regarding the type of text used to convey their ideas... Our surviving ancient Greek and Latin technical texts reflect a range of *authorial decisions*.[11]

Thorsten Fögen emphasizes the concept of self-fashioning when he affirms "im Vordergrund steht die Frage nach der *Selbstdarstellung* der jeweiligen Verfasser gegenüber ihrem Lesepublikum und der dazu von ihnen verwendeten rhetorischen Strategien."[12] The attention devoted to authorial voices and self-fashioning combines a more traditional interest in the prefatory parts of technical and scientific texts, usually taken into consideration by literary scholars precisely because these parts have a particularly high degree of rhetoricity and the personality of the authors is more openly displayed.[13] However, prefaces and prologs are tricky to consider as *the* place where literary qualities are more visible: their more elegant rhetorical style is not sufficient to demonstrate that technical texts *are* indeed literature. This is a methodologically difficult point that merits much more attention. For a discussion of the textuality of these ancient works, it is important to observe whether and to what extent a connection between prefatorial parts and the body of the text exists, especially from the perspective of subsequent reception.[14] For example, in a discussion of Vitruvius' second preface

---

[9] Asper 2012, 4.
[10] See in particular the collective volumes edited by Doody and Taub 2009 and Asper 2012, and the monograph by Fögen 2009.
[11] Doody and Taub 2009, 7 (emphasis added).   [12] Fögen 2009, 7.
[13] Prefaces, prologs and proems of Latin technical and scientific writings are collected in three volumes by Santini and Scivoletto 1990–2 and Santini, Scivoletto and Zurli 1998. Prologs are at the center of among others Formisano 2001 and 2013, and Fögen 2009. For a general treatment of prologs in Latin literature, see the classical study by Janson 1964.
[14] A Latin text which has given the occasion to this kind of discussion in Vitruvius' *De architectura*. See Romano 1987, Gros 1997 and Novara 2005.

I have pointed to a combination of a disjunctive or anti-unitary style, an internal discontinuity of content, and a set of destabilizing meanings as indices of the text's literariness.[15]

As in discussions of *Wissensvermittlung*, the focus on authorial choices and strategies of self-fashioning commits to the literary nature of ancient technical texts only to a certain degree. For these approaches are firmly anchored in the surface of texts, and, by shedding light on modalities of systematization and on authorial intentions, the effect of this kind of scholarship is *de facto* to keep ancient technical and scientific texts distinct from other genres. At the same time, this approach places emphasis more on the peculiarity of the content of these texts (a field of knowledge which needs to be systematized and presented to a more or less larger audience by means of specific authorial strategies and rhetorical devices) and less on their inherently and self-reflexively literary qualities stemming from the very fact that they are texts beyond the fact that they are instruments apt for trasmitting knowledge.

Another strand of scholarship investigates the interrelations between poetry and technical and scientific prose from a number of perspectives. In an important article Heinrich von Staden shows the complexity of this relationship within the Galenic corpus. He focuses on (didactic) poetry as a medium for transferring medical knowledge that is particularly suited to the clear transmission of exact instructions on how to compose and use certain remedies. Galen in this case seems to prefer poetry to prose not for aesthetic reasons but precisely because of its inherent utility.[16] But, at the same time, Galen repeatedly expresses a negative judgment both on poetry and on the written text in general precisely because of their intrinsic ambiguity: the written word always bears an instability of meaning, which is contrary to a scientific way of reasoning and is particularly dangerous in medical practice.[17] Particularly intriguing is that Galen's writings are extremely sophisticated, displaying precisely the kinds of literary complexity against which they argue. Ralph Rosen takes a different approach to this contradictory attitude by pointing out that, even if "not an especially sensitive or insightful reader of poetry," Galen took advantage of the authority of poetic language in order to argue within the "context of scientific argument, which privileged a very different form of authority, with different standards of inference and truth."[18] Gregory Hutchinson discusses, on a

---

[15] Formisano 2016b. Disjunction and discontinuity have for instance been identified as a central motor of a poetic text such as Ovid's *Fasti* by Newlands 2000.
[16] Von Staden 1998, 76.   [17] Von Staden 1998, 86.   [18] Rosen 2012, 188.

more general level, the interrelation between ancient didactic poetry and technical writing, which he calls "didactic prose."[19] His goal is to re-orient the discussion in a more nuanced way towards technical prose texts, usually considered as mere containers from which didactic poets take raw materials for their own texts. As Hutchinson puts it, "the traffic between poetry and prose is two-way: prose can draw on poetry as well as the reverse; poetry can seek to evoke or appropriate characteristics of prose."[20] Hutchinson's analysis sheds a new and different light on the interaction between didactic prose and poetry, especially because it shows that technical prose, just like poetry, can bear a high degree of rhetorical complexity and, more importantly, can be as intertextual as poetry.[21]

The attention devoted to the relationship between poetry and technical texts reveals that the latter can be seen as more complex literary works than scholars of ancient poetry have acknowledged. At the same time, however, this approach confirms the common opinion according to which "literariness" is a factor *external* to texts, consisting for instance in a higher stylistic level, a sophisticated use of rhetorical strategies in order to transmit knowledge, or intertextual practices, which traditionally represent the high point of the analysis of ancient poetic texts. But all of this seems to ignore the possibility that knowledge, when it becomes *text*, has its own poetics, a poetics sometimes in accordance, but sometimes in contrast, with *Wissensvermittlung*, i.e. with the transmission of knowledge itself. This brings us back to Bakhtin's distinction between primary and secondary speech genres discussed above. I shall return to this important point later.

The question mark in the title of the first part of the introduction – *From Words to Acts?* – matters a great deal. This is not a rhetorical question, but is intended to open up a genuine and nuanced discussion of this fundamental topic. In particular, this volume aims at shedding light on what is usually believed to be the most technical aspect of the literature of knowledge, its applicability, and at showing how precisely this aspect can be considered as a textual phenomenon. In fact, I would argue that precisely the technicality of any applicable instruction contained in those texts is a textual device successfully disguised as a gesture towards the extra-textual reality. I shall give two very different examples in order to make the point.

---

[19] Hutchinson 2009, 196.   [20] Hutchinson 2009, 196.
[21] Hutchinson 2009, 198: "The passages will also expose to view an involved intertextuality in both poetry and prose."

In recent decades there has been a great deal of insistence on a specific form of representation that lies at the border between textual and extra-textual reality: diagrams and illustrations. These elements, as is well known, have almost without exception not been continuously transmitted from antiquity in the manuscript tradition, but subsequently reconstructed by later scribes and authors in the Middle Ages and Renaissance.[22] Yet some ancient texts do make explicit references to such diagrams and illustrations. This point has usually been taken as an indicator of how vigorously important the principle of applicability was in the intention of ancient authors. But are these illustrations in technical works exclusively instruments for the transfer of knowledge? In order to make this point clear, I now move on to a text conceived, written and illustrated in a very different age, one which, though its distance from ancient textuality, or perhaps precisely because of this distance, clearly shows how images can work in technical and scientific texts. Roland Barthes dedicated an essay to the *Plates* of the eighteenth-century *Encyclopédie*, which, for their formal perfection and precision, have long been perceived as one of the most accomplished examples of instructional illustration. In his essay, Barthes sheds light on the poetics ruling the logic of the composition of these illustrations. Independently of the original intention of Diderot and D'Alembert, the fathers of the *Encyclopédie*, these *Planches* can have an oneiric quality in our eyes precisely because of the distance in time (an argument which can be easily applied also to ancient texts). Barthes says: "There is a *depth* in the Encyclopedic image, the very depth of time which transforms the object into myth."[23] The objects presented are literally encyclopedic, i.e. they cover the entire sphere of substances shaped by man: clothes, vehicles, tools, weapons, furniture etc. "The *Encyclopédie* identifies the simple, the elementary, the essential, and the causal."[24] Within the *Encyclopédie*, technology is simple because it is reduced to a space of two terms: the human being and the object. There is an aesthetic of bareness: huge, empty, well-lit rooms, in which man cohabits alone with his work: a space without parasites, walls bare, tables cleared. The machines here are anything but enormous relays: man is at one extreme, the object at the other. Between the two lies an architectural milieu, consisting of beams, ropes and gears, through which, like a ray of light, human strength is simultaneously developed, refined, focused and enlarged: hence in the gauze-loom, a little man in a jacket, sitting at the keyboard of a huge wooden machine, produces an extremely fine web, as if he were playing music. As can be seen within the *Planches*,

---

[22] See Stückelberger 1994.   [23] Barthes 1980, 34.   [24] Barthes 1980, 26.

Introduction: The Poetics of Knowledge   21

encyclopedic poetics are always defined as a certain *unrealism*. It is the ambition of the *Encyclopédie* to be both a didactic work (thus based on a severe demand for objectivity, for reality) and a poetic work, in which the real is constantly overcome by some other thing – and that "other" is the sign of all mysteries. This enacts a kind of subversion of reality which is precisely the opposite of the kind of intention we might have attributed to these *Planches*. In a general way the *Encyclopédie* is fascinated by the insistence of reason (*la raison*) on the wrong side of things: it cross-sections, it amputates, it turns inside out, it tries to get behind nature. In most cases the image compels the recomposition of an object that is strictly *unreasonable*.[25]

Barthes' essay is of fundamental importance here, since it clearly shows how even the *Planches* of the *Encyclopédie*, normally considered the most "unliterary" element of a strictly technical work aimed at giving practical instruction, can be read and interpreted (not least because of the temporal distance we experience when we observe them) as sophisticated images that magnificently express the aesthetic of an entire period. Barthes' treatment of the most original work of the European Enlightenment is representative in many ways, among other things being useful for the understanding of ancient and pre-modern scientific texts. The *Encyclopédie* programmatically promoted a variety of mechanical arts which until then had been excluded from standard educational canons, and launched a new and more comprehensive concept of culture. By doing so, it opened up its contemporary cultural system to applied knowledge, which was significantly devalued in the classical ideal of education. But, as Barthes convincingly shows, despite the passionate defence of the values brought up by technology and mechanical arts, that powerful construction of rationalism, the *Encyclopédie* – and precisely in the *Planches*, meant to offer the most apt instrument for the applicability of technical knowledge by privileging images over words – implicitly proclaims the destabilization of the "applicability" of both words and images, since these unavoidably end up referring to themselves.

The other example I shall mention here in order to show how the literary dimension of the technical text not only emerges at the level of style and rhetoric, but also shapes the technical discourse itself, sometimes even destabilizing it, is an ancient text: the *Stratēgikos*, a treatise on the art of war written by the Greek theorist Onasander around AD 49.[26] The "art of

---

[25] I summarize Barthes' essay.
[26] I report here some of the arguments presented in Formisano 2011.

war," i.e. the literary genre devoted to the treatment of various aspects of warfare, for obvious reasons represents a privileged field in order to observe how and to what extent applicability matters.[27] In an important study bearing the title *Teach Yourself How to Be a General*, Brian Campbell aims to reconstruct the impact of military treatises and, as he calls them, "manuals," in the context of contemporary military reality: the guiding question is whether and to what extent these books were useful instruments for war. The answer to this question cannot be anything but frustrating, since "it is impossible to estimate the general availability of military textbooks and to discover how often the texts we have were actually used."[28] That said, Campbell does not want to fully deny *utilitas* to these works: he sees the utility of their content, which consists mainly of historical *exempla* and ethical precepts, in the moral guidance they gave to those in military command rather than in any strictly practical instruction. Elsewhere Campbell points out: "The important point is perhaps not that generals actually read military handbooks just as they went to battle, but the whole *genre* is typical of a mindset in society about the role of a commander, his responsibilities in battle, and perhaps also the qualities of character essential for dealing with matters of life and death."[29] This perspective, which I endorse, needs to be expanded by directing more attention to the voice of the texts themselves, i.e. to the particular textuality *of* these books. The obsession typical of military history, old and new, with the question of whether military tracts were actually useful on the battlefield contains some strikingly paradoxical aspects, especially noticeable if we put the ancient tradition in relationship with its modern reception in later authors such as Machiavelli and Clausewitz.[30]

The *Stratēgikos* is a peculiar type of text. Given the importance normally – but probably excessively – assigned to war in antiquity, we might have expected to find far more of these texts, and to find them in continuity with each other, as is the case many other literary genres. But, as Maurice Lenoir has shown in a seminal article, the presence of military literature in Latin is surprisingly scanty in view of Rome's lasting reputation for military success.[31] A major topic in the scholarship on Onasander – as that on

---

[27] Since within this volume only one chapter (Kaiser's – Chapter 5) discusses the art of war, it is useful to discuss it here at some length, given the relevance of this discipline precisely within a discussion on practical applicability.
[28] Campbell 1987, 27.   [29] Campbell 2004, 17.
[30] See Formisano and Böhme 2011; Formisano 2014 and 2016.
[31] Lenoir 1996, 83 ff.

the other military writers – has traditionally been an attempt at establishing the contribution his text aims to make to the practical improvement of the army. But a closer look at the text of the *Stratēgikos* reveals that it is concerned with various topics: Roman *virtus*, the role of fortune or τύχη, the concepts of φόβος, ζῆλος and φθόνος, the idea of *bellum iustum* and, perhaps most importantly, the psychological qualities of the (ideal) general.[32] Many of these topics derive from Xenophon, Aineias the Tactician, Plato and Polybius: Onasander seems to be reproposing Greek moral and technical norms for the benefit of Romans, among other things in order to establish continuity between Greece and Rome.[33] This point, too, is certainly valid, but it is worth noting that the *Stratēgikos* does not in fact refer to any predecessors by name or contain any reference to contemporary military technology or techniques: its norms and rules have a moral and ahistorical significance.[34] Onasander presents them as valid *in the abstract*; and precisely this intellectual quality to his art of war makes his text especially relevant to the tradition of military writing culminating with Clausewitz. More relevantly for the discussion in this volume, I would argue that the *leitmotiv* of this text, clearly thematized in its preface, is the tension between theory and practice, and that the text provides a perhaps surprising answer to the questions raised by that tension. The *Stratēgikos* is characterized by an absence of historical *exempla*, an absence that is rather astonishing for this genre, and by an absolutely generic treatment of military matters without any reference to the historical moment. Several terms in the text point to the notion of testimony through *sight*, hinting at the image of the commander's view over the battlefield, to external observation rather than action and direct involvement. As it has been observed that he:

> offers us a treatise on military science that is essentially philosophical... Onasander gives us a moral and ethical discourse which touches on the nature of the good general, and of the essence of military endeavour, which, it turns out (and here is the twist), *lies less in action than in dissembling*. Onasander's ideal campaign would be completely bloodless on both sides; two perfect matched armies would employ so many feints and counterfeints that they would never come to blows![35]

The attention given to "dissembling," which in Smith's opinion turns out to be paradoxical since it would eventually prevent any action, ends

---

[32] For a list of topics, see Ambaglio 1981, 355 ff. and Campbell 1987, 13–14.
[33] Ambaglio 1981, 362 ff.   [34] See Petrocelli 2008, 15.   [35] Smith 1998, 165 (emphasis added).

up being tightly connected with the *visual paradigm* governing the culture and literature of the first century of the Empire.[36] This visual paradigm emerges at several points throughout the text, becoming particularly relevant when Onasander supposedly gives "practical" advice to the general. I would argue that the contemplative and *theoretical* nature declared by Onasander at the opening of his work, which is in accordance with the "scopic paradigm" as described by Bartsch, represents much more than a rhetorical gesture. Vision (*opsis*) and contemplation become a substantive approach to strategy. But strategy is normally associated with action, while its opposite, contemplation, characterizes literary or philosophical knowledge. By substituting action through contemplation Onasander's text implicitly undermines its very subject matter, which is meant to give practical guidance for the conduct of war. Here Bakhtin's distinction between primary and secondary speech genres is particularly relevant: once a certain norm which is meant to support an extra-textual action enters the text, it unavoidably produces other meanings, which not only cannot be converted into practice, but which resist or even undermine application itself. If we take the role of vision in this text seriously, it leads us to a reconsideration of the generic classification so often applied by classicists. The *discrimen* between "literature" and "technical writing" becomes much more subtle, nuanced to the point that it vanishes, or only remains as a hint of scholarly preoccupations. In this regard at least, the scholarly principle of separating out the "literary" aspects of military texts as "intended to amuse and delight,"[37] as opposed to the technical, applicable core which needs to be taken *seriously*, oversimplifies the connection between literature and action, i.e. the principle of applicability in ancient technical texts. Onasander's book, like many other ancient texts, is today too easily classified as a "technical" text or manual, and yet not only its literary style but even its most technical content – its discussions of how to conduct war – in fact participate in a discourse whose nature is highly textual and literary.

Both Roland Barthes' essay on the *Planches* and Onasander's treatise on the ideal general illustrate how the rhetoric of practical applicability is more than a simple, literally interpretable textual device, but something which deserves closer attention from a theoretical perspective, since it often subverts rather than supports what we call "practice."

---

[36] See Bartsch 2006, 16. [37] Campbell 2004, 17.

REFERENCES

D. Ambaglio 1981, 'Il trattato "Sul comandante" di Onasandro', *Athenaeum* 59: 353–77

M. Asper 2007, *Griechische Wissenschaftstexte: Formen, Funktionen, Differenzierungsgeschichten* (Stuttgart)

(ed.) 2012, *Writing Science: Medical and Mathematical Authorship in Ancient Greece* (Berlin, Boston)

M. Bakhtin 1986, *Speech Genres and Other Late Essays*, trans. V. M. McGee (Austin)

R. Barthes 1980, *New Critical Essays*, trans. by R. Howard (Los Angeles)

S. Bartsch 2006, *The Mirror of the Self: Sexuality, Self-Knowledge and the Gaze in the Early Roman Empire* (Chicago)

B. Campbell 1987, 'Teach Yourself How to Be a General', *Journal of Roman Studies*, 77: 13–29

2004, *Greek and Roman Military Writers* (London, New York)

A. Doody and L. Taub (eds.) 2009, *Authorial Voices in Greco-Roman Technical Writing* (Trier)

T. Fögen 2009, *Wissen, Kommunikation und Selbstdarstellung: Zur Struktur und Charakteristik römischer Fachtexte der frühen Kaiserzeit* (Munich)

M. Formisano 2001, *Tecnica e scrittura: Le letterature tecnico-scientifiche nello spazio letterario tardo-latino* (Rome)

2011, 'The Stratēgikós of Onasander: Taking Military Texts Seriously', *Technai*, 2: 39–52

2013, 'Late Latin Encyclopaedism: Towards a New Paradigm of Practical Knowledge' in J. König and G. Wolff (eds.), *Encyclopaedism from Antiquity to Renaissance* (Cambridge), 197–215

2014, 'Kriegskunst' in M. Landfester (ed.), *Renaissance-Humanismus, Der Neue Pauly*, Supplemente 9 (Stuttgart)

2016a, 'Arte della guerra e rivoluzioni militari' in V. Ilari (ed.), *Future Wars: Storia della distopia militare* (Milan), 131–44

2016b, 'Reading Dismemberment: Vitruvius, Dinocrates and the Macrotext', *Arethusa* 49, 2: 145–59

M. Formisano and H. Böhme (eds.) 2011, *War in Words: Transformations of War from Antiquity to Clausewitz* (Berlin, Boston)

P. Gros 1997, 'Vitruvio e il suo tempo' in P. Gros (ed.), *Vitruvio, De architectura* (Turin), ix–lxxvii

M. Horster and C. Reitz (eds.) 2003, *Antike Fachschriftsteller: Literarischer Diskurs und sozialer Kontext* (Stuttgart)

2005, *Wissensvermittlung in dichterischer Gestalt* (Stuttgart)

G. Hutchinson 2009, 'Read the Instructions: Didactic Poetry and Didactic Prose', *Classical Quarterly*, 59: 175–90

T. Janson 1964, *Latin Prose Prefaces* (Stockholm)

J. König and T. Whitmarsh (eds.) 2007, *Ordering Knowledge in the Roman Empire* (Cambridge)

W. Kullmann and J. Althoff (eds.) 1998, *Vermittlung und Tradierung von Wissen in der griechischen Kultur* (Tübingen)

M. Lenoir 1996, 'La Littérature de re militari' in C. Nicolet (ed.), *Les Littératures techniques dans l'antiquité romaine: Statut, public et destination, tradition* (Geneva), 77–107

B. Meißner 1999, *Die technologische Fachliteratur der Antike: Struktur, Überlieferung und Wirkung technischen Wissens in der Antike* (Berlin)

C. Newlands 2000, 'Connecting the Disconnected: Reading Ovid's *Fasti*' in A. Sharrock and H. Morales (eds.), *Intratextuality: Greek and Roman Textual Relations* (Oxford), 171–202

A. Novara 2005, *Auctor in Bibliotheca: Essai sur les textes préfaciels de Vitruve et une philosophie latine du livre* (Leuven)

D. Paniagua 2006, *El panorama litterario tecnico-científico en Roma (siglos I–II d. C.): Et docere et delectare* (Salamanca)

C. Petrocelli 2008, *Onasandro. Il generale: Manuale per l'esercizio del comando* (Bari)

E. Romano 1987, *La capanna e il tempio: Vitruvio e l'architettura* (Palermo)

R. Rosen 2012, 'Galen on Poetic Testimony' in M. Asper (ed.), *Writing Science: Medical and Mathematical Authorship in Ancient Greece* (Berlin, Boston), 1177–89

C. Santini, N. Scivoletto and L. Zurli (eds.) 1990–8, *Prefazioni, prologhi, proemi di opere tecnico-scientifiche latine*, vol. III (Rome)

C. Smith 1998, 'Onasander on How to Be a General' in M. Austin (ed.), *Modus Operandi: Essays in Honour of Geoffrey Rickman* (London), 151–66

A. Stückelberger 1994, *Bild und Wort: Das illustrierte Fachbuch in der antiken Naturwissenschaft, Medizin und Technik* (Mainz)

P. J. van der Eijk 1997, 'Towards a Rhetoric of Ancient Scientific Discourse: Some Formal Characteristics of Greek Medical and Philosophical Texts' in E. J. Bakker (ed.), *Grammar as Interpretation: Greek Literature in Its Linguistic Contexts* (Leiden), 77–129

H. von Staden 1998, 'Gattung und Gedächtnis: Galen über Wahrheit und Lehrdichtung' in W. Kullmann and J. Althoff (eds.), *Vermittlung und Tradierung von Wissen in der griechischen Kultur* (Tübingen), 65–94

CHAPTER 3

Machines on Paper
From Words to Acts in Ancient Mechanics

Markus Asper

Peter Damerow in memoriam

Machines are artifacts.* As such, they result from an 'act', namely from a process of construction. At the same time, machines are usually the result of 'words', the product of a traditional, that is, verbally transmitted and usually institutionally protected body of knowledge that regulates how to build such machines and what to do with them. Therefore, one assumes, writings on such bodies of knowledge may seem to be perfect candidates for an investigation of application, for the precarious process that turns words into acts. As far as ancient mechanics is concerned, 'acts' have enjoyed much more public recognition than 'words'. The so-called 'Antikythera device', for example, has by now been the focus of numberless exhibitions and conferences, while hardly anybody reads Hero or Ps.-Aristotle on mechanics any more, let alone Biton's or Philo's treatises on catapult-making. Across the body of mechanical writing, application has usually been taken for granted instead of being recognized as the problem that it actually is. For if one browses through the body of extant mechanical writings, it is far from evident what exactly constitutes an 'act' in this field. What application?, one might ask. Let me begin with a rare passage that points to a certain range of what mechanical writing can do. Athenaeus Mechanicus, probably a first-century BC author of a rather strange treatise on siege-machinery, while brushing aside the writings of his predecessors (he mentions, among others, Archytas and Aristotle), distinguishes between learning and doing things:[1]

---

* Thanks to Philip van der Eijk and Marco Formisano for making me write on mechanics, to Joyce van Leeuwen for generously providing a revised text and diagram of Ps.-Aristotle's *Mechanics*, Probl. 17, to Oliver Overwien for helping me with the Arabic version of Hero's *Mechanics*, to Serafina Cuomo for encouragement and helpful remarks, and especially to Peter Damerow and Jürgen Renn who introduced me to the field more than fifteen years ago.

[1] Athenaeus Mechanicus, Περὶ μηχανημάτων (*Mech.*) pp. 4–5 W. = p. 44 W.-B. I quote from the following editions: F. Hultsch, *Pappi Alexandrini collectionis quae supersunt*, 3 vols. (Berlin 1878)

27

Νεωτέροις μὲν γὰρ φιλομαθοῦσιν οὐκ ἄχρηστα εἴη <πρὸς ἕξιν> τοῦ στοιχειωθῆναι· τοῖς δὲ βουλομένοις ἤδη τι πράττειν μακρὰν παντελῶς ἂν εἴη καὶ ἀπηρτημένα τῆς πραγματικῆς θεωρίας.

Admittedly, for young men who strive for knowledge, (those texts) might not be useless as far as ?introductory teaching? is concerned; for those, however, who already want to do something, (those texts) would be completely irrelevant and detached from practical investigation.

Athenaeus reads his famous predecessors' texts basically as introductory writing, while he himself wants to contribute to *doing* something. He thus distinguishes two 'acts' of doing mechanics, learning about it and turning it to practical effects, and constructs them, to a certain extent, as polar and serial: there is a fixed biographical order for both dealings with mechanical knowledge, which means, for example, that one cannot do both at the same time.[2] Already here, the reader forms the impression that 'doing' is much more important to Athenaeus and his addressee, probably M. Claudius Marcellus,[3] than learning. Later on, when describing a 'tortoise' (χελώνη) that carries a ram (κριός), he mentions one more act:[4]

Τοῦ γὰρ ἔργου καλῶς διατυπουμένου τούτου, τοῦ ἀρχιτέκτονος εὑρίσκεται εὐδοξία· κατὰ δὲ λόγον ἐκτιθεμένου, τὰ συντάγματα μέγιστον ἕξει κλέος ἐν τοῖς ὑπομνήμασιν.

When this piece of work is well devised, a great reputation of its builder will emerge; but when someone explicates it in words, the description will have most renown in technical writings.

Athenaeus again pits 'doing something' against 'writing something', although this time writing is not for instruction but geared towards building a reputation.[5] Athenaeus seems to make a sarcastic point, namely

---

(repr. Amsterdam 1965 = H.); E. W. Marsden, *Greek and Roman Artillery: Technical Treatises* (Oxford 1971) (= M.); L. Nix, 'Herons von Alexandria Mechanik in der arabischen Übersetzung des Kosta ben Luka' in Heron Alexandrinus, *Opera*, vol. II (Leipzig, 1900), 1–299 (= N.); W. Schmidt, 'Herons von Alexandria Druckwerke und Automatentheater' in Heron Alexandrinus, *Opera*, vol. I (Leipzig 1899), 1–453 (= S.); M. Thévenot, *Veterum mathematicorum Athenaei, Apollodori, Philonis, Bitonis, Heronis et aliorum opera* (Paris 1693) (= Th.); C. Wescher, *La Poliorcétique des grecs* (Paris 1867) (= W.); D. Whitehead and P. H. Blyth, *Athenaeus Mechanicus, On Machines* (Stuttgart 2004) (= W.-B.). Unless otherwise noted, all translations are my own.

[2] The search for acts in the field of mechanical writings shows some resemblance to the long-established functional analysis of genre (see Asper 2007a, 18–23). Compare, for example, Netz 2009, 108 on three types of mathematical literature which he calls 'ludic', 'survey' and 'pedagogic'.

[3] See the discussion in 15–20 W.-B.   [4] Athenaeus, *Mech.* (as in 1), 15 W. = 50 W.-B.

[5] It is unclear to me why Whitehead and Blyth in their recent edition decide to bracket this sentence. The fact that an equivalent in Vitruvius is missing and that therefore the sentence is possibly not from Agesistratus does not at all prove that it cannot be from Athenaeus either.

that, unlike in constructing a real machine, machines on paper[6] bury the mechanical knowledge involved in obscure writings, and thus the reputation following from it is nothing, a waste of effort.[7] In any case, Athenaeus has apparently several acts in mind, to which mechanical words may lead.

In modern science writing (unlike in treatises on ethics, cooking or religion), the response that the author usually aims at is not really an act, unless we call understanding an 'act'.[8] Theoretical authors, ancient and modern, usually aim at explanation and consensus. Ancient mechanical writing, however, by no means a unified body of knowledge, occupies a shifting position between theory and practice. 'Mechanics' (τὰ μηχανικά or ἡ μηχανική (namely, τέχνη)) covers everything from geometrical proof of the law of the lever to the details of siege-craft. Accordingly, in the group of texts treated in this chapter, there seem to be several ways to understand 'act', i.e. the activity the author is aiming for in his readers, what he wants them to do. One can identify about four such 'acts', arranged in the order of decreasing focus on practical matters: (i) the practical construction of machines; (ii) comprehensive and systematic knowledge about the construction of machines; (iii) decision-making with regard to purchasing machines and paying engineers; and (iv) the proper 'scientific' explanation of how and why machines work, entailing quasi-mathematical proof. I shall have to leave open the question of whether or not all four are forms of application at all. (In a certain sense, however, explanation is some form of application, too, insofar as it transforms a certain piece of knowledge into something else.) The chapter will illustrate these four formats and their peculiar means of communication with some examples. I begin with the most obvious.

Constructing Machines

Mechanical knowledge solves certain practical problems, for example, how to extract oil from olives or how to lift great weights, or responds to certain social needs, for example, the construction of seemingly self-moving automats or of impressive siege-engines. One way in which such

---

[6] I borrow the expression 'machine(s) on paper' from James W. Goedert's brilliant art show at the Antenna Gallery, New Orleans (8 January – 5 February 2011), whose machines are real and do, admittedly, things very different from the ones I shall discuss in this chapter.

[7] Although ὑπόμνημα has a wide range of meanings, the common denominator seems to be 'writings that do not circulate widely, are for internal use only', etc. Compare Asper 2007a, 28 n. 130, 53 f.

[8] Understanding and 'act' are not necessarily unrelated: see Aristotle's so-called 'practical syllogism' (e.g. *Eth. Nic.* VII.4–5, 1146 b 35–1147 b 19), on which see Broadie 1968, Welch 1991.

knowledge becomes explicit[9] is transmission and, perhaps, competition among experts. The media of such explicit knowledge are, mostly, words, occasionally accompanied by diagrams, numbers and such. In these cases, words serve as a means of storage in order to preserve and to distribute such forms of knowledge. At the same time, mechanical knowledge can become a little unwieldy in writing (an especially convincing application of Plato's arguments in the *Phaedrus* against writing down serious things).[10] In many passages of writing about mechanics the precise way of construction remains unclear to the reader, perhaps intentionally, as some modern readers have claimed with respect to siege technology, or perhaps because these treatises address an expert audience. I shall touch upon two or three issues here: the status of numbers and diagrams, and structure.

Those mechanical treatises that aim at the act of construction are full of numbers specifying the dimensions of certain machines and their parts. The following text from Cato's handbook on agriculture may suffice as an introductory example. Cato (*Agr.* 18.1–2), after listing the materials needed (12), describes how to put together an efficient press. The following excerpt presents approximately one-third of the description:

> Torcularium si aedificare voles quadrinis vasis, uti contra ora sient, ad hunc modum vasa conponito: arbores crassas P. II, altas P. VIIII cum cardinibus, foramina longa P. III S exculpta digit. 6, ab solo foramen primum P. I S, inter arbores et et parietes P. II, in II arbores P. I, arbores ad stipitem primum derectas P. XVI, stipites crassos P. II, altos cum cardinibus P. X, suculam praeter cardines P. VIIII, prelum longum P. XXV, inibi lingulam P. II S.

> If you wish to build an olive press-room for four presses, make them face in alternate directions. Arrange them in this way: 'Trees' two feet thick, nine feet high including tenons, with sockets cut out 3¾ feet long and 6 fingers wide beginning 1½ feet from the ground. 2 feet between the 'trees' and the walls; 1 foot between the two 'trees'. 16 feet at right angle from the 'trees' to the nearest of the posts. Posts 2 feet thick, 10 feet high including tenons. Windlass 9 feet long plus tenons. Press-beam 25 feet long, including tongue 2½ feet. (trans. A. Dalby)

This is actually a recipe, that is, a series of precise instructions without any general explanation or description. Cato never gives us an explanation

---

[9] The machines on paper discussed in this chapter are too complex for putting much emphasis on a 'tacit' level of knowledge à la Polanyi (see the discussion in Polanyi 1966, Collins 2010). Orally or in writing, the knowledge needed, for example, in order to put together an effective siege-engine cannot ever have been tacit or implicit. For an exception, see 'Understanding machines' below (pp. 44–8).

[10] See the account in Kullmann 1991, 1–21.

of why it is that the prescribed dimensions are the best or that the recommended materials are superior to others. Nor does he explain how it actually works. It seems that Cato expects his addressee to already know the mechanical principle of such presses, which is why he can concentrate on dimensions and materials.[11] Modern historians of technology would classify Cato's press as a lever-and-drum press, as opposed to various types of screw presses that were invented later.[12] Such terms of classification would instantly supply the reader with the information needed in order to visualize the machine. Cato's reader, however, knows what machine Cato is talking about and, apparently, knows only one kind of such machines. Thus, nowhere in Cato's text does one find an abstract qualification of how in general this machine is supposed to work. Let us move on by about 250 years to Neronian times and compare Hero's introductory passage on the same machine, from the third book of his *Mechanics*, transmitted only in Arabic:[13]

> Now agricultural machines [al-ālātu = lit. 'tools'], namely the ones one presses wine and oil with, are quite close to the applications of levers [muḫlun = μόχλος] which we have discussed above... The beam called 'mountain' [ǧabalun = ὄρος] which others call 'press' ['uṣṣāratun][14] is nothing but a lever and its fulcrum [ḥaǧaru lladī taḥta l-muḫli = lit. 'the stone that is below the lever' = ὑπομόχλιον].[15]

Hero classifies the press in analytic terms, by which I mean that for him such a press is a complex machine that can be analysed as being put together from elements like levers, a windlass, etc. that he had defined above as 'simple machines'. From such a perspective, the press is an application of a lever, perhaps combined with a windlass. Once this analytic step is taken, all that remains is to identify which parts correspond to which element of the simple machine. Matters of size and dimensions, material, etc. do not play a role in this text, which is not about construction but about classification (at least in this passage) and, perhaps, explanation. What follows is an analytic description, not a 'recipe'. It is doubtful that Hero's intention is to guide his readers to actually construct such presses. Both texts are, however, similar in that they do not bother to refer to a diagram.[16]

---

[11] For technical details, see Schürmann 1991, 129–32.
[12] According to Pliny, *HN* XVIII 317, who distinguishes 'ancient' from more recent 'Greek' presses. See Curtis 2008, 382.
[13] *Mech.* III 13, vol. II, p. 226 f. N.-S.
[14] As conjectured by Oliver Overwien (manuscript reading: 'uṣṣār).
[15] My English translation is not based on the Arabic, as it should be, but on Nix's German.
[16] Which one may nonetheless find in the manuscripts of Hero.

These two texts are already sufficient in order to demonstrate how different approaches can be taken towards the same machine on paper and that ancient mechanical writers aim for more than just one 'act' when dealing with these machines. For now, however, let us stay with our first act, construction. These presses are, in principle, quite simple; they were used widely and had already been used for a long time when Cato was writing about them.

In both respects, Hellenistic artillery was different. The technology was fairly new, at least when compared with beam presses for olives or grapes;[17] it was not something everybody could observe or even contemplate from close quarters, when in action; it was technologically quite complicated; and the knowledge concerned must have been subject to some form of confinement, e.g. to a royal institution at Alexandria or Pergamum. For all these reasons, Hellenistic or early imperial texts on artillery display a very special sound, rich in numbers and attempts at technical visualization, but also tending to appear opaque to the un-initiated. One example may suffice: this is how Biton, a second-century BC author whose work poses some challenges,[18] begins to describe the base of Zopyrus' mountain belly-bow:[19]

> ἔστιν ἄρα βάσις ἡ Α, ἧς τὸ μῆκος ποδῶν [ι]ε΄, τὸ δὲ πλάτος ποδῶν γσ΄, ὕψος δὲ ποδὸς α΄· ἐπάνω αὐτῆς κιλλίβαντες οἱ ΧΧ, ὧν ὁ μὲν ἐλάσσων, ὁ δὲ μείζων, καὶ ὁ μὲν ἐλάσσων ὁ Ι, ὁ δὲ μείζων ὁ Θ. ἔστι δὲ τοῦ ἐλάσσονος τὸ ὕψος πόδες γ΄, τοῦ δὲ μείζονος πόδες ε΄.

> There is a base, A, the length of which is 5 ft., the breadth 3½ ft., the height 1 ft. Above it are the trestles, XX, of which one is smaller, one larger, and the smaller is I, the larger Θ. The height of the smaller one is 3 ft., of the larger one, 5 ft. (trans. M.)

As is clear from the letters and as it is also explicitly stated by the author,[20] there was a diagram illustrating the construction which is now lost. This fact certainly accounts partly for the difficulties modern readers have with Biton. On the other hand, even with the diagrams imagined, Biton decides

---

[17] Compare Cuomo 2007, 41–67 on the 'shot-gun model'. As for the age of beam presses, a late sixth-century *skyphos* from Attica is sometimes taken to provide a *terminus ante quem* for mechanical presses.
[18] See Lewis 1999 on Biton's date and the question of why he does not cover torsion artillery at a time when this technology was the only one used.
[19] Biton, Κατασκευαί πολεμικῶν ὀργάνων καὶ καταπαλτικῶν p. 76 M. = p. 65 W.
[20] M. p. 76 = p. 67 W.: τὸ δὲ σχῆμα οἷόν ἐστι ὑπογέγραπται. Almost every machine that Biton describes was also illustrated with diagrams.

not to give any abstract account of the machine, that is, an explanation of how it works, what the principles of its construction are, etc. For the expert mechanic, probably either the name of the machine (γαστραφέτης) or the one of its inventor (Zopyrus of Tarentum) or of the place of its first construction (ἐν Κύμῃ τῇ κατ' Ἰταλίαν) carried some information about such details. As in the case of Cato, the machine on paper, instead of trying to present a perfect mimesis of the real machine, comes closer to being a supplement to the knowledge about that machine that the reader already brings to the text. Naturally, the supplement gives details and supplies a sort of knowledge that contemplation cannot disclose and that memory cannot accurately store, that is, numbers and, thus, proportions.

Authors like Biton, Philo and Athenaeus present the numbers that describe the machines in question as being absolute perhaps because, in the case of individual inventions, as, e.g., Zopyrus' belly-bow mentioned above, they wish to claim to have had access to the original dimensions. Thus, the more precise the array of as many numbers as possible, the greater is the benefit to their authority as authors. There is, however, another possibility to explain the habit of giving just one set of numbers: just as in Babylonian or Greek practical mathematics,[21] the numbers may have exemplary status, that is, they demonstrate that one model with such-and-such dimensions actually works. The apprentice mechanic is perhaps supposed to learn how to construct a certain machine by building it the first couple of times with precisely these numbers, but later goes on to extract from these their actual proportions by himself and, by doing that, learns the principles of the machine on paper that were never actually mentioned by its authors. Such is the method by which recipes work, even today.

One of the most important points about artillery, at least whenever it is constructed on site, is its reach in proportion to the weight hurled. Since both primarily depend upon the dimensions of the ballistic machine, the problem of proportion, although not addressed in the descriptions of constructions themselves, must nonetheless have been vital in construction, as both Biton and Philo attest:[22]

> Biton: ἐάν τε γὰρ βούλῃ μείζονα κατασκευάζειν, ἐπιτέλει, ἐάν τε ἐλάσσονα· μόνον πειρῶ τὴν ἀναλογίαν φυλάττειν.
>
> If you wish to construct larger or smaller machines, go ahead; just try to preserve the proportion.

---

[21] E.g., Asper 2009, 108–14.
[22] Biton, *Catap.* p. 67 f. W. = p. 76 f. M.; Philo, *Bel.* p. 50 Th. = p. 106 M.

Philo: διό φημι δεῖν προσέχοντας μεταφέρειν τὴν ἀπὸ τῶν ἐπιτετευγμένων ὀργάνων σύνταξιν ἐπὶ τὴν ἰδίαν κατασκευήν, μάλιστα δέ, ὅταν τις εἰς μεῖζον μέγεθος αὔξων τοῦτο βούληται ποιεῖν καὶ ὅταν εἰς ἔλασσον συναιρῶν.

Therefore, I insist that one must be careful when one adapts the construction of successful machines to one's own construction, especially when one wants to enlarge or diminish the scale.

In this context, Philo even presents a carefully narrated story of how the problem was first dealt with experimentally, then transformed into a mathematical one and eventually solved only because of the generous funding of experimental science by the Ptolemies.[23] He then sets out in detail a complicated algorithm that promises to effectively solve the problem, that is, it relates the weight of the stone hurled to the diameter of the hole that receives the spring of the catapult (Philo p. 51 Th. = p. 108 f. M.) and which in turn determines the dimensions of the whole catapult.

Diagrams are at least as crucial as numbers, but have fared with even less luck in transmission. Some manuscript writers apparently left out all diagrams they might have found in their sources; in other cases, modern editors decided to provide visualizations according to their own taste.[24] Only recently have diagrams embarked on a career in classics, mostly thanks to the work of Reviel Netz and Ken Saito; on the other hand, recent art history has taken an interest in technical visualization ancient and modern.[25]

While in theoretical mathematics the combination of letters and diagrams, the so-called 'lettered diagram', has been hailed as a Greek invention and as a superb way to utilize the possibilities of paper in order to minimize the contingencies of transmission in writing, I wonder whether one could not contemplate the priority of the technological sketch with respect to the mathematical[26] (after all, people must have been in urgent need of complex machines quite early). It is quite evident that complex machines on paper have to rely upon diagrams.[27] Consider, for example, the following passage from Hero's manual on artillery:[28]

Ἡ δὲ χοινικὶς γίνεται τόνδε τὸν τρόπον. ἐμβολέα δεῖ κατασκευάσαι ὅμοιον τῷ ΑΒΓΔΕΖ ὑπογεγραμμένῳ, ἔχοντι τὰς μὲν ΑΕ, ΒΖ περιφερείας, τὰς δὲ

---

[23] On this passage, see Asper 2013; and Keyser 2013.
[24] Compare the alarming remark of Wilhelm Schmidt in his 1899 preface to Heron's *Automata*, li, but see the same on *Pneumatika (Spir.)*, xxvi.
[25] See, e.g., Netz 2009, 5; Saito 2006, and Saito and Sidoli 2012; for ancient technical diagram in modern art history, see Bogen 2013.
[26] For a similar argument with respect to architecture, see Asper 2007b, 82 f. See p. 62 M. on Biton's und Vitruvius' *scenographia*.
[27] See Philo, *Bel.* p. 62.15 Th. = p. 126 M.; Athenaeus, *Mech.* p. 39 W. = p. 60 W.-B.
[28] Hero, *Bel.* p. 96 W. = p. 30 M.

ΕΓ, ΖΔ εὐθείας, τὴν δὲ ΑΒ ἴσην τῇ τοῦ τρήματος διαμέτρῳ, καὶ πρὸς τοῦτον ἐκτορνεύσασθαι τὴν χοινικίδα.

The washer is made in the following way. You construct a template like ΑΒΓΔΕΖ as illustrated, with ΑΕ and ΒΖ curved, ΕΓ and ΖΔ straight, and ΑΒ equal to the diameter of the whole. Following this template, chisel out the washer.

The diagram is, as usual, lost in transmission, but imagine the mass of text that Hero would have had to insert here, had he not decided to give an illustration. Without such an illustration, the transmission is problematic (which is precisely what it is now). Note also that there is nothing theoretical in this description.

As these few quotations have shown, mechanical authors also provide an unorthodox range of specimens for research on diagrams. In the context of constructing complicated and costly machines, a diagram's function goes beyond just making sure that construction will take place in the way intended. 'Machines on paper' must always work, and thus the diagrams partly aim for an illusion, namely to convince the reader that the machine works smoothly. From such a perspective, diagrams turn, quite surprisingly, into 'fantasies of power'.[29] In that respect, diagrams become part of what has been called 'literary technologies', that is, literary tools meant to enhance the writer's authority.[30]

Sometimes in our mechanical treatises the language that accompanies the diagrams closely resembles the language of theoretical mathematics ('Let there be a such-and-such ABC...'), and sometimes the Platonizing tendency to erase the authorial voice even leads to certain absurdities. For example, Biton says at the beginning of his description of a belly-bow:[31]

> Ἐχομένως δὲ <ταύτης> τῶν καταπαλτικῶν γαστραφέτου σοι ἀρχιτεκτόνευμα προκεχείρισμαι ἀναγράψαι. ἔχει δὲ τόνδε τὸν τρόπον. ἔστω γὰρ ὁ ὑποκείμενος γαστραφέτης ὃν ἠρχιτεκτόνευσε Ζώπυρος ὁ Ταραντῖνος ἐν Μιλήτῳ.

> Next after this, I have undertaken to write up for you the design of one of the catapults, a belly-bow. It has the following arrangement. Let the proposed belly-bow be the one which Zopyrus of Tarentum designed at Miletus. (trans. E. Marsden)

Here, the 'mathematization' of language has been taken to extremes: Biton decides to present the individual invention of Zopyrus, realized at a certain

---

[29] For diagrams in technical treatises as 'fantasies of power', see Bogen 2013.
[30] Shapin 1984 uses the expression in discussing Robert Boyle's technical writings.
[31] Biton, *Catap.* p. 61 f. W. = p. 75 M.

place and time (which probably means during a certain and identifiable siege), in terms of the Platonizing mathematician who pretends that he has nothing to do with the events in space the implications of which he investigates ('let there be two lines AB and CD that cross each other in E . . .'). More such absurdly mathematized diagrams and descriptions will be discussed below under 'Understanding machines' (pp. 44–8). For now, however, all I want to show is how crucial, next to numbers (and a precise idea of what they are supposed to mean), diagrams are for a successful attempt to get from words to acts with respect to machines, i.e. to get one's machines off the paper.

Codifying Knowledge about Machines

Besides constructing machines, managing the knowledge that one needs in order to successfully construct those machines (i.e. acquiring, storing, transmitting and keeping accessible that knowledge) is a different problem and, depending on how complex the knowledge concerned is, a difficult one. Certainly, displaying mechanical knowledge as a systematic body aims for a different act from that of describing the construction of a given, individual machine. Tentatively, I understand this act as an attempt at codifying a body of knowledge, that is, giving an authoritative sketch of what belongs to the *technē* in question.

Among the literature about machines, Vitruvius' *On Architecture (De arch.)* IX and Hero's *Mechanika* qualify, I believe, as such mechanical *technai*.[32] Far from discussing these accounts in detail, I shall highlight only one feature, namely their 'systematicity'. The concept of systematicity has been much debated in recent years by philosophers of language and cognition.[33] In this context, the concept is of interest, because any

---

[32] Both base their accounts and perhaps their overall structure on Hellenistic predecessors, probably Philo of Byzantium's *Syntaxis* (in nine books – still extant are Book IV on belopoeics, V on pneumatics, VII on *paraskeuastika* and VIII on poliorcetics) and/or Ctesibius of Alexandria, to whose texts the same would apply. Pappus of Alexandria's synthesis of mechanics (*Collection* VIII), on the other hand, seems to be based entirely on Hero's *Mechanics*.

[33] Cummins et al. 2001, 167:

> The current debate over systematicity concerns the formal conditions a scheme of mental representation must satisfy in order to explain the systematicity of thought. The systematicity of thought is assumed to be a pervasive property of minds, and can be characterized (roughly) as follows: anyone who can think T can think systematic variants of T, where the systematic variants of T are found by permuting T's constituents. So, for example, it is an alleged fact that anyone who can think the thought that *John loves Mary* can think the thought that *Mary loves John*, where the latter thought is a systematic variant of the former.

codified body of knowledge must be a representation of decontextualized knowledge which emerged in non-representative, context-determined acts, namely actual construction. For managing such knowledge properly, that is, keeping it accessible over large distances of time and space, the representation of it as systematic has proved successful.[34] Authors display their knowledge in a way that does not mainly consider application, but constructs knowledge-internal structures. In addition, the display not only makes these structures explicit, but tries to mirror them with textual features. Practitioners do not typically need 'systematicity', at least not always, and certainly not in order to build a certain machine or to solve a practical problem. Systematicity of knowledge emerges, however, perhaps in contexts of public display, when success in a competitive situation is at stake, and certainly in contexts of tradition and transmission.[35]

Such systematic representation follows the procedure often observed in Greek theoretical literature, beginning with Pre-Socratic thought and actually discovered as a rational practice that crosses and constructs fields by Fuhrmann.[36] The main features are: (a) to define a field of knowledge and several subfields within it; (b) to identify, name and define elements of analysis to which all objects of discourse can be reduced; and (c) to arrange these elements in such a way that causal structures become cognizable. In this respect, the beginning of Vitruvius' book on mechanics is worth quoting (*De arch*. X 1):

> Machina est continens e materia coniunctio maximas ad onerum motus habens virtutes. Ea movetur ex arte circulorum rotundationibus, quam Graeci κυκλικὴν κίνησιν appellant.
>
> A machine is a constant construction from wood that exerts the greatest forces towards moving weights. It moves by the revolutions of circles, which the Greeks call circular movement.

Vitruvius begins by giving a definition of what a *machina* is. His next sentence identifies, very much in the spirit of Ps.-Aristotle, the *archē* of its ability to move weights: it all has to do with circles. That is, Vitruvius

---

[34] With respect to mechanics, see already Vernant 2006, 303.
[35] I have described 'systematicity' as the hallmark of the Greco-Roman *technē* as follows (Asper 2015): 'Across the range of knowledge concerned, the unifying feature of these accounts is "systematicity", i.e., a desire to identify conceptual structures and patterns in the body of knowledge concerned and to use these patterns in order to provide texts that formally mirror the knowledge which they are meant to convey.'
[36] As a feature of the proper *logos* already in Diogenes of Apollonia (end of fifth century BC) fr. 64 B 1 Diels and Kranz. Fuhrmann 1960 was the first to describe the features of such systematic codification of knowledge across various fields.

defines mechanics as a subfield of physics, namely the physics concerned with circular movements,[37] to the analysis of which (with respect to straight lines and linear movements) he will devote an entire chapter (X 3). In a third step, three classes of machines are differentiated and defined, namely the *genus scansorium* (ἀκροβατικόν), the *genus spirabile* (πνευματικόν) and the *genus tractorium* (βαροῦλκον).[38] According to practical importance, Vitruvius devotes the bulk of what follows to the last class of machines. At the end of the third chapter (X 3.9), all βαροῦλκα are found to be based on the combination of straight and circular elements and movements. This combination is precisely what fuels the explanation of the following classes of machines, namely machines drawing or manipulating water (chapters 4–8), ballistic machines, siege-engines (chapters 10–15) and presses.[39] That is, all machines of practical importance can, according to Vitruvius, be analysed into these elements and thus both understood and constructed. To be sure, the actual description of certain machines is rich in detail, and sometimes even contains narrative passages that explain the invention of machines (e.g. X 13.1–3), or even complex tales (e.g. 16.3–8)[40] that add social-normative and methodological dimensions to systematic knowledge.

Vitruvius wrote, at least partly, with the intention of seeking patronage from an emperor and presented his codifying account of mechanical knowledge within a concept of *architectura* as a master-discipline. Hero, the great mechanic of the Neronian age who is heir to a long line of Alexandrian mechanics, and who probably works within a well-established institution devoted to such knowledge, can afford to be more accurate and even more systematic, because neither the hierarchy of fields nor patronage needs to concern him. Nonetheless, with respect to the basic intention of highlighting 'systematicity', both are comparable.

---

[37] An account of which in Peripatetic natural philosophy one finds in Bodnár 2011, 448–53. This would presumably be the theoretical background of Vitruvius, too.

[38] As the terms show, Vitruvius follows a Greek system here, in which the first two terms smack of theoretical discourse (note the -ικος suffix, for which see Kenneth Dover's introduction to his 1993 commentary on Aristophanes' *Frogs*, p. 30 f.). Cf. v. Hesberg 1987, 54 on Hero's use of Hellenistic philosophical dihaeresis in *Aut.* I, 1–3.

[39] Only the taxameter described in chapter 9 and the treatment of defence strategies interrupts the order that grosso modo unfolds as Vitruvius has indicated. He decides not to cover presses, though, perhaps as unworthy of an emperor's attention.

[40] This narrative reports the rise, fall and rise of the Rhodian architect Diognetus and contains, besides the main moral about the relation of mechanical knowledge and common sense in siege, a second-order moral about the instable public recognition of architects (!), even a mini-treatise on the applicability of models (X 16.5). This narrative certainly functions as a frame-tale of book X, that is, an embedded narrative which nonetheless implicitly unfolds potential contexts and interpretations of the knowledge it is embedded in (see my essay on frame tales, Asper 2011).

Hero analyses the field of mechanics into five basic machines that he calls δυνάμεις (windlass, lever, pulley, wedge and screw).[41] He spends his whole second book on explaining these five basic machines, reduces their functions to and proves them by the Ps.-Aristotelian theorem of the law of the lever, conceptualizing two levers as the radii of two concentric circles (chapter 7, pp. 111–14 N. – see below), even reduces some machines to others (the screw is a sort of wedge, chapter 17) and then proceeds by looking at typical combinations of these elements (chapter 21 ff.). All this happens in the second book, though.[42] The first chooses another approach, namely it takes its cue from a generalization of the most often posited problem, that is, moving a given weight by a given force, which Pappus attributes anecdotally to Archimedes (p. 1060.2 f. H.). This problem is solved by a model machine, an ideal combination of these δυνάμεις, the so-called βαροῦλκος, which is a system of four cogwheels, one upon which the force is exerted and one attached to the weight. By varying the relative proportions of the cogwheels, any given force can, in theory, move any given weight (although there are practical limits, duly discussed by Hero in chapter II 32, p. 171 N.). Interestingly, from the presentation of the *baroulkos* at the very beginning of the extant *Mechanics*, Hero works backwards, that is, towards more general problems of a mathematical nature, arriving at definitions of congruence and similarity (I 12 f., p. 27 ff. N.) and then moving on to basic principles of mechanics in geometrized fashion, e.g. cylinders on inclined planes (chapter I 23, p. 61 ff. N.), ending up with principles of statics and equilibrium. The third book, however, talks about practical matters, including all kinds of presses. Thus, the system of mechanical knowledge as presented in the *Mechanics* does not cover all fields identified by Vitruvius, e.g. artillery and siege-engines or the taxameter (on both topics, however, there exist separate texts by Hero); nor does it proceed in a parallel way. Nonetheless, by beginning with a generalized problem, then working 'up' towards the determining factors of utmost generality, presenting and explaining the five elements of all machines, and then working 'down' on practical matters of just one field, it creates a systematic map of mechanical knowledge that applies to *all* subfields of mechanics. In both accounts, what I called 'elements' a, b and c in my own systematic account are not only present, but also determine the overall structure of the exposition, which, to be sure, in addition contains lots of practically

---

[41] The Arabic is *quwan* (sing. *quwwa*); Pappus preserves the Greek term δύναμις (*Coll.* VIII 11, vol. III, p. 1060.7 H.).
[42] The arrangement of the three books cannot be the original one, since in chapter I 1, vol. II, pp. 4.13–15 N., Hero refers to chapters II 8ff., p. 113 ff. N., as something preceding.

relevant details. It is precisely the systematic character of these accounts, however, that guarantees their communicative success,[43] that makes us, e.g., understand and remember the main structures of mechanical knowledge, that is, makes us accept this body of knowledge as codified. Note, however, how elements a, b and c are rather remote from practical concerns, that is, from the first act the words of mechanics can aim at.[44]

Deciding about Machines

As I claimed in the beginning, I believe that constructing machines is, perhaps at first glance surprisingly, not the only act ancient mechanical writing is aiming for. Some treatises contain so much material that does not contribute to actually constructing machines that something else must be on their authors' and readers' minds. It directly concerns application. One might, e.g., consider a section in Athenaeus' strange text that his most recent editors have dubbed 'bad practice' and that is replete with stories like the following: Athenaeus talks about 'sambucas', that is, scaling devices operated from ships (αἱ ἐκ τῶν πλοίων μηχαναί), and how they are often constructed badly. As an example, he relates the following little story:[45]

> Οἱ γὰρ ἐν τῇ περὶ Χίον πολιορκίᾳ, ἀστοχήσαντες καὶ μείζονας τῶν πύργων τὰς σαμβύκας κατασκευάσαντες, ἐποίησαν τοὺς ἀναβάντας ἐπ' αὐτὰς ὑπὸ τοῦ πυρὸς ἀπολέσθαι μὴ δυναμένους ἐπιβῆναι ἐπὶ τοὺς πύργους, χαλάσαι τε οὐκ ἦν οὐδενὶ τρόπῳ αὐτάς·... Διόπερ, μετὰ καὶ τῶν ἄλλων τοὺς μηχαναῖς μέλλοντας χρᾶσθαι τεχνίτας δεῖ μὴ ἀπείρους εἶναι τῶν ὀπτικῶν.

> The ones in the siege of Chios, having taken aim wrongly and having constructed sambucas that were higher than the towers, they caused the ones who went up on them to die from fire because they could not get to the towers and it was not possible in any way to lower them [i.e. the sambucas]... Therefore, among other fields, the technician who wants to use machines must not be inexperienced in optics.

The topic is familiar from, among others, Polybius, Biton and Hero. The first relates how Philip's army intended to make an early morning surprise attack on the town of Meliteia – and then discovered that the ladders they

---

[43] In an excellent paper, Tybjerg 2005 compares Hero with Philo and makes clear that Hero's ambition, at least partly, is less to describe inventions than to provide an integrated outlook on mechanics as a field (especially 211–13).
[44] That Hero has fared so badly with pioneering scholars such as Drachmann, Heiberg or Diels, who called him 'a real philistine' in 1893 ('einen reinen Banausen', quoted from Tybjerg 2005, 207), has simply to do with the fact that they only respected, in my terminology, acts one and four.
[45] Athenaeus, *Mech.* p. 27 f. W. = p. 56 W.-B.

had brought were much too short.⁴⁶ The second, in describing the *helepolis* of Posidonius the Macedonian, hints at the kind of knowledge one needs in order to safely avoid such deadly blunders.⁴⁷ The third mentions that only he who knows how to construct and operate his *dioptra* device will avoid such failures.⁴⁸ From Polybius' ninth book we know that, according to him, the *stratēgos* needs all kinds of non-military knowledge, especially geometry, astronomy, etc.⁴⁹ – knowledge that Polybius himself apparently does not have. What all three have in common is that the knowledge concerns not the act of constructing such machines but decisions about how to use them. Anecdotes are a natural means of such lessons about faulty constructions.

In his short section illustrating bad practice, Athenaeus also describes a strange device that he describes to the famous Ctesibius: a movable giant tube, mounted on a wagon, that can be raised to the height of any wall and inside which men can walk or climb, protected from the defenders' missiles. Athenaeus passes a predictably harsh, but interesting judgement on Ctesibius' invention:⁵⁰

> Γενναίου δὲ τοῦτο ἄξιον οὐθενός, ἀλλ' ἐκ θαυμάτων τὸ μηχάνημα συγκείμενον καὶ μάλιστα τὸν τεχνίτην τὸ θαυμάσαι.
>
> This machine does not have any genuine value, but is put together from marvels and mostly (in order) to marvel at the technician.

Who is supposed to marvel at the technician Ctesibius? It is probably not the city under siege (although one might compare Plutarch on the Roman general Marcellus, who allegedly marvelled at his opponent Archimedes).⁵¹ The notion of the technological 'marvel', despite its possible use in military matters, points towards a completely different area of ancient mechanical thought, namely the well-known branch of *automata* (and, partly, of *pneumatika*) and the social context of patronage and court culture.⁵² The main point of constructing automata is to impress the observer by suggesting

---

⁴⁶ Polybius, *Hist.* V 97.5 f. I have treated the story in context in Asper 2012.
⁴⁷ Similarly Biton, *Catap.* p. 52 W. = p. 70 M., stresses that in the construction of ἑλεπόλεις it is vital to master an optical method to make sure that the siege-tower has the right height.
⁴⁸ *Dioptra*, p. 190.14–21 in the edition by H. Schöne, *Heronis Alexandrini opera quae supersunt omnia*, vol. III (Leipzig), 1903.
⁴⁹ Polybius, *Hist.* IX 12–19. It is also interesting (and perhaps relevant to the ongoing discussion on the date of Biton) that Polybius does not seem to know any 'optical' method.
⁵⁰ Athenaeus, *Mech.* p. 31 W. = p. 58 W.-B.
⁵¹ The context, however, makes it quite clear that the ability to marvel at one's enemy is evidence for Marcellus' great character and thus is an exceptional stance.
⁵² Cf. Hero, *Spir.* I introd., vol. I, p. 2 ἐκπληκτικόν τινα θαυμασμόν. See the brief remarks of v. Hesberg 1987, 55.

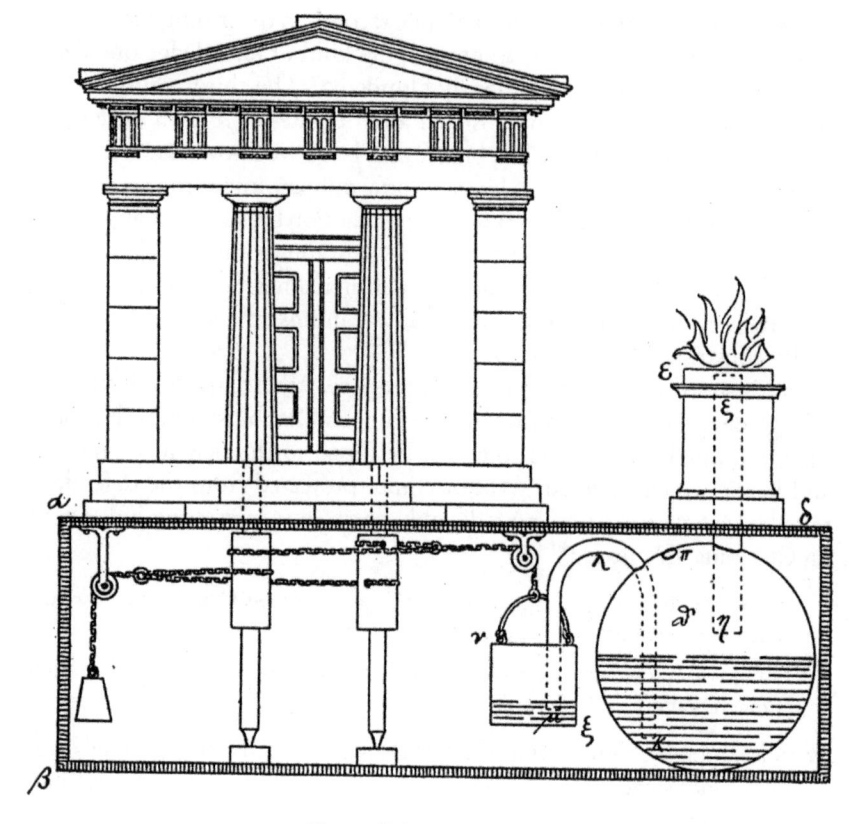

Fig. 3.1 (Diagram p. 176 S.)

that technological apparatuses can do something unexplainable – move by themselves or, in the case of Hero's theatrical arrangements (as described in his *Περὶ αὐτοματοποιητικῆς*), even perform little scenes from well-known tragedies. Many of these 'automatic performances' depict elements of public ritual or religious practice connected with large-scale symposia (sacrifice, etc.). The following is a typical arrangement, formatted as a mathematical problem:[53]

> Ναΐσκου κατασκευή, ὥστε θυσίας γινομένης τὰς θύρας αὐτομάτως ἀνοίγεσθαι, σβεσθείσης δὲ τῆς θυσίας πάλιν κλείεσθαι. Ἔστω ὁ προειρημένος ναΐσκος ἐπὶ βάσεως τῆς ΑΒΓΔ, ἐφ' ἧς ἐπικείσθω βωμίσκος ὁ ΕΔ· διὰ δὲ τοῦ βωμίσκου διώσθω σωλὴν ὁ ΗΖ.

[53] Hero, *Spir.* I 38, vol. I, p. 174 S.

> Construction of a little sanctuary, in such a way that, when sacrificial fire is lit, the doors open by themselves, and when the fire is extinguished, they close again. Let the mentioned sanctuary be mounted on a base ΑΒΓΔ, on which there sits also a little altar ΕΔ. Through the little altar there should be pushed a pipe ΗΖ.

Hero explains that when the fire is lit, the air in the pipe warms up and expands, and thus makes, through a transmitting mechanism that involves water, corresponding vessels, chains and axles, the doors of the sanctuary to turn. Clearly, the main point of the whole apparatus, which is probably of rather modest dimensions,[54] is not to *do* anything. Its only function, by hiding the purely mechanical causes of the unexpected movements, is to make the observer marvel at the sudden epiphany of a divine presence that opens the door when the ritual is at its peak. We know such scenes from Hellenistic hymns, especially from Callimachus' 'mimetic' hymns, where the divine presence is announced by such acts without visible agents.[55] Since both mechanical marvels and Alexandrian hymn point to the same courtly context, this coincidence cannot be chance. As has been stated before, we should thus understand such *automata* in the context of the representative practice of Hellenistic courts.[56] Some such marvels were, for example, part of the *pompē* of Ptolemy Philadelphus in 278 BC, famously described by Callixeinus, which procession also featured an automatic goddess Nysa.[57] Typically, these apparatuses performed elements of actions related either to symposiastic or religious aspects of festivals, both realms of courtly culture and its typical ways of searching for and displaying symbols of power.

While the purely military use of machines with which I started this chapter is much closer to our understanding of 'act', we should realize that both *automata* and *belopoiika* probably address the same circle of powerful people. These have an interest in two aspects of technology, and they are two sides of the same coin: the same power that is gained or defended by *belopoiika* one can represent within one's court circle by *automata*.

Obviously, the members of such circles will not themselves construct machines. They can, however, decide to allocate the means to have the latest catapults constructed for them; and, even more importantly, they

---

[54] See v. Hesberg 1987, 66–9. [55] See Callimachus, *Hymn.* 2, 1–4; 5, 2 f.
[56] Schürmann 1991, 249 f. Apollonius of Rhodes' description of mechanical devices in Aeetes' palace provides an interesting parallel (*Argon.* III 228–31, and v. Hesberg 1987, 59–65). I thus disagree with understanding *automata* in an educational context – compare Wilson 2008, 361; Tybjerg 2003, 449–51.
[57] Kallixeinos (FGrHist 627 F 2, quoted by Athenaeus); see v. Hesberg 1987, 50–2; Schürmann 1991, 37; on the procession itself, see Huß 2001, 321–3.

probably have to choose from a large group of available experts. They might, however, want to have some guidance on what machines to choose for what purposes, and how to generally decide matters that concern machinery, either in court contexts or in military matters. Taken in that way, the frequent practice of dedicating writing on mechanics to powerful figures makes perfect sense.[58] On the one hand, the expert authors of these treatises are, of course, in search of patronage. On the other, the specific 'act' the 'words' of such treatises are geared towards is less the construction of certain machines as qualified *decision-making* about such constructions and their uses. This is, for example, quite obvious in the case of the strategic manual of Aeneas the tactician,[59] and it is also implied by Polybius, whose advice on technological matters addresses not technicians but people with the power to decide. A similar point could be made with respect to medical writing that is often addressed to influential persons, e.g., by Galen. In the case of mechanical writers, the rhetorical 'mathematization' of their knowledge might be part of a corresponding strategy to secure a position higher up the social ladder[60] and thus closer to the centre of power and knowledge than mechanics used to be. Modern historians of technology have sometimes passed severe judgements on some of our authors,[61] at least partly because these modern readers did not realize that construction is not the only act the words of mechanical writing can aim at. To some extent, what they are actually aiming at is the act of decision-making about artillery at the highest political level.

Understanding Machines

While act one (construction of complex machines) and act three (helping powerful people to decide technical matters) hardly come as surprises, what I perceive to be the fourth act is certainly odd. Essentially, this is the attempt to *theoretically* understand machines that everybody uses in daily practice. Even more radically, the writers who produce the words that are supposed

---

[58] E.g. Philo dedicates his work to 'Aristo', Biton to 'king Attalus', Athenaeus to 'Marcellus', Vitruvius to 'imperator Caesar'. Although in some of these cases the addressee's identity is disputed, there is no question but that they have all been highest-ranking individuals. Compare, e.g., Galen's addresses to high Roman officials like Boethus.

[59] Compare, e.g., Aineias' advice on the disposition of troops (chapter 1), the selection of gate-keepers (Chapter 5), the pronouncement of κηρύγματα (chapter 10.3), or how to deal with mercenaries (chapter 13). See Whitehead 2001, 39–42, especially 41: 'the people, *whoever they were*, upon whom the polis' survival chiefly depended' (Whitehead's italics).

[60] Compare Plutarch on the relative merits of Archimedes' theoretical and practical, i.e. mechanical, achievements (*Vit. Marc.* 17, 307 C–D).

[61] E.g., Diels on Athenaeus' alleged 'Rokokocharacter' (1893, III; quoted at 17 W.-B.).

to lead readers to that act attempt to analyse and understand especially the most simple tools, such as a wedge, or the most ubiquitous tool-aided techniques, such as rowing, in theoretical terms, that is, by deduction from a general law and its application. What strikes us as odd, in comparative terms, is the desire to understand or to explain the principles of devices one uses every day.

The oldest extant treatise on mechanics, the pseudo-Aristotelian *Mechanics*, is the textual result of such an act of understanding. Despite its unclear authorship and date,[62] it is a fascinating piece for several reasons: first, it formulates a non- and probably pre-Archimedean 'law of the lever', and follows its applications in various tools and situations. Second, although at first glance its structure is nothing but a series of physical problems in the format 'Why is it that x is the case? Perhaps because x is y and y is the case' – that is, it exhibits the typical structure of so-called problem literature, a Peripatetic genre that perhaps goes back to the sophists and that gives explanations in the form of question and answer, and exhibits a syllogistic structure[63] – at a second glance, however, the text emerges as a rigorous deductive piece that attempts to derive all solutions of physical problems from one complicated mathematical-physical demonstration of how circles and their radiuses behave.[64] Third, as with many Peripatetic texts, there is a tangible desire at play to explain just everything in the world, in this case movements παρὰ φύσιν, and especially what is closest and best-known to us. Explanation here means deductive proof, based on more general truths – for example, the law of the lever. From the perspective of readers primarily concerned with acts one and three, this treatise will appear close to useless: it does not help the reader construct anything; nor is the knowledge concerned of interest to the wealthy and powerful. This is theory. Accordingly, the treatise stands out among mechanical writings (only Hero in his *Mechanics* and in general the writings discussed with respect to act two show some points of contact).

In order to illustrate the spirit and the structure of the text, a look at just one problem will be sufficient. Have we not always wondered why wedges work the way they do? This is why:[65]

Διὰ τί τῷ σφηνὶ ὄντι μικρῷ μεγάλα βάρη διίσταται καὶ μεγέθη σωμάτων, καὶ θλῖψις ἰσχυρὰ γίνεται; ἢ διότι ὁ σφὴν δύο μοχλοί εἰσιν ἐναντίοι ἀλλήλοις,

---

[62] Nobody, to the best of my knowledge, has suggested a compelling identification. Nonetheless, most modern critics take the text to be from the early post-Aristotelian Peripatos, perhaps written by Strato of Lampsacus. See Meißner 1999, 45 n. 49; Asper 2007a, 74 f.; skeptically, Bodnár 2011, 454.
[63] Asper 2007b, 74 f.   [64] See Schiefsky 2009, 50–3 on what he calls the 'reduction program'.
[65] Ps.-Aristotle, *Mech.* 17, 853 a 19–32.

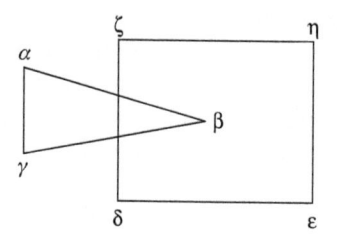

Fig. 3.2 (Diagram © Joyce van Leeuwen)

ἔχει δὲ ἑκάτερος τὸ μὲν βάρος τὸ δὲ ὑπομόχλιον, ὃ καὶ ἀνασπᾷ ἢ πιέζει. ἔτι δὲ ἡ τῆς πληγῆς φορὰ τὸ βάρος, ὃ τύπτει καὶ κινεῖ, ποιεῖ μέγα· καὶ διὰ τὸ κινούμενον κινεῖν τῇ ταχυτῆτι ἰσχύει ἔτι πλέον. μικρῷ δὲ ὄντι μεγάλαι δυνάμεις ἀκολουθοῦσι· διὸ λανθάνει κινῶν παρὰ τὴν ἀξίαν τοῦ μεγέθους. ἔστω σφὴν ἐφ' ᾧ ΑΒΓ, τὸ δὲ σφηνούμενον ΔΕΗΖ. μοχλὸς δὴ γίνεται ἡ ΑΒ, βάρος δὲ τὸ τοῦ Β κάτωθεν, ὑπομόχλιον δὲ τὸ ΖΔ. ἐναντίος δὲ τούτῳ μοχλὸς τὸ ΒΓ. ἡ δὲ ΑΓ κοπτομένη ἑκατέρᾳ τούτων χρῆται μοχλῷ· ἀνασπᾷ γὰρ τὸ Β.

Why do great weights and sizes of bodies split apart when a small wedge is applied, and why do they produce strong pressure? Probably because the wedge is two levers that work against each other, and each of the two has both weight and fulcrum which turns around and presses. In addition, the force of the blow makes a great weight that hits and moves. And because it moves what is moved fast, it is even more powerful. Although it is small, great forces emerge; thus, its movements escape our notice despite its sizable impact. Let there be a wedge ΑΒΓ and what the wedge is exerted upon, ΔΕΗΖ. So there is a lever ΑΒ, the weight is below Β, the fulcrum is ΖΔ. On the opposite side of this one, there is the lever ΒΓ. And ΑΓ upon which force is exerted, uses a lever on each of these two sides. For Β is the turning point (the fulcrum).

This passage explains how wedges work by analysing them as a co-ordinated pair of levers. At the beginning of the treatise, levers were, via balances, explained as an application of moving radiuses in concentric circles. The author demonstrates at the beginning (1, 848 b 3–5, 849 b 19–21), in a complicated, pseudo-mathematical proof, that, when two concentric circles are given, turning with the same torsional moment, a point on the circumference of the bigger one moves faster than a point on the circumference of the smaller, and that the difference in velocity is the same as the difference in the distance of these points from the centre of the circle. From this proposition, by introducing the notions of balance and lever, the author deduces a law of the lever in a dynamic, that is, non-Archimedean form.⁶⁶ Of the following thirty-two problems, sixteen analyse mechanical problems as applications of levers according to this law, that is, in the form of deductions. In many of them we find this strange pseudo-mathematical conceptualization of the bodies or tools involved: 'Let AB be the lever, C the *hypomochlion*', etc. – pointing to a diagram that we, in most cases, neither have nor need.

So these are the words – but where is the act? The author's ambition primarily focuses on subjecting phenomena to rigorous explanation. The rigour is deductive in the sense that Euclidean exposition is. Everybody, however, knows the phenomena and uses the devices explained. Nobody really needs an explanation of them, let alone a mathematical-physical one, at least not for practical reasons (as in act one). Steermen steer ships without wondering whether the rudder could be analysed as a lever (probl. 5). All of us have broken dry wood into smaller pieces using our knees (probl. 14) without ever worrying about levers, and will continue to do so. Dentists pull teeth with prongs not thinking of those prongs as crossed levers (probl. 21). After we have understood the explanation of the wedge, the reader cannot do anything she could not before, when she was using wedges all the time without ever dreaming of analysing them *more geometrico*. Especially with respect to the two most reputable fields of mechanics, that is, siege-engines and artillery and the construction of clever *automata*, the treatise would probably have provoked derision from the experts.⁶⁷ However, one might see a parallel between acts three and four in that they both rely on unveiling what was hidden:⁶⁸ the marvel only fulfils its proper function when spectators discover and understand the craft of the illusionist mechanic, the theoretical treatise which focuses on explanation, unveils the omnipresence of levers in our world which we use all the time without ever thinking about them and without even knowing theoretically the concept of levers.

This text follows Aristotelian concepts both of how to explain nature and of how mechanics and geometry are related.⁶⁹ Unlike the texts discussed with respect to acts one to three, here 'application' has acquired a completely

---

⁶⁶ *Mech.* 3, 850 a 39–b2: ὃ οὖν τὸ κινούμενον βάρος πρὸς τὸ κινοῦν, τὸ μῆκος πρὸς τὸ μῆκος ἀντιπέπονθεν. Compare Archimedes, *Equil. plan.* I 6, vol. I, p. 152.11–13 Heiberg: τὰ σύμμετρα μεγέθεα ἰσορροπέοντι ἀπὸ μακέων ἀντιπεπονθότως τὸν αὐτὸν λόγον ἐχόντων τοῖς βάρεσιν (prop. 7 is the same, but for ἀσύμμετρα μεγέθεα).

⁶⁷ See the anecdote about the Peripatetic Phormio lecturing *de omni re militari* to Hannibal, who responded *multos se deliros senes saepe vidisse, sed qui magis quam Phormio deliraret vidisse neminem* (Cicero, *De or.* II 18, 75).

⁶⁸ I am grateful to Brooke Holmes for suggesting this parallel.

⁶⁹ See, e.g., *An. post.* I 9, 76 a 23–5; I 13, 78 b 35–9.

different meaning: it means the instantiation of a general proposition concerning physics, we would say – of a natural law, in everyday contexts. This act of application is quite special because it means learning to see the world with the eyes of a physicist, or to analyse and explain the world around us. Understanding the principle of the lever and seeing it operative around us does not actually help us in any practical respect – just as knowing the four ways of logical opposition or understanding time does not usually help us in practical ways, to cite two typical Aristotelian examples.[70] Nor could one understand the 'act' the *Mechanics* is aiming for as didactic (unless one wishes to understand the *Categories* or the *Physics* as didactic). I have tried to describe above with act two how a useful didactic treatise on mechanics would usually be structured and carried out. There, to be sure, fundamentally theoretical knowledge, that is, insight into first principles, comes into play as the welcome peg on which an author can hang the whole structure of the knowledge concerned. Nonetheless, these principles are, in those treatises, worth knowing not for themselves, but because they lead to the construction of wonderful machines, such as the water-lifting screw or Hegetor's turtle (Vitruvius, *De arch.* X 6.1–4 and 15.2–7, respectively).

The reader finds nothing of this in the Ps.-Aristotelian *Mechanics*, which seems to aim for knowledge out of a fascination with insight *per se*. Perhaps this last act could be defined as 'analytic'. It owes its existence to the fact that it is simply a pleasure to understand the world around us, an ever increasing part of which are machines.

Instead of a Conclusion: Further Questions

I have distinguished four acts that the words of ancient Greek mechanical knowledge appear to be aiming at. For each act, I have tried to identify texts that are devoted mostly to this one. Admittedly, especially when reading the more comprehensive treatises of mechanics, first of all, Vitruvius, *De arch.* X, perhaps also Pappus, *Coll.* VIII, one might consider them as working all four acts, or at least more than just one of the four, into a whole. Nonetheless, even in Vitruvius one ends up with the impression that there is a hierarchy of acts intended.

These four acts are certainly not specific to mechanical literature. What I said about Ps.-Aristotle also applies, I believe, to most genuine Aristotelian works. Act two, on the other hand, is the defining feature of a whole genre of literature, the *technē*, that begins probably with the sophists, is in full

---

[70] *Cat.* 11 b 17–19 (although I am ready to admit, as Stephen Menn reminds me, that Aristotle makes use of that theory within his dialectics, e.g. in the *Topics*); *Phys.* IV 10, 217 b 29–224 a 17.

swing in imperial times and still dominates modern notions of handbooks, introductions, etc. Nonetheless, in mechanics, and especially in its subfields artillery and siege-craft, the problem of application is more serious than in most other fields, perhaps with the exception of medicine.

When we consider the social reality of ancient mechanical knowledge in practice, which must have consisted of groups of migrating experts concentrated at courts, powerful cities, etc., not unlike the *sophistai* lining up at the court of Croesus (Herodotus I 29.1) or fifth-century sophists somehow always ending up at Athens,[71] perhaps the path from words to acts is less of an issue than the fact that there are so many words at all. The main practice of ancient mechanics cannot have been writing. Accordingly, most of the inventors' names given in Athenaeus or Vitruvius do not refer to them as authors. Intense competition among Greek expert culture has, however, as in so many other fields, resorted to texts as vehicles of authorial success over distances of place and time.[72] At least with respect to act one, the texts would have been supplementary in comparison with constructive practice; when it came to matters of reputation, patronage and competition, the respective role of the two might have been different. Perhaps Hero of Alexandria hints at that fact when he describes his personal perspective on the literary history of writings about machines:[73]

> Ἐπεὶ οὖν οἱ πρὸ ἡμῶν πλείστας μὲν ἀναγραφὰς περὶ βελοποιϊκῶν ἐποιήσαντο, μέτρα καὶ διαθέσεις ἀναγραψάμενοι, οὐδὲ εἷς δὲ αὐτῶν οὔτε τὰς κατασκευὰς τῶν ὀργάνων ἐκτίθεται κατὰ τρόπον οὔτε τὰς τούτων χρήσεις, ἀλλ᾽ ὥσπερ γινώσκουσι πᾶσι τὴν ἀναγραφὴν ἐποιήσαντο, καλῶς ἔχειν ὑπολαμβάνομεν ἐξ αὐτῶν τε ἀναλαβεῖν καὶ ἐμφανίσαι περὶ τῶν ὀργάνων τῶν ἐν τῇ βελοποιΐᾳ, ὡς μηδὲ ἴσως ὑπαρχόντων, ὅπως πᾶσιν εὐπαρακολούθητος γένηται ἡ παράδοσις.

> My predecessors have composed a great number of detailed treatises on artillery, including measurements and designs, but not one of them sets out the construction of machines in order, nor do they set out their uses, but it is as if they have written down the details for all the readers who already know [about artillery]. Therefore I think it appropriate to take over from them and give a clear description of machines in artillery, perhaps even of those that do not exist any more, in such a way that my account may be easy for everyone to follow.[74]

---

[71] Again, the above-quoted story in Vitruvius (X 16.3–8) provides some clues: Diognetus and Callias publicly compete in Rhodos for the position of chief architect. The latter wins with an epideixis involving models which are later confuted by dire reality.

[72] At least occasionally Vitruvius refers to authors as his sources (e.g. X 13.3 Diades). See Asper 2007a, 35–42 in general, 160–73 with respect to the somewhat untypical field of mathematics, 351–6 for medicine, especially Galen.

[73] Hero, *Bel.* p. 73.6–11 W. = p. 18 M.   [74] Translation follows closely 19 M.

In Hero's writings, the account is apparently meant to be non-supplementary and thus not address expert readers; at least, this is what Hero says *pro domo*.

Nonetheless, even with such a narrative in mind, taking mechanics from professional experts to ruling πεπαιδευμένοι, some problems remain. How are we supposed to understand the detailed references to certain war machines by the name of the engineer and often even by historical context in many texts? Take, for instance, the *sambuca* of Damius of Colophon, the *gastraphetes* of Zopyrus of Tarentum, the stone-thrower of Charon of Magnesia, etc., all described by our authors with detailed references to dimension, materials, construction and even appearance. Where did this information come from? How did these acts turn into words or numbers at all? Perhaps there were groups of connoisseurs who would remember principles of construction or even individual solutions to general problems by name and occasion, such as, among modern gun-lovers, the names 'AK-47' (*Awtomat Kalaschnikowa, obrasza 47*) or 'Colt Government' (Colt M1911) evoke not only a certain artifact and its historical context and effects, but an individual technical solution of general problems. All of this must, unfortunately, remain speculation.

REFERENCES

M. Asper 2007a, *Griechische Wissenschaftstexte: Formen, Funktionen, Differenzierungsgeschichten* (Stuttgart)
 2007b, 'Medienwechsel und kultureller Kontext: Die Entstehung der griechischen Sachprosa' in J. Althoff (ed.), *Philosophie und Dichtung im antiken Griechenland* (Stuttgart), 67–102
 2009, 'The Two Mathematical Cultures of Ancient Greece' in E. Robson and J. Stedall (eds.), *The Oxford Handbook of the History of Mathematics* (Oxford), 107–32
 2011, '"Frame Tales" in Ancient Greek Science Writing' in K.-H. Pohl and G. Wöhrle (eds.), *Form und Gehalt in Texten der griechischen und der chinesischen Philosophie* (Stuttgart), 91–112
 2012, '"True" and "False" Errors in Ancient (Greek) Computation' in M. Geller and K. Geus (eds.), *Productive Errors: Scientific Concepts in Antiquity*. Max Planck Institute for the History of Science (Berlin), 47–66
 2013, 'Making up Progress – in Ancient Greek Science Writing' in M. Asper (ed.), *Writing Science: Medical and Mathematical Authorship in Ancient Greece* (Berlin), 411–30
 2015, 'Explicit Knowledge' in M. Hose and D. Schenker (eds.), *A Companion to Greek Literature* (London), 401–14

I. Bodnár 2011, 'The Pseudo-Aristotelian Mechanics: The Attribution to Strato' in M.-L. Desclos and W. W. Fortenbaugh (eds.), *Strato of Lampsacus: Text, Translation, and Discussion* (New Brunswick), 443–55

S. Bogen 2013, 'Diagrammatic Reasoning: The Foundations of Mechanics' in M. Asper (ed.), *Writing Science: Medical and Mathematical Authorship in Ancient Greece* (Berlin), 279–98

A. Broadie 1968, 'The Practical Syllogism', *Analysis*, 29: 26–8

H. M. Collins 2010, *Tacit and Explicit Knowledge* (Chicago, London)

R. Cummins, et al. 2001, 'Systematicity and the Cognition of Structured Domains', *Journal of Philosophy*, 98: 167–85

S. Cuomo 2007, *Technology and Culture in Greek and Roman Antiquity* (Cambridge)

R. I. Curtis 2008, 'Food Processing and Preparation' in J. P. Oleson (ed.), *The Oxford Handbook of Engineering and Technology in the Classical World* (Oxford), 369–92

F. Fuhrmann 1960, *Das systematische Lehrbuch: Ein Beitrag zur Geschichte der Wissenschaften in der Antike* (Göttingen)

H. V. Hesberg 1987, 'Mechanische Kunstwerke und ihre Bedeutung für die höfische Kunst des frühen Hellenismus' in *Marburger Winckelmann-Programm*, 47–72

W. Huß 2001, *Ägypten in hellenistischer Zeit: 332–30 v. Chr.* (Munich)

P. Keyser 2013, 'The Name and Nature of Science: Authorship in Social and Evolutionary Context' in M. Asper (ed.), *Writing Science: Medical and Mathematical Authorship in Ancient Greece* (Berlin), 17–62

W. Kullmann 1991, 'Platons Schriftkritik', *Hermes*, 119: 1–21

M. J. T. Lewis 1999, 'When Was Biton?', *Mnemosyne*, 52: 159–68

B. Meißner 1999, *Die technologische Fachliteratur der Antike: Struktur, Überlieferung und Wirkung technischen Wissens in der Antike* (Berlin)

R. Netz 2009, *Ludic Proof: Greek Mathematics and the Alexandrian Aesthetic* (Cambridge University Press)

M. Polanyi 1966, *The Tacit Dimension* (Chicago) (2nd edn 2009 with a foreword by A. Sen)

K. Saito 2006, 'A Preliminary Study in the Critical Assessment of Diagrams in Greek Mathematical Works', *SCIAMVS*, 7: 81–144

K. Saito and N. Sidoli 2012, 'Diagrams and Arguments in Ancient Greek Mathematics: Lessons Drawn from Comparisons of the Manuscript Diagrams with Those in Modern Critical Editions' in K. Chemla (ed.), *The History of Mathematical Proof in Ancient Traditions* (Cambridge University Press), 135–62

M. Schiefsky 2009, 'Structures of Argument and Concepts of Force in the Aristotelian Mechanical Problems', *Early Science and Medicine*, 14: 43–67

A. Schürmann 1991, *Griechische Mechanik und antike Gesellschaft: Studien zur staatlichen Förderung einer technischen Wissenschaft* (Stuttgart)

S. Shapin 1984, 'Pump and Circumstance: Robert Boyle's Literary Technology', *Social Studies of Science*, 14: 481–520
K. Tybjerg 2003, 'Wonder-Making and Philosophical Wonder in Hero of Alexandria', *Studies in History and Philosophy of Science*, 34: 443–66
   2005, 'Hero of Alexandria's Mechanical Treatises: Between Theory and Practice' in A. Schürmann (ed.), *Geschichte der Mathematik und der Naturwissenschaften in der Antike. Vol.* III: *Physik/Mechanik* (Stuttgart), 204–26
J.-P. Vernant 2006, 'Some Remarks on the Forms and Limitations of Technological Thought among the Greeks' in J.-P. Vernant, *Myth and Thought among the Greeks*. Trans. J. Lloyd. (New York) (orig. title *Mythe et pensée chez les Grecs*, Paris 1965), 299–318
J. R. Welch 1991, 'Reconstructing Aristotle: The Practical Syllogism', *Philosophia*, 21: 69–88
D. Whitehead 2001, *Aineias the Tactician: How to Survive under Siege*. 2nd edn. (Bristol)
A. I. Wilson 2008, 'Machines in Greek and Roman Technology' in J. P. Oleson (ed.), *The Oxford Handbook of Engineering and Technology in the Classical World* (Oxford), 337–66

CHAPTER 4

Si qui voluerit
Vitruvius on Architecture as 'the Art of the Possible'

Elisa Romano

In an essay many years ago I drew attention to the fact that Vitruvius, in his *De architectura*, represents architecture's 'birth' as distinct from its 'origin'. 'Birth' refers to the abstract process of theoretical foundation (*unde architectura nascatur*), while the 'origin' is the actual process of historical beginnings (*unde origines aedificiorum sint institutae*).[1] Vitruvius was thus well aware of the double aspect of his discipline as both practice and theory, and explicit formulations in his treatise warrant modern attempts to reconstruct in it two different 'archaeologies' of architecture, one historical and one theoretical. By an 'historical archaeology' I mean the first developments of architecture as a set of building techniques, whereas a 'theoretical archaeology' is the process of founding architecture as an intellectual practice. We may approach this latter, theoretical analysis by treating Vitruvius' work as thought about the theoretical possibilities of architecture.[2] Some of the questions posed by the editors of this volume may find a first possible answer in the space that separates these two aspects of architecture: this space can create distance or, on the contrary, it can be a site of overlapping and intersection.

Architecture between Practice and Theory: From Imitation to the Text

Viewed historically, building techniques find their first impulse and achievements in the ability to imitate first nature, then the actions of other men. The hut, symbol of architecture in its primal stages, is the result of imitation that has been made possible by aptitude for learning (*docilitas*):

> It was then that some of them from these first groups began to make shelters of foliage, others to dig caves at the foot of mountains and yet others to build

[1] Vitruvius 2,1,8–9.   [2] Romano 1987, 183–5.

refuges of mud and branches in which to shelter in imitation [*imitantes*] of the nests of swallows and their way of building. Next, by observing [*observantes*] each other's shelters and incorporating the innovations of others in their own thinking about them [*adicientes suis cogitationibus res novas*], they built better kinds of huts day by day. Since men were naturally imitative and quick to learn [*imitabili docilique natura*], they would show each other the results of their building, proud of their own inventions, and so, sharpening their wits in competition [*exercentes ingenia certationibus*], became more competent technically every day. (2,1,2–3)[3]

On the other hand, viewed theoretically, architecture can be seen as the application of written rules that have been fixed in a text, in a *corpus* of precepts.[4] However, this theoretical aspect does not exhaust the intellectual engagement of the architect, as is shown by the phrases *adicientes suis cogitationibus res novas* and *exercentes ingenia certationibus* in the passage above, referring to the application of intelligence to practical, historical problems.

In comparison with other fields of knowledge that in late Republican times were on their way to definition as autonomous intellectual practices, such as law or oratory, architecture has a peculiar status, for it is not limited to a theoretical level or to the pragmatic level of political action, as is the case with law and oratory, but inescapably involves a translation onto the level of manual and material execution. In the programmatic first chapter of Book 1, devoted to defining the architect's knowledge, Vitruvius underlines the indissoluble co-existence of a practical and a theoretical component in that knowledge (*fabrica* and *ratiocinatio*, or elsewhere, with a different wording, *opus* and *ratiocinatio*):[5]

> The practical aspect [*fabrica*] consists in the continued and consummate exercise of experience [*continuata ac trita usus meditatio*], through which a work is manually produced according to a fixed design, while theoretical reflection [*ratiocinatio*] accounts for, and gives demonstration of, what is produced. So architects who have tried to achieve manual ability without an education [*sine litteris*] were not successful; by contrast, those who have relied only on theory and book-learning [*ratiocinationibus et litteris solis*] have

---

[3] The English translation of all the passages of the *De architectura* I shall quote in the footnotes of the present chapter is taken from Schofield 2009, except for 1, 1, 1–2 and 1, 1, 16. On the importance of *docilitas*, cf. Gros 1997, xxxviii.
[4] From an historical point of view, this is a later development, marking the shift from a stage in which practical action is regulated by what J. Assmann has defined as 'mimetic memory' to the stage of its textual codification: Assmann 1997, xv f.
[5] See Vitruvius 1,1,15.

Architecture and the Art of the Possible 55

pursued a shadow rather than reality; and only those who have mastered both aspects have reached their goal with authority. (1,1,1–2)[6]

Vitruvius' model is the Hellenistic architect Hermogenes, the prototype of the architect who both created buildings and gave later generations of architects written works from which to learn the principles of the discipline: 'It is clear that Hermogenes created spectacular buildings with impressive and penetrating ingenuity and has left written sources from which posterity could derive the theory underlying these disciplines' (3,3,9).[7] To what extent did Vitruvius himself succeed in achieving the model he proposed, i.e. in combining theoretical ability with technical skill? To ask this is also to ask: what relationship between 'words' and 'acts' emerges from the only ancient treatise on architecture that has survived to the modern age?

The answer scholars tend to offer is mostly negative. In the field of Vitruvian studies, even the most recent ones, this relationship has commonly been understood, and therefore investigated, as the possible correlation between the treatise and our archaeological evidence. As a consequence, the stress has usually been put on the assumed distance between the *De architectura* and the contemporary reality of architecture. Both in its descriptive and in its normative component, the treatise has mainly been interpreted as the anachronistic reflection of an architectural stage that was already in the past at the time of writing, and its author as a conservative closed to the innovations of his own age.[8] According to this critical point of view, Vitruvius relies on book learning, and offers ideal and abstract models

---

[6] The constitution of the text and its interpretation are controversial; I give here my reading and translation. See also Romano in Gros et al. 1997, 65–7, Romano 2005, 88–90. For a recent analysis, based on philosophical categories, of the so-called 'couples philosophiques' in Vitruvius (e.g. *fabrica/ratiocinatio*), cf. Viola 2006, 119–36.

[7] Hermogenes, one of the masters of Eastern Hellenism, who flourished between the end of the third century and the first quarter of the second century BCE, is one of Vitruvius' main models; it was long wrongly assumed that he was his only source, but this hypothesis has been completely refuted by the thorough study of Gros 1978.

[8] Vitruvius took almost no part in the process of transforming Rome: this fact is now beyond dispute, and there are no more attempts to downplay it or evade it by assuming a higher chronology for the treatise. On the contrary, analysis of Books 3 and 4, which have the most references to contemporary Roman buildings, allows us to date these two books securely to between 27 and 23 BCE: cf. Gros 1976, esp. 15–52, Gros in Callebat et al. 1999, xlix–lx. This fact has been linked to Vitruvius' implicit disapproval of the innovations used in the building activity of the Principate, due not to ideological traditionalism, but to the fact that they diverged from the rules of Hellenistic architecture that he took as a model. His limited participation in the new monumentalization of Rome is matched by the pre-Augustan image presented by the *De architectura*, which is anachronistic in the context of the years in which the treatise was composed: cf. Gros 1990, lxxiv–lxxxv. Wallace-Hadrill 2008, 144–210 has reacted against the common portrayal of Vitruvius as a conservative, due to his not being identifiable as a spokesman of the Augustan 'revolution' as we usually understand it, and has stressed the theoretical significance of his project, which, following Augustan cultural guidelines, gave importance to the Italic tradition. On Vitruvius' 'conservatism', see also König 2009, 40–1

that have no correspondence to actual reality. The most evident contradiction that has been noted pertains to Vitruvius' own autobiographical experience. When in Book 5 he presents the basilica in its ideal features (5,1,4–5), he seems to split his personality between himself as a writer and as a designer–builder. He proposes different solutions from those which he himself had carried out in the basilica of Fanum and which he will illustrate a little later,[9] creating an apparent inconsistency in the text.

The basilica in Fanum departs from Vitruvius' set of rules about basilicas, and this dichotomy between the normative solutions prescribed for basilicas and the ones carried out in fact has been seen as a divergence between 'words' and 'acts'.[10] However, this is to compare two non-homogeneous series: on the one hand, the text, and, on the other, a series of (attested or reconstructed) objects. Given the complexity of such a text as the *De architectura*, so stratified and rich in inconsistencies and internal oscillations, such a reading cannot yield unequivocal conclusions.

Instead of this sort of 'external' point of view, it seems more interesting to adopt, as it were, an 'internal' one. Such an internal point of view is not much concerned with possible 'objective' correspondences between the treatise and external reality, but instead involves the way Vitruvius regards the relationship between, on the one hand, the conceptual moment of research and writing and, on the other, the practical moment of execution in the field of architecture; that is, instead of merely noting the divergence between words and acts, between descriptions/rules and actual buildings, the goal is to trace the relation between them.

From 'Words' to 'Acts' within the Text of the *De architectura*

In a comprehensive – and so far unsurpassed – essay published in 1997, Pierre Gros underlined the delicate and complex balance between the normative and the descriptive aspects of the *De architectura*:

> the normative aspects of Vitruvius' discourse, on which so much stress is laid in the seven books about the art of building (*aedificatio*), do not aim

---

and n. 31, who tends to reduce the significance of this aspect in comparison with the author's self-legitimating strategy, because, according to her, 'Vitruvius is interested not so much in supporting Augustus and the project of Roman imperialism as he is in elevating himself'.

[9] Vitruvius 5,1,6.

[10] As far as concerns this dichotomy, Corso in Gros et al. 1997, 541 speaks of a 'dicotomia tra dire e fare'. These divergences are listed and discussed in Corso in Gros et al. 1997, 531, Saliou 2009, l–li, 140–1. Cf. also Pizzigoni 2010. It is interesting to note how Vitruvius himself had, in a sense, foreseen the charges of detachment from contemporary architectural reality that might be addressed to him. In 5,5,7, for example, he anticipates the objection of an imaginary interlocutor about the absence in Rome of theatres equipped with voice resonators identical to the ones he describes: 'perhaps someone will say [*dicet aliquis forte*] that plenty of theatres are built in Rome every year in which no consideration at all has been taken of these rules'.

so much at outlining a theory of architecture, as at promoting a correct practice: Vitruvius' purpose is to express what he calls *potestas artis*, i.e. all the potentials contained in the builder's *techne*, at the levels of both carrying out (*opus*) and planning (*ratiocinatio*).[11]

From the first, programmatic chapters of Book 1, Vitruvius repeatedly indicates *opus* as the ultimate aim of the architect's culture itself, and as the touchstone for testing the technical articulations of architecture. The architect's education is aimed at practice, *ad faciendum*, as is explicitly stated in 1,1,16: 'to undertake those works that are to be carried out through manual ability and the exercise of technique is for those who have been trained in only one technical discipline aimed at practice [*ad faciendum sunt instituti*]'. It may also be noted that, in order to stress the crucial importance of practical execution, Vitruvius resorts to linguistic signs that are very simple and, consequently, all the sharper: consider the frequency of the noun *opus* and the verb *facere*, which mark a clear thread in the treatise from its first lines. Vitruvius, indeed, ends the preface to Book 1 by offering Augustus 'well-defined rules collected together', so that the *princeps*, by paying attention (*adtendens*),[12] 'can personally get to know the qualities of the works (*opera*) that have been already built and those that are still to be built' (1, *praef.* 3). In chapter 1,1, which deals with the architect's education, the necessity of possessing many different kinds of knowledge is demonstrated in relationship to the actual execution of the *opus*: from drawing, which will be useful for depicting the *opus* in the planning stage, to knowledge of *historiae*, which helps when explaining the origin of figurative subjects represented *in operibus*, up to ethical philosophy, which teaches the moral integrity without which no *opus* can be carried out:

> Then he must have the expertise in drawing which will enable him to represent more easily the appearance of the work he wishes to design [*quam velit operis speciem*] with painted models... he should also have a wide knowledge of history because architects often devise a great deal of ornament for their buildings [*in operibus*], the meaning of which they must be able to explain to those who ask why they have made them... philosophy in fact makes the architect high-minded and ensures that he will not be arrogant, but rather flexible, fair and trustworthy without greed, which is the most important quality, since no work [*nullum opus*] can be carried out satisfactorily without loyalty or integrity. (1,1,4–5; 7)

The *opus* is also the point of reference for the definition of the categories of architecture: *ordinatio* consists in adapting the individual elements of an

---

[11] Gros 1997, xix.
[12] On the frequent usage of the participle *adtendens* by Vitruvius and its meaning, cf. Gros 1997, xxv.

opus to the right measure, *dispositio* is the appropriate placing of elements and the elegant execution of the *opus* (1,2,2), and so on.

The connection between text and *opus* is also forged at a more subtle level. Vitruvius more than once underlines the link between his treatise and operational action. In 2,8,8, for example, after illustrating the different kinds of masonry and their advantages and flaws as regards resistance and durability, he observes that 'if someone, on the basis of this treatise [here defined as *commentarii*], wants [*si qui voluerit*] to choose a kind of masonry, he will be able to judge its durability'. The phrase *si qui voluerit* is a formula that Vitruvius uses to address his reader while assigning him undefined features by means of an indefinite pronoun.[13] Elsewhere the definition of the addressees remains generic but becomes more specific, as in the case of the *aedificantes*, who will be able to choose their building material on the basis of Vitruvius' instructions: *de copiis quae sunt necessariae in aedificiorum comparationibus . . . uti non sint ignota aedificantibus exposui* (2,10,3). In the preface to Book 6 the identity of these builder–addressees is further specified: they are *patres familiarum*, family men, who, because of the general incompetence of architects, choose to rely on a handbook (this is the meaning of the expression *litteratura confirmati*) and to direct the construction of their own houses in person:

> When I see that a knowledge of such a fundamental discipline is boasted of by the uneducated and unskilled, and by those who are not only ignorant of architecture but know absolutely nothing of construction, I cannot but praise the heads of families who are so confident in what they have read about architecture that they build for themselves [*litteraturae fiducia confirmati per se aedificantes*], reckoning that if they must trust someone unskilled, then they themselves are more entitled to spend sums of money to satisfy their own, rather than other people's, desires. (6, *praef.* 6)

It has often been pointed out that the *De architectura* has several levels of address,[14] but, whether it addresses Augustus or an unspecified reader, whether generic builders or family men, the same pattern of literary communication can be discerned: a triangle whose three sides link the author (and/or his text), the addressee[15] and the *opus*, the work to be built on the

---

[13] Cf. also 4,3,3, quoted below. Vitruvius uses the conditional link *si qui* (recurring sixteen times, plus two instances of *ne qui*) more often than *si quis*, only twice attested. On this preference, probably reflecting a usage of colloquial language, cf. Romano 1997, lxxxix.

[14] On the addressees of the *De architectura*, cf. Romano 1987, 174–83, Fleury 1990, xxxiii–xxxvi, Gros 1997, xxiv–xxvii, Fögen 2009, 113 and n. 20.

[15] Vitruvius (like, a few decades later, Celsus in his *De medicina*) avoids using the second person in speaking to his addressee: on this 'authorial choice' in the context of the construction of a self-image, cf. Hine 2009, especially 28–30.

basis of the written text, of the *litteratura*. An undefined addressee, once more hidden behind the formulaic protasis *si qui voluerit*, appears again in a very interesting passage of the *De architectura*: in Book 4, after dealing with the Ionic order in Book 3 and the Corinthian in 4,1, Vitruvius goes on to treat the Doric order and declares that theorists of Hellenistic architecture have long refused to build temples in the Doric order. He nonetheless states that he will explain its rules, addressing them to 'anyone who wants to build' sacred buildings of that order in future:

> Some ancient architects denied that temples should be built with the Doric order because false and unpleasant modular systems were generated in such buildings: that is what Tarchesius said, but also Pytheos, and especially Hermogenes... However, as the sequence of our argument requires, we shall explain this system here as we have learnt about it from our masters, so that anyone wishing [*si qui voluerit*] to proceed following these rules in this way will have at his disposal a full explanation of the proportions with which he can bring sacred temples in the Doric style to completion correctly and without errors. (4,3,1; 3)[16]

This is a unique case in the treatise: although the Doric order was in decline, Vitruvius describes its proportions and the rules regulating its ratios, striving to solve the modular problems that had compromised their applicability. He does so in order to meet his builder–reader halfway, and so to help his addressee, who in no other passage of the work plays such a decisive role as here.[17] On the one hand, he is responding to his addressee's possible desire to renew a prestigious, if declining, style; on the other hand, he is declaring that, thanks to the teachings of his *praeceptores*, he is capable of overcoming the difficulties posed by the Doric order (to be precise, he proposes to solve the so-called 'Doric corner conflict' in the Doric frieze by means of the half-metope, 4,3,5).[18] This illuminating example – a truly key one – is of special importance: it shows that the *De architectura* is not just a description and classification of attested monuments but is intended as a theoretical contribution to future buildings, and that, in this planning dimension, the addressee plays a central role.

Pierre Gros, while stressing 'the prevalence of the expository system over real-life experience', has nonetheless observed that 'it happens, more often than one might think on a first consideration, that Vitruvius proposes his

---

[16] On the decline of the Doric order, cf. Gros 1992, xx–xxi, Corso in Gros et al. 1997, 459.
[17] My observations about this passage (Romano 1987, 175–6) have been shared by Gros 1992, xx–xxi, 126.
[18] On this complex problem, cf. Gros 1992, xxi, 134–7.

own ideas and extends normative systems that he seems to have worked out on his own to edifices that did not traditionally partake in them'. The latter is true of Vitruvius' treatment of the Doric order (4,3,1–3), but one might also consider how he treats *opus reticulatum*, to take just one other example. He raises queries about the resistance and durability of this building technique, denouncing its flaws and noting contra-indications of its use, and in this he anticipates a phenomenon that will actually take place: *opus reticulatum* was in fact abandoned in the following decades, and it has been suggested that this process may have been affected by the theoretical reflections of the last decades of the first century BCE, and in particular by Vitruvius' body of precepts.[19]

Applying the Text: Translating Knowledge beyond the Workshop

The text of the *De architectura* thus offers some concrete clues that Vitruvius' aims include the possible application of his precepts. We should now ask ourselves: a) what form does the idea of applicability take in the text?; and b) what means does Vitruvius offer his addressee (identified as reader–builder) so that the rules he sets out can be applied?

A first answer could be found in a traceable compositional layer within the treatise. Here I am thinking of those passages in which the author gives detailed sequences of clear-cut explanations, collected together in a paratactic and enumerative style, with in some cases the implied presence of drawings at the bottom of the roll (*volumen*) that was to contain each of the ten books of the work.[20] We can define these sets of directions, and the diagrams that accompanied them, as a layer of 'operating instructions' that hint at a workshop – or better, in this case, building-site – experience.[21] I shall not treat this topic at length, not only because it is very well known, but also because on this issue an important insight has already been achieved (here I am thinking, once more, of Gros' contributions). In constructing his own identity as an 'expert' who places his knowledge at others' disposal, Vitruvius aims to go beyond the sphere of the workshop (or building-site). In this – convincing – interpretation, he tends to avoid

---

[19] Cf. Gros in Callebat et al. 1999, lii–lv, 109–13.
[20] Cf. e.g. Vitruvius 3,3,13: *quae apud Graecos entasis appellatur, in extremo libro erit forma . . . subscripta*; 3,4,5: *item in extremo libro forma et demonstratio erit descripta*; 3,5,8: *de volutarum descriptione . . . in extremo libro forma et ratio earum erit subscripta*; 5,5,6: *haec autem si qui voluerit ad perfectum facile perducere animadvertat in extremo libro diagramma musica ratione designatum*; see also 5,4,1; 8,5,3; 9, *praef.* 5; 8.
[21] Cf. e.g. Vitruvius 1,6,6 and 3,3,7.

the multiplication of prescriptions, because that would be to fall back into the restricted sphere of what we could call, in a more modern term, the 'secrets' of the corporation. He would lose touch with his interlocutors, who were those new figures who, in the last decades of the first century BCE, were taking charge of building activities as individuals (politicians, though also private citizens). The typological sample-book that Vitruvius is trying to collect in the form of his *corpus* represents precisely the attempt to go *beyond* the stage of prescription.[22]

This compositional layer thus exhibits a residual character: it by no means exhausts Vitruvius' interest in the applicability of his own precepts. We could class it as a basic level at which his normative system can be translated into application, a level that merely concerns the execution of instructions.

A second possible answer looks to a more advanced level of 'translatability', one that demands the performance of all the potentials of the architect's *ars*. Applicability, at this level, requires a theoretical frame of reference, in which, above all, the epistemological value of mistakes emerges.[23]

Using an argument pattern that is commonplace in didactic literature (including didactic poetry), which points out others' mistakes and warns the addressee how to avoid them, Vitruvius disapproves of architects who have adopted a number of columns on the sides double that at the front of a peripteral temple:[24]

> But in peripteral temples the columns should be placed so that there are twice as many intercolumniations down the sides as there are on the front; in that way the length of the building will be twice its breadth. For those who have simply doubled the number of columns have clearly gone wrong [*erravisse*] because the length is evidently greater than it should be by one intercolumniation. (3,4,3)

To take just one other example, he suggests how to prevent the inconveniences caused by the fragility of concrete masonry through the expedient of metal reinforcements:

> But anyone who wants to avoid this pitfall [*si qui noluerit in id vitium incidere*] should leave a cavity in the middle behind the outer facing-stones,

---

[22] The general role played by sketches and diagrams in the treatise will also, according to this view, need to be put back into perspective. Cf. Gros 1996.
[23] As Sharrock 2005 stresses with regard to Ovid's *Ars amatoria*, the acknowledgment of one's own mistakes is extremely rare among ancient didactic writers. In the tradition of didactic prose, too, a mistake is usually recognized only when it has been made by others; Sharrock cites Vitruvius' passage about the city of Mytilene as an example.
[24] On this critical assertion by Vitruvius, cf. Corso in Gros et al. 1997, 325.

and build walls two feet thick with courses of red dressed stone, fired brick or ordinary hard stone in the cavity: then the outer faces should be bound to these with iron clamps and lead. In this way the wall, which has not been built with just a pile of material but in courses, will last indefinitely without defect. (2,8,4)

There is an extreme type of mistake, namely those that derive from the attempt to carry out the impossible, illustrated by the 'impossible cities' described in the treatise. Salapia, a city of Apulia, had been founded in marshland and therefore had an unhealthy climate; its inhabitants ultimately requested a better site for their city (1,4,12). Mytilene, on the island of Lesbos, is too exposed to the winds, which reveals the first mistake to be avoided in planning the road network of a city:

> So it seems best to avoid this fault and make sure that what often happens [*vitandum videtur hoc vitium et advertendum ne fiat*] in a number of cities does not recur. For example, Mytilene, on the island of Lesbos, is a town that was built magnificently and elegantly but orientated unwisely. When the south wind blows in the city, men fall ill: when the north-west wind blows, they return to good health but cannot stand about in the side streets and avenues because of the biting cold. (1,6,1)

A further example is the project of a city on Mount Athos, whose plan the architect Dinocrates had proposed to Alexander the Great. Dinocrates had designed a city in the form of a male statue, with the city walls in the left hand, and in the right a bowl that would collect the water of the streams on Mount Athos, so that it would from there be poured into the sea. However, it would have been impossible to supply with food except through difficult shipments by sea, leading Alexander to criticize the plan on grounds of judgement: *Dinocrates, inquit, adtendo egregiam formae compositionem et ea delector, sed animadverto si qui deduxerit eo loci coloniam fore ut iudicium eius vituperetur.* Thus the reader learns that the architect must use good judgement (*iudicium*) or he will be subject to criticism.[25]

---

[25] 2, *praef.*, 3:

> Alexander, entranced by the idea behind the project, immediately asked if there were any fields in the vicinity which could maintain the city with a regular supply of corn. When he discovered that it would be impossible without transporting it across the sea, he said, 'Dinocrates, I appreciate the ingenious design of the project and am delighted with it, but I note that the judgement of someone who was to found a colony there would be heavily criticized. For just as a newborn baby cannot grow without the nurse's milk nor be led up the steps of its growing life, so a city without fields and their produce arriving inside the walls cannot expand or support a large number of inhabitants without an abundance of food, or maintain its population without a good supply. Therefore, though I believe your design has much to recommend it, at the same time I regard the site as unsuitable.

Architecture as the Art of the Possible

While the unfeasible is a source of mistakes to be avoided, Vitruvius also encourages his readers to judge what is possible, to consider which work could actually be achieved, even if it has not yet been implemented or, as in our key example concerning the Doric temple at 4,3,3, is no longer implemented,[26] or to consider kinds of buildings that are usually built in one place but not (or not yet) in others. A case in point is the eustyle temple, recommended by its usefulness, beauty and solidity, qualities duly underlined by Vitruvius, even though there are no instances of that sort of temple in Rome:

> Now we must give an explanation of the principles of the eustyle temple, which is much the most praiseworthy and incorporates a system of design which is excellent with respect to function, appearance and stability... We have no example of this type in Rome, but at Teos in Asia there is the hexastyle Temple of Father Liber. (3,3,6; 8)

Or palaestrae, which the author describes even though they are not part of Italian tradition: 'Now, despite the fact that the construction of gymnastic complexes is not an Italian custom, I think it appropriate to discuss them and show how they are organized by the Greeks, since the method of construction has been handed down to us' (5,11,1). The potential space for new developments in architecture, either for restoring unusual forms or for importing techniques employed outside Rome and Italy, is thus found in the gap between what is, or is lacking, in the present context, and what might be, or exists elsewhere. In passing, it is interesting that this gap often coincides with the divergence between Greek and Roman practices and hints at the 'us/them opposition' which Andrew Wallace-Hadrill has recognized as one of the main lines of Vitruvius' cultural project;[27] to the examples of palaestrae and the eustyle temple more could be added.

However, the architect's ability to intervene is not limitless: the gap cannot always be filled. An instance of this is *opus latericium*, or brick masonry, whose many advantages Vitruvius praises: such advantages are attested by, among other things, its use in royal buildings, such as Croesus' palace or the Mausoleum of Halicarnassus, where it was preferred to more precious materials such as marble. In Rome, however, the height of buildings and the laws limiting the thickness of walls to less than what is required by brick walls make the use of *opus latericium* impossible:

> But I shall explain why this kind of structure should not be built by Romans in the City, not forgetting the causes and reasons for it. Public laws forbid

[26] Cf. p. 59 above.   [27] Wallace-Hadrill 2008, 150–1.

that walls thicker than a foot and a half be built on public land; other walls should be constructed of the same thickness lest the interior spaces become too narrow. But brick walls a foot and a half thick cannot support more than one storey, yet they can if they are two or three bricks thick. Yet given the enormous size of the City and the great density of its population, it is essential to provide innumerable houses. Accordingly, since one-storey housing could not provide living-space for such a mass of people, this very fact forces us to resort to high-rise buildings. So tall buildings constructed with stone piers, walls of fired brick and rubble party walls, and tied together by a great number of wooden floors, provide partitions of great practical use. This is why the Roman people have excellent housing because of the upward proliferation of many storeys of various configurations without obstruction within the city walls. (2,8,16–17)

The architect will thus have to adapt himself to necessity and choose concrete structures (*opus caementicium*). In my opinion, this ability to adapt oneself to circumstances constitutes the main criterion that determines the implementation of Vitruvian precepts: adaptation to the nature of the soil and climate of the place when planning to build, adaptation to the resources available in the area, adaptation to the needs of those for whom edifices are built.

Sometimes, for instance, necessity demands something that should be rejected at the theoretical level, such as the use of wattle walls, i.e. walls with an easily inflammable wooden framework, which it would be better not to use. Nonetheless, as one is often compelled to use it for the sake of speed or due to lack of funds or the limitations imposed by the site, Vitruvius explains how to carry it out:

> But I wish that half-timbering had never been invented: despite its advantages with respect to speed of construction and space-saving, it is equally prone to cause devastating and widespread disasters since it is predisposed to burn like torches... But since lack of time or money or the need for partition-walls in spaces without structural support force many people to use half-timbering, it should be made like this. (2,8,20)

I shall give some more examples. The architect will have to consider the inconveniences of the pycnostyle temple, in particular the lack of comfort caused by the placing of the columns so close to each other that women are compelled to walk in single file by the narrow space:

> Both these kinds [pycnostyle and systyle] present practical difficulties because mothers ascending the stairs to make supplications cannot pass through the intercolumniations arm in arm but must go single file: and again the view towards the folding doors is obstructed and the cult statues

themselves are hidden because the columns are so close together: also, walks around the temple are impeded by the narrowness of the spacings. (3,3,3)

The choice of the site for a theatre is made according to the health of the audience, and the architect's ingenuity is needed in choosing it:

> Once the forum has been laid out, one must then select the healthiest possible site for the theatre for the spectacles of the games on the festival days of the immortal gods... This is because people sitting with their wives and children for some time through the games are captivated with pleasure, and their bodies, remaining immobile with enjoyment, present open pores into which wafts of air penetrate, which would infuse their bodies with noxious vapours if they come from marshy or other unhealthy areas. And so if the site of the theatre is chosen very carefully, problems like this will be avoided [*Itaque si curiosius eligetur locus theatro, vitabuntur vitia*]. We must also be careful that it is not exposed to the south... This is why one must at all costs avoid areas which have disadvantages like this, and select healthy ones. (5,3,1–2)

In the orchestra of the Roman theatre one will have to plan a space for the senators' seats, not contemplated in Greek theatres: 'in the orchestra are the places reserved for the seats of the senators' (5,6,2). Moreover, porticoes will have to be built behind the *scaena*, so that the audience can seek shelter in case of sudden rain. Here again it is a professional quality of the architect that is needed to solve the problem:

> Porticoes must be built behind the stage-set so that when sudden showers interrupt performances, people will have somewhere to shelter away from the theatre and there will be plenty of space for the preparation of the stage-equipment; such, for example, are the porticoes of Pompey, and again, those of Eumenes in Athens... and in other cities, which had very punctilious architects [*diligentiores habuerunt architectos*], there are porticoes and open walkways around the theatres. (5,9,1).[28]

In 5,6,7, there is a clear statement that the necessity to adapt, and to be aware that rules cannot have universal application, is a professional skill of the architect:

> These modular systems, however, cannot cater for the requirements and desired effects in all theatres, but the architect must be alert [*sed oportet architectum animadvertere*] to the proportions he should follow to conform to the modular system and those which should be modified to suit the character of the site or the size of the building [*ad loci naturam aut magnitudinem*

---

[28] Cf. also 4,5,1–2; 4,8,6.

operis temperari]. For there are some components which, for practical purposes, must be made the same size in both small and large theatres, such as the steps, the transverse passageways, the parapets, the passages, the ramps of stairs, the actors' platform, the boxes for the magistrates and anything else that crops up which necessarily requires some divergence from modularity lest the functioning of the theatre should be interfered with. Again, if some shortage of materials occurs [*exiguitate copiarum*] during construction, for example, of marble, wood or any other supplies, it will not be inappropriate to add or subtract something so long as this is not done too rashly, but with common sense. This will happen if the architect has practical experience and also is not devoid of mental flexibility and technical skill [*si architectus erit usu peritus, praeterea ingenio mobili sollertiaque non fuerit viduatus*].

This paragraph summarizes the factors that can lead to adaptation on the architect's part: the character of the site (*loci natura*), the size of the building (*magnitudo operis*), the possible shortage of materials (*exiguitas copiarum*). At the same time, it contains another crucial implication: it is in this adaptability that the architect proves both his practical experience and his intelligence, technical skill and intellectual qualities. In other words, adaptability to 'what is possible' achieves the blending of theoretical aptitude and practical ability, of *ratiocinatio* and *fabrica*, that Vitruvius has set himself as a model, and so is one of the theoretical requirements of the architect's *scientia*.

In sum, we can say that architecture too, just like politics according to a famous definition, is 'the art of the possible'.

REFERENCES

J. Assmann 1997, *La memoria culturale: Scrittura, ricordo e identità politica nelle grandi civiltà antiche* (Turin) (orig. title *Das kulturelle Gedächtnis: Schrift, Erinnerung und politische Identität in frühen Hochkulturen*, Munich 1992)

L. Callebat, P. Gros and C. Jacquemard 1999, *Vitruve: De l'architecture livre II*, texte établi par L. Callebat, introduit et commenté par P. Gros; recherche sur les manuscrits et apparat critique par C. Jacquemard (Paris)

Ph. Fleury 1990, *Vitruve: De l'architecture livre I*, texte établi, traduit et commenté par Ph. Fleury (Paris)

Th. Fögen 2009, *Wissen, Kommunikation und Selbstdarstellung. Zur Struktur und Charakteristik römischer Fachtexte der frühen Kaiserzeit* (Munich)

P. Gros 1976, *Aurea templa: Recherches sur l'architecture religieuse de Rome à l'époque d'Auguste* (Rome)

1978, 'Le Dossier vitruvien d'Hermogénès', *MEFRA*, 90: 687–703

1990, *Vitruve: De l'architecture livre III*, texte établi, traduit et commenté par P. Gros (Paris)

1992, *Vitruve: De l'architecture livre IV*, texte établi, traduit et commenté par P. Gros (Paris)

1996, 'Les Illustrations du *De architectura* de Vitruve: Histoire d'un malentendu', in P. Gros et al. (eds.), *Les Littératures techniques dans l'Antiquité romaine. Entretiens sur l'Antiquité classique* 42 (Vandoeuvres-Genève), 19–44

1997, 'Vitruvio e il suo tempo', in P. Gros, A. Corso and E. Romano (eds.), *Vitruvio, De Architectura* (Turin), ix–lxxvii

P. Gros. A. Corso and E. Romano (eds.) 1997, *Vitruvio, De Architectura* (Turin)

H. M. Hine 2009, 'Subjectivity and Objectivity in Latin Scientific and Technical Literature' in L. Taub and A. Doody (eds.), *Authorial Voices in Greco-Roman Technical Writing* (Trier), 13–30

A. König 2009, 'From Architect to Imperator: Vitruvius and his Addressee in the *De Architectura*', in L. Taub and A. Doody (eds.), *Authorial Voices in Greco-Roman Technical Writing* (Trier), 31–52

V. Pizzigoni 2010, 'La basilica di Fano: la sola architettura nota di Vitruvio' in H. Burns, F. P. Di Teodoro and G. Bacci (eds.), *Saggi di Letteratura Architettonica, da Vitruvio a Winckelmann*, vol. III (Florence), 251–69

E. Romano 1987, *La capanna e il tempio: Vitruvio o dell'architettura* (Palermo)

1997, 'Fra astratto e concreto: la lingua di Vitruvio', in P. Gros, A. Corso and E. Romano (eds.), *Vitruvio, De Architectura* (Turin), lxxix–xcv

2005, 'Il difficile rapporto fra teoria e pratica nella cultura romana', in F. Bessone and E. Malaspina (eds.), *Politica e cultura in Roma antica* (Bologna), 81–99

C. Saliou 2009, *Vitruve: De l'architecture livre V*, texte établi, traduit et commenté par C. Saliou (Paris)

R. Schofield 2009, *Vitruvius, On Architecture*, trans. R. Schofield with an introduction by R. Tavernor (London, New York)

A. Sharrock 2005, 'Those Who Can, Teach: Ovid's *Ars Amatoria* and Contemporary Instructional Writing' in C. Reitz and M. Horstner (eds.) 2005, *Wissensvermittlung in dichterischer Gestalt* (Wiesbaden), 243–263

A. Viola 2006, *Vitruve: Le savoir de l'architecte* (Paris)

A. Wallace-Hadrill 2008, *Rome's Cultural Revolution* (Cambridge University Press)

CHAPTER 5

Caesar's Rhine Bridge and Its Feasibility in Giovanni Giocondo's Expositio pontis *(1513)*

Ronny Kaiser

Prefatory Notes and Considerations

It has been claimed that Caesar described the construction of the Rhine bridge so accurately in Book IV, Chapter 17 of his *Commentarii de bello Gallico* that a model of it might easily be built.[1,2] This is a curious proposition, as it suggests that Caesar supplies a kind of technical instruction for how the (wooden) bridge over the Rhine was constructed. That the about 160 words contained in the chapter do not justify this conclusion is not at issue here, although such should be pointed out.[3] Nevertheless, it is an interesting position, in that it raises the question of just what function and effect Caesar intended this chapter to have. This problem was also confronted at the beginning of the sixteenth century by the humanist Giovanni Giocondo, whose interpretation of the bridge chapter constitutes the focus of this chapter.

In order to understand Giocondo's treatment of the bridge chapter properly it will first be necessary to place Caesar's description of the Rhine bridge as well his *Commentarii* in their historical context. Caesar's stay in Gaul fell within the 50s BC and was generally characterized by military conflicts with the Gallic and Germanic tribes. His *Commentarii*, of which he himself wrote the first seven books, Aulus Hirtius the last, essentially describe the domestic political disturbances in Gaul and Caesar's military activity there. The Roman commander Caesar, who time and again puts

---

[1] The treatise was originally published in Giocondo 1513b. I have unfortunately not been able to see this edition and thus quote instead from the 1544 Basle edition of Heinrich Glarean: Giocondo 1544b in Glarean 1544a.
[2] See Maurach 2003, 104: 'Die Konstruktion der Holzbrücke über den Rhein beschreibt er [Caesar] so genau, daß man sie im Modell recht leicht nachbauen kann.'
[3] In general, scholars have identified two reasons for Caesar's crossing of the Rhine: on the one hand, the will to power of the future dictator, who went to Gaul specifically in order to conquer it (see Holmes 1913, 80–6, 138–68); and, on the other hand, as a reflex of current Roman politics (see Mensching 1975, especially 12; Maier 1978, 55–75).

his own military and strategic accomplishments in relief, aims above all to describe his military measures as necessary for the maintenance and protection of Roman rule.

The discussion of Caesar's authorial intention necessarily raises the nearly insoluble problem of identifying the audience he envisioned for his *Commentarii*. Officers and magistrates should probably be considered first,[4] as such readers were keenly interested in geopolitical conditions, the disposition of masses of material and people, and the tactical deployment of military units. Yet with their tendentious contents the *Commentarii* certainly also served Caesar's self-fashioning needs. Thus it can be assumed that he had in mind his own partisans as well as his political opponents,[5] both of whom he sought to present with a carefully sculpted image of his military and strategic accomplishments. Directly connected with the question of audience is the problem, often discussed in scholarship, that no certainty can be achieved regarding the *Commentarii*'s precise date of composition, their mode of publication or the possibility that they were ever publicly read.[6] However that may be, it is likely that Caesar intended his work to be received by the leading political actors in the city of Rome, including both friends and enemies. Indeed, scholars tend to view the *Commentarii* as less a piece of literature than of contemporary political propaganda. Decisive is the perspective of the Roman proconsul and general, whose deeds secure for him a privileged position in the account. In the narration of these deeds Caesar shrinks from no adornment, exaggeration or abridgement.[7] His opponents he portrays as overly mighty, conniving or menacing, in order to portray himself as an even greater and more glorious conqueror. In light of these considerations it can be assumed that Caesar sought to influence politics in the city of Rome with his *Commentarii*.

In addition to his two expeditions to Britain in 55 and 54 BC, Caesar's two crossings of the Rhine were a novelty in Roman military history; according to the few sources we have for the Republican period, he was the first Roman general east of the river.[8] By crossing the Rhine, Caesar left the (Greco-Roman) ecumene[9] and entered completely new territory. This is a decisive point, as the Rhine crossing must be interpreted against the backdrop of Roman politics. The triumvirate founded by Pompey, Crassus and Caesar in 60 BC threatened to break down soon thereafter,

---

[4] See Rüpke 1992, 211.   [5] See Rüpke 1992, 223.   [6] See Rüpke 1992, 201 ff.
[7] See Urban 1989, 244.   [8] See Mensching 1975, 12.
[9] In antiquity 'ecumene' was used as a geographico-social term referring to the known, inhabited world (see Schmitt 2000).

thus occasioning the pact of Lucca in 56 BC. This agreement provided for the renewal of the coalition through the determination of common political goals and the safeguard of the balance of power between the three parties.[10] Caesar emerged the strongest force. As the only one to hold both an office and a military command he saw to it that Crassus and Pompey, both now without an army, would enjoy the election victory they desired. Thus both were chosen consuls for the year 55 BC.[11] This electoral victory provides the political background for Caesar's first crossing of the Rhine. Caesar was clearly determined to remain a strong presence in Roman politics. An extraordinary deed – such as crossing the Rhine – was the perfect opportunity. Although we have no contemporary sources for the reaction the Rhine crossing elicited in Rome, we can surmise that Caesar had the ways and means to communicate the event's cultural and ideological import in Roman politics and to translate it into political power. It can thus be assumed that the exploit won great respect in Rome and that Caesar's political presence continued to be felt despite his physical absence.

The description of the first bridge's construction appears in the fourth book of the *Bellum Gallicum* (and is thus chronologically located in 55 BC). It forms part of the narration of Caesar's first Rhine crossing, which can be divided into four parts: Caesar's arguments for his enterprise (*BGall.* IV, 16), a description of the bridge's construction (*BGall.* IV, 17), Caesar's safeguards in the occupied territory east of the Rhine (*BGall.* IV, 18) and the Roman retreat back to Gaul (*BGall.* IV, 19). First, Caesar justifies the crossing of the Rhine: *cum videret Germanos tam facile impelli ut in Galliam venirent*.[12] At the same time he seeks to demonstrate the capabilities and the daring of the Roman people.[13] In doing so, he also highlights his own capabilities and daring, thus setting himself apart in particular from his antagonist in Rome, Pompey. Furthermore, he relates his efforts to make the Sicambri surrender the cavalry of the Usipetes and the Tencteri, which had retreated to Sicambrian territory.[14] To justify their rejection of this request, the Sicambri claim the River Rhine as the border of the Roman Empire's sphere of influence.[15] In this way Caesar legitimizes his actions in

---

[10] See Oppermann 1968, 92. [11] See Maier 1978, 56 f. [12] Caesar, *BGall.* IV, 16, 1.
[13] Cf. ibid. [14] Cf. ibid., 2–4.
[15] Ibid., 4: *Ad quos cum Caesar nuntios misisset, qui postularent eos qui sibi Galliaeque bellum intulissent sibi dederent, responderunt: populi Romani imperium Rhenum finire; si se invito Germanos in Galliam transire non aequum existimaret, cur sui quicquam esse imperii aut potestatis trans Rhenum postularet?* 'To them Caesar sent envoys to demand the surrender of the men who had made war upon himself and Gaul. They replied that the Rhine marked the limit of the Roman empire: if he thought it unfair that the Germans should cross into Gaul against his will, why did he claim any imperial power across the Rhine?' (trans. taken from Edwards in Caesar 1994, 199).

Gaul thus far and presents the Roman reader with the new borders of the Roman Empire, which he himself has extended. In addition, he portrays himself as a dutiful Roman proconsul who has taken pains to rid Gaul of the dangers caused by Germanic incursions, required the surrender of enemies and brought succour to Roman allies, namely the Ubii, who had given hostages.[16]

Immediately before describing the construction of the first bridge Caesar considers the Ubii's suggestion of crossing the Rhine with pontoons, ultimately reaching the conclusion that *navibus transire neque satis tutum esse... neque suae neque populi Romani dignitatis esse*.[17] It is likely that Caesar intends here an implicit comparison with other great commanders known to his intended audience.[18] Ancient literature reports famous instances of large and important rivers having been crossed with bridges of interconnected boats, thus offering a point of contrast to Caesar's statement. For example, Herodotus (484–425 BC) narrates how the Persian King Darius I (550–486 BC) made a pontoon bridge over the Bosphorus out of Achaemenid boats in order to cross with his army against the Greeks in 490 BC.[19] Similarly, according to Diodorus Siculus, Alexander the Great crossed the Hellespont by connecting ships.[20]

Thus two considerations are central for understanding Caesar's passage. First, great commanders prior to Caesar used bridges made of connected boats to cross into foreign cultural areas. By consciously and explicitly rejecting such a course of action as incommensurate with Roman and his own *dignitas*, Caesar contrasts himself with predecessors like Darius and Alexander and, in this respect, sets himself above them. Second, Darius and Alexander left their own respective cultural zones and entered another one by crossing the Bosphorus and the Hellespont. Hence Caesar is implicitly in a similar situation when on the verge of crossing the Rhine, also a civilizational border. But Caesar would like to mark the act of crossing into foreign cultural territory with a grave and prestigious deed indicative of the cultural superiority of the Romans, as well as of Roman and – of course – Caesarian dignity. Therefore he highlights the difficulties involved in constructing the bridge as a strategy to enhance the glory of his own achievements. Despite all dangers and difficulties he considers the construction of a solid bridge the only respectable option for crossing the Rhine.[21]

---

[16] See Trzaska-Richter 1991, 122. [17] Caesar, *BGall.* IV, 17, 1.
[18] See Rüpke 1992. [19] Cf. Herodotus IV, 85.
[20] Cf. Diodorus Siculus, *Univ. Hist.* XVII, 17, 1–3. [21] Cf. Caesar, *BGall.* IV, 17, 2–10.

At the same time, Caesar gives proof of his excellence as a commander by citing safety reasons for his decision not to cross the Rhine by means of a pontoon bridge (*BGall.* IV, 17, 1). Caesar's own reputation is equally at stake, if not more so, in the second argument he makes against pontoons: as we have seen, he states quite explicitly that this solution accords with neither his own nor the Roman people's dignity. Thus from the outset Caesar moves himself into the focus of his narration: the construction of a sturdy bridge would increase the glory of the Roman people to be sure, but above all – and this is directly announced to the Roman reader – it would increase Caesar's own glory. His dignity and military capabilities are set into relief by his swift construction of a wooden bridge. The description of the bridge's construction serves, therefore, to highlight his own virtues, as well as to give the impression that Roman interests are commensurate with his own.

The actual description is not as detailed as we might like, perhaps because Caesar presupposed that the construction of such a bridge for military purposes was familiar to his readers.[22] It is possible that he therefore describes only what is new in his method. Whatever the case may be, he employs the technical language of the architects of his time. Since there are no other extant accounts of bridge construction from classical or late antiquity, Caesar's short description is difficult for modern readers to understand and has led to various theories about the bridge's appearance.[23]

Perhaps the passage has seemed even more impressive to the modern reader on account of its technical vocabulary and compressed syntax, which make a quick reading and translation quite difficult. Yet it is doubtful that Caesar describes the bridge so accurately that it could easily be recreated, as Maurach suggests, considering that the description itself does not exceed the length of one short chapter. The form and content of the text and the chapter's place in the larger literary context suggest that Caesar did not

---

[22] Especially because a detailed description of a well-known technique would have been unsuitable for the character of his *Commentarii*.

[23] H. J. Edwards has also drawn attention to this problem:

> If it were reckoned along the river-bed, it would mean that the successive pairs of balks would incline inwards at different angles as the bridge approached mid-stream from each bank: for it is not to be supposed that the width of the roadway varied. We are not told at what angle the balks inclined, nor at what height above the waterline the transoms were, nor how far each pier or trestle was from the next. It is highly probable that, in a military bridge of this kind, for the sake of rapidity in construction, a uniform profile was followed for the part of each pier or trestle visible above the water-line. (Caesar 1994, 627)

Perhaps the best way to understand Caesar's description of his bridge is in comparison with the wooden bridges depicted on Trajan's Column (see Deissmann 1991, n. 241, 597).

envision practical applications for his description. Instead, he sketches for the reader a literary outline of certain aspects of the bridge's construction. His goal in doing so is to demonstrate the technical superiority of the Roman pioneers,[24] to bring to the fore his own glory and dignity, and even to depict the triumph of discipline and civilization over (Germanic) barbarism.[25] Thus the description intends primarily to reduce deeds to words and was probably not aimed at prompting further deeds.

In the short treatise *Expositio pontis*, however, the Renaissance architect Giovanni Giocondo reverses this textual polarity, treating Caesar's work as if its words could form the basis for further deeds. In this manner the treatise, which was part of Giocondo's larger edition of Caesar's *Opera omnia* (published by Aldus Manutius in Venice in 1513), effects a practical and theoretical transformation of Caesar's chapter dealing with the Rhine bridge. In what follows I shall describe this treatise and identify the various strategies of transformation it employs, i.e. the means by which Giocondo adapts the ancient reference text to his own contemporary circumstances, thereby inverting the original relationship between theory and practice.

I use the term 'transformation' here with reference to a model for exploring processes of intercultural change that has been developed in the Collaborative Research Center 644 'Transformations of Antiquity' at the Humboldt-Universität zu Berlin. Eschewing the concept of passive reception, the transformation paradigm focuses especially on the productive reciprocity between a reference area (in this case Roman antiquity) and a – for lack of a better word – reception area that adopts, adapts, appropriates, assimilates or otherwise integrates the reference area into its own cultural matrix. This productive reciprocity is manifested in specific processes of transformation and is governed by (personal or non-personal) agents and certain media. The name given to the productive reciprocity of cultural transformation is 'allelopoiesis'.[26]

Humanistic editions and commentaries on ancient authors are prime examples of the media in and through which transformations of antiquity are initiated and conducted. Their annotations and paratexts transform ancient texts according to the cultural and intellectual interests of the reception area, adapting ancient knowledge to the reading habits and the needs of the new circle of recipients. In this way ancient knowledge is reorganized, negotiated and developed in novel directions. Humanistic commentaries perform a very important function in the context of cultural

---

[24] See Maurach 2003, 103.   [25] See Koutroubas 1972, 13.
[26] See Bergemann et al. 2011, 43–5. For the concept of 'transformation', see also Böhme 2011.

transformations of antiquity, as they explicitly seek to couple ancient texts with the state of knowledge and the intellectual exigencies of the reception area.[27]

In light of its graphical and textual arrangement, Giocondo's treatise on Caesar's Rhine bridge can be seen as a kind of commentary whose purpose is to explicate and structure the ancient knowledge found in the edition of Caesar's works according to a very specific interest. This kind of commentary, which is typical of how humanists appropriate and transform ancient texts, is the subject of what follows. The main focus is on the question of how Caesar's chapter on the Rhine bridge is read and transformed by Giocondo. To this end, I shall above all explore the visual and textual appropriation strategies through which the ancient text is removed from its original context and read within a humanist discourse of theory and practice. In this way, Giocondo's reading of Caesar's chapter will emerge not only as an act of reception and transformation but also – and more relevantly for the subject of this volume – as an important example of an ancient text being applied in such a way that the relationship between words and deeds is clearly affected.

Giovanni Giocondo as Architect and Humanist Scholar

Although Fra Giovanni Giocondo (1433/5–1515)[28] was one of the most famous architects of his time, we have surprisingly little information about his life. Like his older contemporary, Leon Battista Alberti, Giocondo may have come to architecture by way of his attempts at interpreting the technical writers of antiquity, especially Vitruvius.[29] For the first part of his life we only have information about his humanistic studies.[30] And these studies seem to have followed him for the rest of his life, as his editions of ancient authors after 1500 attest.

In 1489–93 Giocondo was active in Naples as an architect for the duke of Calabria, who would become King Alfonso II of Naples in 1494. The duke often sent him to different parts of the realm to inspect and assess ancient remains; some drawings made during these missions have survived.

---

[27] Scholarship on the meaning and transformative function of humanistic editions and commentaries on ancient literature is still in its infancy. For humanistic editions of ancient literature, see Holtz, Schirrmeister and Schlelein 2012; for commentaries as media of transformation, see Parker 1993; Most 1999; Pade 2005; Hafner and Völkel 2006; Weichenhan 2011a and Weichenhan 2011b; Kaiser 2013; Enenkel and Nellen 2013; Enenkel 2014; therein: Kaiser 2014a; Kaiser 2014b.
[28] Unsurpassed and still fundamental for a biographical overview is Brenzoni 1960.
[29] See Willich 1992, 64. He also points out (65) that Giocondo's precise religious affiliation is uncertain: in some sources he is described as a Dominican, in others as a Franciscan.
[30] See Willich 1992, 65.

In years previous, Giocondo seems to have pursued his humanistic studies throughout Italy, as is suggested by the collection of Roman inscriptions he dedicated to Lorenzo Magnifico de' Medici in 1489.[31] Furthermore, he was Greek and Latin tutor to the young Julius Caesar Scaliger (1484–1558). Between 1496 and 1499 Giocondo was invited to France by Louis XII. He spent part of this period at the Château d'Amboise on the Loire,[32] where a large colony of artists and artisans from Italy had gathered. There his salary was equivalent to that of an illuminator, thus making him one of the best remunerated of his countrymen. His unusual title 'deviseur des bastiments' suggests that the term 'architect', as Alberti had established it in Italy, was new to France.[33]

Uncertainty reigns about the ten years Giocondo spent in France. His best-attested work is probably the bridge of Notre-Dame in Paris; the construction documents show that he provided the design, supervised the stonework and was responsible for the levelling.[34] Giocondo also worked in the Republic of Venice. In 1506 he was appointed to build fortifications, which work influenced his contemporaries.[35] In addition, Giocondo is known to have devised a plan for the Rialto bridge. It was probably rejected for reasons of cost, and the bridge was built according to the simpler design of Scarpagnino instead. In 1513 Giocondo had the honour of succeeding the deceased Bramante as one of the builders of the new St Peter's. The choice of the scholar-theorist shows the high value that was accorded a humanistic education. Almost certainly the fame he earned with his *editio princeps* of Vitruvius (first published in Venice in 1511) was decisive. In Rome he and Raphael (1483–1520) worked as equal *capomaestri*.[36]

Scholarship has tended to concentrate on Giocondo's architectural achievements. In contrast, I shall focus in what follows on his humanistic studies, which have largely been ignored.[37] These resulted, for example, in his outstanding editions of Vitruvius, Pliny the Younger's letters (1508), Sallust's *Opera* (1509), Frontinus' *De aquaeductibus* (together with Vitruvius, 1513), Julius Caesar's works (1513), and Cato, Varro, Columella and

---

[31] See Willich 1992, 65.
[32] Miron Mislin indicates the period from 1497 to 1498 (see Mislin 2006, 2225).
[33] See Willich 1992, 65.
[34] See especially Mislin 2006, who reconstructs Giocondo's architectural efforts in Paris. Cf. also Willich 1992, 66. Of the many palaces and residences that are associated with Giocondo's name, only a few can be attributed to him with any probability.
[35] See Willich 1992, 66.   [36] See Willich 1992, 66.
[37] To my knowledge, only Brenzoni summarizes Giocondo's work as an editor, but without ascribing any importance to it (see Brenzoni 1960: L'attività umanistica di Fra Giocondo, 75–101). This seems to be a serious lacuna in the scholarship.

Palladio (all together in 1514).³⁸ Giocondo's humanistic studies, which are mirrored precisely in these editions, clearly reflect a particular interest in those ancient texts that have a historical and technical focus.

Giocondo's Treatise on Caesar's Rhine Bridge

The edition of Caesar which Giocondo published in 1513 spans about 300 folio pages and contains Caesar's *Commentaries on the Gallic War* as well as the *Commentaries on the Civil War*. A dedicatory letter to Giuliano de' Medici II (1479–1516) is prefixed to the edition, in which Giocondo complains in particular about the insufficient recognition and recompense accorded to textual editors.³⁹ The whole preface gives voice to the general difficulties of the editor and, especially, to Giocondo's own difficulties in procuring and dealing with the manuscripts relevant for the present edition. To claim glory by highlighting these difficulties,⁴⁰ he admits, is a standard component of humanistic rhetoric and self-presentation; on the other hand, it is necessary precisely because the contribution of editors is not widely recognized.⁴¹ Giocondo understands his edition of Caesar's *Opera* as an approximation of the ancient original; he seems to be aware of his inability to restore the ancient text definitively.⁴²

In this respect his focus is on the philological reconstruction of the text and thus on the revival of antiquity. The *praefatio* is followed by other paratexts placed before the works of Caesar, each supplemented with preceding or succeeding images: the edition includes maps of and texts on ancient Gaul and Spain, images of and short texts on the Rhine bridge, Avaricum, Alesia, Uxellodunum and Massilia and *indices* of names of Gallic and Spanish cities, places and peoples.⁴³ It is noteworthy that in

---

³⁸ Giocondo 1508, 1509, 1511, 1513a, 1514.
³⁹ Giocondo 1544a, a 2ʳ–a 4ʳ, here: a 2ʳ: *Si diligentius quis consideret, Iuliane illustriſſime, quod is, qui corrupta antiquorum scripta, ut emendata in manus hominum exeant, curas, labores exhauriat: quam vero nullius, vel perexiguae admodum apud plurimos laudis particeps fiat, admiretur profecto, cur sibi quisquam id oneris assumat, quo in perferendo: quum maxime enitendum sit, minimam tamen mercedem consequatur*. 'Illustrious Giuliano, considering closely that whoever devotes care and labour to making corrupt works of ancient authors available to the public in improved versions gains little or no fame, it is a wonder why anyone would take this burden upon himself; although the greatest effort must be expended, only the slightest payment is received in return.'
⁴⁰ See Giocondo 1544a, a 3ᵛ.
⁴¹ Giocondo insists that the *utilitas* of editing makes editors deserving of *laus*: Giocondo 1544a, a 3ʳ.
⁴² See Giocondo 1544a, a 3ᵛ.
⁴³ The arrangement is as follows: a map of Gaul (Glarean 1544a, a 4ᵛ–a 5ʳ); a short *Galliae Divisio* (ibid., a 5ᵛ–a 6ʳ); an image of the Rhine bridge (ibid., a 6ᵛ); the *Expositio Pontis* (ibid., a 7ʳ–a 8ᵛ); a short text on Avaricum (ibid., a 8ᵛ); an image of Avaricum (ibid., bʳ); an image of Alesia (ibid., bᵛ); a short text on Alesia (ibid., b 2ʳ–b 2ᵛ); a short text on Uxellodunum (ibid., b 2ᵛ); an image of

the illustrations and texts relating to the Rhine bridge, Avaricum, Alesia, Uxellodunum and Massilia certain technical details are marked with letters of the alphabet, thus ensuring their systematic recording. The arrangement of these paratexts and images mainly follows the narrative structure of the Caesarian reference text. Yet they also act as a filter, distilling very specific information from the reference text and thereby focusing the interest of the reader. The paratexts mainly treat geographical, ethnographical and technical information, and these categories endow the text with a specific structure and system that it did not necessarily have in antiquity. In this way Caesar's *Commentarii* are adapted to the reading habits of humanist readers, whereas these paratextual strategies themselves are expressions of such reading habits.

A second preface by Aldus Manutius rounds out the paratexts.[44] Aldus very briefly treats the historical context, the conditions of creation and the question of the authorship of the various, stylistically quite diverse, works passed down in the *Corpus Caesarianum*.[45] At the end of the preface Aldus argues that Caesar's *Commentarii* should be considered truthful accounts, adducing as evidence the letters Cicero wrote to Atticus reporting Caesar's latest deeds in Gaul. This, according to Aldus, is the clearest proof that the contents of Caesar's *Commentarii* are certainly true.[46]

The treatise dealing with the construction of the Rhine bridge is part of the paratextual ensemble that precedes Caesar's works and moves certain of their technical aspects into the foreground. The purpose of this erudite ensemble is not only to explain the ancient text to the reader, but also to adapt it to the reading habits and interests of its new humanistic recipients, as well as to exhibit Giocondo's own humanistic skills. Thus the paratexts do not simply elucidate the ancient text as completely as possible: they also fragment it. They generally aim at organizing the ancient text along certain thematic lines and thus at establishing new spaces for interpretation

---

Uxellodunum (ibid., b 3ʳ); an image of Massilia (ibid., b 3ᵛ); a short text on Massilia (ibid., b 4ʳ); a map of Spain (ibid., b 4ᵛ–b 5ʳ); a very short *Hispaniae Descriptio* (ibid., b 5ᵛ); an index of Gallic places, cities and peoples (ibid., b 5ᵛ–b 7ʳ); a short index of Spanish places and cities (ibid., b 7ᵛ).

[44] See Aldus 1544.
[45] See Aldus 1544, b 8ᵛ.
[46] Cf. Aldus 1544, b 8ᵛ: *Illud insuper addederim, bonam partem eorum quae his commentarijs scripta sunt, haberi etiam in epistolis M. Tullij ad Atticum, in quibus cum quid novi gestum erat a Caesare statim Atticum, Quintumue Ciceronem fratrem aut M. Brutum certiorem faciebat. Id quod maximo testimonio est, uerissima esse, quaecunque his commentarijs scripta habentur.* 'Moreover, I should add that a significant portion of what is written in these *Commentarii* is also found in the letters of Marcus Tullius to Atticus. In these letters he was wont to inform Atticus, his own brother Quintus Cicero or Marcus Brutus immediately whenever Caesar had done something new. This is the greatest piece of evidence for the truthfulness of whatever is written in these *Commentarii*.'

in which the ancient reference text is subjected to very specific questions. Such selective commentary on ancient knowledge can be understood as an act of appropriation, aimed at adapting the ancient reference text to new intellectual, cultural and practical exigencies. At the same time the ancient text exhibits a certain resistance. That is, it is difficult if not impossible for it to be transformed entirely arbitrarily.[47] Instead, the substance of the text provides certain docking sites that can be occupied and exploited by the reception area according to its own needs.

Giocondo's treatise *Expositio pontis* interacts with one such site, elucidating a specific passage of Caesar's text while ignoring its context. The main focus is on technical issues. The image of the Rhine bridge preceding the treatise illustrates only a certain view of the finished construction, making no attempt to capture the full length or complexity of the bridge. It is an architectural cross-section labelled with single capital letters from A to G, listed and explicated in the treatise with corresponding lower-case letters.[48] The layout of the letters in alphabetical order correlates with the letters as they appear in the treatise and basically follows the logic of the bridge's construction. The ancient reference text is focused selectively through these letters, both in the image and in the treatise, and in this way the technical information contained therein is disentangled. This highlighting of specific individual elements of the bridge's construction both decreases and systematizes the totality of information contained in the chapter.

The visualization of ancient knowledge is a medial transformation strategy: the textual boundaries of the ancient reference object are transgressed and its contents are enhanced creatively in order to facilitate the comprehension of a quite difficult passage – in this case, the description of the bridge's construction. On its own, the image of the bridge gives only an impression and is hardly sufficient to make the whole structure, as described by Caesar, completely understandable. The picture is only made comprehensible by its position in the edition and its direct relation to the treatise. Thus there is a close connection between the image and the treatise. The picture is probably provided in support of both Caesar's description of the bridge and the theoretical and technical content of the treatise, and all the parts work in tandem to help the reader understand the passage. In addition, the image gives the impression that the bridge could be reconstructed from the contents of the reference text.

---

[47] For some considerations of the idea of resistance in transformation processes, see Kaiser 2014b.
[48] See the image of the Rhine bridge (Glarean 1544a, a 6^v) and Giocondo 1544b, a 7^r.

The *Expositio pontis* itself can be structurally divided into four sections of various length which are connected to and remain in permanent argumentative relation with the reference text, reinterpreting and transforming it. In the first, very short section Giocondo mentions Caesar's two Rhine bridges. In the second section the capital letters shown in the figure are briefly explained. The third section deals with grammatical difficulties in the reference text, the fourth with mechanical aspects which are illustrated for the reader on the basis of selected examples.

In the first section Giocondo postulates that Caesar *pontem eadem forma & ratione bis fecit*.[49] By claiming that Caesar's description was realized twice in the exact same manner, Giocondo implicitly moves the semantic-discursive differential of theory and practice[50] into the focus of his treatise. In this way the reference text is already transformed, refracted through the issue of theory and practice. The specific interest of the reader and commentator determines the interpretation of the reference text and overwrites the actual intention of its author.

The second section, in which the capital letters are explained, is remarkable. Giocondo emends the bridge chapter by changing the syntax, thus simplifying the description of the bridge's construction provided by Caesar.[51] Whereas the point of Caesar's text is to explain how and that the bridge is sunk into the bed of the river, Giocondo is interested rather in the way the entire structure was put together. His explanation of the bridge's structure aims at simplifying and systematizing the ancient information. He resolves the narrative structure of the reference text, highlighting important aspects and ignoring less relevant ones. Giocondo does not discuss this syntactic and structural reorganization. Nor does he have to, as his summary of the main points of the bridge's construction makes a

---

[49] Ibid., a 7ʳ.   [50] Cf. Scharloth 2005.
[51] See Giocondo 1544b, a 7ʳ: A: *Tigna bina sesquipedalia paulum ab imo praecuta, dimensa ad altitudinem fluminis, &c.* – pairs of one-and-a-half-foot-thick balks sharpened a little way from the base and measured to suit the depth of the river; B: *Trabes bipedales immissae super utraque tigna, quae binis utrinque fibulis ab extrema parte distinebantur* – two-foot transoms set onto the pairs of balks, which were kept apart by a pair of braces on the outer side at each end; C: *Fibulae quae disclusae destinent bipedales trabes* – braces which, kept asunder, keep the two-foot transoms apart; D: *Vbi fibulae disclusae in contrariam partem reuinciuntur* – there the braces, kept asunder, are fastened to the opposite side; E: *Materia directa, quae iniecta supra bipedales trabes totum opus contexebat* – timber in right angles, laid atop the two-foot transoms, interconnected the whole structure; F: *Sublicae obliquae ad inferiorem partem fluminis adactae, quae pro ariete subiectae, & cum omni opere coniunctae, vim fluminis exciperent* – piles driven in aslant on the side facing downstream, thrust out below like a buttress and close joined with the whole structure, so as to take the force of the river; G: *Fistuca, qua adigebantur tigna in flumine* – a rammer, by means of which the balks were driven into the river.

knowledge of the reference text unnecessary for understanding this section. His main concern seems to be to distil the essential steps of the bridge's construction and to provide readers with this information. Thus the issue of theory and practice is emphasized. In this descriptive section the value of ancient technical knowledge, as passed down by Caesar, is reinforced by being condensed and newly systematized according to the essential elements of the bridge construction.

Giocondo introduces the issue of theory and practice into the first two sections by submitting the reference text to a question that Caesar implicitly provides but does not actually treat. This question is continued in the third section, too, which turns on the immediate relationship between the reference text and its feasibility. At issue is a problematic grammatical construction:

> Haec utraque, insuper bipedalibus trabibus immißis. Hunc locum sic corrigendum puto: Haec utraque insuper bipedales trabes immissae: hac ratione, ut insuper sit praepositio, & haec utraque sit accusandi casus. Quod si duriusculum hoc quisquam existimarit, sciat Caesarem ipsum simili usum constructione, in secundo de Bello ciuili, in expugnatione Massiliae... Quare si sic, ut puto, perseuerat corruptus librariorum vitio locus, neque sensus constabit, neque constructio, nisi implexa & litigiosa grammaticis. Sed ut utraque constent, sensus scilicet, & constructio, tam ingeniosis, quam grammaticis, & operi verba sint conformia, & opus verbis: animadvertendum est, quod postquam Caesar descripsit modum figendi, & adigendi tigna in fundo fluminis, ex qua adactione magnam stabilitatem, & firmitatem assecuta sunt, vertit se ad bipedales trabes, quae transuersam totius pontis latitudinem perficiebant, & qua ratione immitti possent, & quo modo sustinerentur, docet.[52]

> These pairs of balks had two-foot transoms inserted from above. I think that this passage should be corrected as follows: Into these pairs of balks two-foot transoms had let atop. This way *insuper* is a preposition and *haec utraque* is accusative. But if anyone should consider this a little awkward, he should know that Caesar himself used a similar construction in the second book on the Civil War, in the description of the conquest of Massilia... In my opinion, therefore, if the passage, corrupted by scribal error, is left in its current form, neither the sense nor the construction will be known, except as intertwined and controversial to the grammarians. But for both to be clear, namely the sense and the construction, for the engineers as well as

---

[52] See also Caesar *BGall.* IV, 17, 6: *Haec utraque insuper bipedalibus trabibus immissis, quantum eorum tignorum iunctura distebat, binis utrimque fibulis ab extrema parte distinebantur.* 'These pairs of balks had two-foot transoms let into them up, filling the interval at which they were coupled, and were kept apart by a pair of braces on the outer side at each end' (trans. taken from Edwards in Caesar 1994, 201).

Caesar's *Rhine Bridge and Giovanni Giocondo's* Expositio pontis 81

for the grammarians, and thus for the words to conform to the work and the work to the words, it has to be considered that, after describing how to connect the beams and how to sink them in the river – which gave them great stability and strength – Caesar turns to the two-foot transoms, which determined the width of the entire bridge, and shows how they can be inserted and how they are held. (Giocondo 1544b, a 7^{r-v})

Leaving us in suspense about how exactly he understands the passage in its unchanged state,[53] Giocondo clearly underscores the fact that a certain formulation of the reference text must be corrected. Thus he changes the syntax in order to clarify its contents. His emendation is contrary to the

---

[53] Not knowing how Giocondo understands Caesar's text, it is very difficult to translate his emended version of it. This is also noted by Heinrich Glarean (1488–1563) in his edition of Caesar, in which he incorporates Giocondo's short treatises as well as his illustrations, supplementing them with his own annotations. Although he praises Giocondo's image of the Rhine bridge, he cannot understand what kind of linguistic problem Giocondo had with the passage in question. See Glarean 1544b, i8^v/144–k^r/145:

> Antequam ad haec Caesaris uerba deveniamus, hoc Lectorem praemonere uoluimus: Placere mihi Iucundi picturam, quae pleraque Caesaris uerba luculenter exprimit, exceptis ijs quae Caesar de sublicis ita describit. Ac nihilo secius sublicae ad inferiorem partem fluminis obliquae adigebantur, quae pro ariete subiectae & cum omni opera coniunctae, uim fluminis exciperent, & aliae item supra pontem, & c. Hodie solent pilis lapideis trabes praefigere, ad excipienda omnia quae temere fert flumen. Quae uero ad inferiorem fluminis partem obliquae adigebantur, mihi non omnino intelliguntur. Et uide num <pro> ariete[s] subiectae dictum sit, pro fulcimento. Nunc ad proposita Caesaris uerba ueniamus, quae in Annotationibus meo quidem iudicio, Iucundus corrumpit, non emendat. Ea enim in Caesaris commentario bene habent. Sed disting<u>endum est post utraque, quod uerbum in nominandi casu dictum est, non accusandi. & construitur cum uerbo distinebantur, & insuper dictum quasi desuper, ut sit hic sensus: haec utraque supple tigna sesquipedalia, paulum ab imo praecura etc. distinebantur ab extrema parte binis utrinque fibulis, insuper bipedalibus trabibus immissis, quantum eorum tignorum iunctura distabat. <u>Nec uideo quid difficultatis uisum sit Iucundo.</u>

> Before coming to Caesar's words, we would like to advise the reader that I find Giocondo's picture pleasing. It expresses most of Caesar's words very well, except what Caesar says about the wooden stakes, which he describes as follows: 'and further, piles were driven in aslant on the side facing downstream, thrust out below like a battering ram and close joined with the whole structure, so as to take the force of the stream; and others likewise at a little distance above the bridge etc.' [trans. largely taken from Edwards in Caesar 1994, 203]. It is common today to erect balks before bridge posts, in order to intercept things flowing down the river. I am baffled, however, as to what was driven in aslant on the side facing downstream. Maybe the term 'thrust out below like a battering ram' was used for 'support'. Let us now come to the text of Caesar in question, which, as far as I am concerned, Giocondo corrupts rather than emends in his *Annotations*. For the passage is found in good condition in the *Commentarius*. It has to be punctuated after *utraque*, which is in the nominative and not in the accusative case; it is constructed with the verb *distinebantur*; and *insuper* is used as *desuper*. The sense is as follows: These both – i.e. the one-and-a-half-foot-thick balks slightly pointed at the ends etc. – were kept apart by a pair of braces on the outer side at each end and had two-foot transoms inserted into them from above, filling the interval at which they were coupled. I cannot see what Giocondo found so difficult.

manuscript tradition, which exhibits no variants for this passage. Giocondo's focus is on the use of *insuper* and the question of the remaining structure of the sentence. In point of fact Caesar uses *insuper* as an adverb, and *haec utraque* is the subject of the verb *distinebantur*. Giocondo, however, believes this passage is corrupt and denies that it makes sense as it stands. It is not only the *sensus*, according to Giocondo, but also the *constructio* that suffers from the supposed corruption. With *constructio* he refers to both the syntax of the passage and to the actual construction of the bridge. For Giocondo the sentence is *implexa & litigiosa* in its syntax, its explanatory value, and its real-world feasibility. The textual corruption is not only linguistic, but also has practical consequences.

Giocondo reads *insuper* as a preposition and *haec utraque* as its accusative object. Having thus eliminated the (original) subject of Caesar's version, Giocondo must convert the ablative construction *bipedalibus trabibus immissis* into the nominative *bipedales trabes immissae*. This emendation suggests either that Giocondo thought that the received text might be misunderstood by readers interested in putting it into practice, or that it was actually difficult for him to understand. At any rate the object of his intervention is to point out that the pairs of balks (*bina tigna*), which were sunk into the river bottom, had two-foot transoms (*bipedales trabes*) inserted into them from above. His grammatical considerations have a direct practical relevance and appeal to the actual construction process. Thus he emends the linguistic structure of the text so that it can vouchsafe for the practical feasibility and functionality of the bridge's construction.

In adopting the prepositional reading of *insuper* (with an accusative object), Giocondo articulates certain practical-theoretical implications of his linguistic-philological considerations. These become particularly clear when he writes that ultimately *& operi verba sint conformia, & opus verbis*. To effect this conformity between words and deeds, he advises the reader to consider (*animadvertendum est*) the way the bridge's construction is described. The upshot is that Giocondo systematizes the content of the reference text. He reads Caesar's chapter solely under the pragmatic aspect of its comprehensibility and feasibility.

In this way the ancient text is read as an articulation of (ancient) theory, theory that is closely allied with Giocondo's practical interests. This reading, this specific form of transformation, is made possible by the highly technical character of Caesar's passage. It is wholly understandable that Giocondo, as an architect and builder, was particularly interested in the practical feasibility of the bridge and in systematizing what the text has to say about building technique.

Caesar's Rhine Bridge and Giovanni Giocondo's Expositio pontis   83

Giocondo buttresses his proposed emendation by citing similar prepositional uses of *insuper* in Caesar,⁵⁴ as well as in Vitruvius.⁵⁵ These references, which I shall not go into in any further detail, serve primarily to confirm his own reading. The specific passages are technical in nature and therefore have to be understood in light of the semantic-discursive differential of theory and practice. Giocondo engages in a kind of a technical intertextuality. For him it is not the common but rather the rarer, prepositional use of *insuper* in specialized texts that is of interest. In addition to clarifying the reference text, the intertextual references serve to demonstrate Giocondo's philological skills, helping him to present himself as a humanist expert on technical texts.

Moreover, the prepositional reading of *insuper* informs Giocondo's understanding of the placement of the transoms and thus has consequences for the nexus of theory and practice. To emphasize again his own understanding of Caesar's text and of the bridge's construction, Giocondo paraphrases his own emended version and clarifies his thought process. In support he cites Leon Battista Alberti (1404–72), who had interpreted Caesar's text similarly.⁵⁶

---

⁵⁴ See Caesar, *BCiv.* II, 9, 2: *Hanc insuper contignationem, quantum tectum plutei ac vinearum passum est, laterculo adstruxerunt.*
⁵⁵ See Vitruvius, *De arch.* V, 12, 4: *Deinde insuper eam exaequationem pila quam magna constituta fuerit, ibi struatur.*
⁵⁶ See Giocondo 1544b, a 7ᵛ–8ʳ:

> ... dicens: Quod super haec utraque, id est super bina tigna, quae & in parte superiori, & ea quae in parte inferiori posita erant bipedales trabes immissae, quantum eorum tignorum iunctura distabat, binis utrinque fibulis ab extrem aparte distinebantur. Quibus disclusis, & in contrariam partem reuinctis, & c. In hanc eandem sententiam mecum venire videtur Leo Baptista Albertus, vir & ingenio & literis clarus, in suo de Architectura, qui eiusdem Caesariani pontis descriptionem repetens non aliter ei uisum fuit potuisse sibi ipsi satisfacere, nisi his verbis: Huiusmodi autem immissae trabes binis utrinque fibulis ab extrema parte distinebantur. Quibus disclusis, & c.
>
> ... saying: that over these two, i.e. over these two pairs of balks which are placed on both the upper and lower ends, the two-foot transoms were inserted into them from above, filling the interval at which they were coupled, and were kept apart by a pair of braces on the outer side at each end. Having been kept apart and bound opposite each other etc. It seems to me that the same view was reached by Leon Baptista Alberti – a man renowned not only for his genius but also for his written works – in his *De architectura*. Repeating the description of the very same Caesarean bridge, he seems to think it could only be described satisfactorily in the following words: these transoms were inserted and were kept apart by a pair of braces on the outer side at each end. Having been kept apart etc.

Giocondo may have used the edition from which I quote: Alberti 1485; therein the chapter *Leonis Baptistae Alberti De universorum opere lib. IIII*, fol. g viiiᵛ–i viiʳ, here: fol. i iiʳ. For an overview of Leon Battista Alberti, his treatise *De re aedificatoria* and its importance in Renaissance humanism, see Biermann 1997; Burns 1999; Choay 1999; Locher 1999; Vickers 1999; Grafton 2001; Wulfram 2001; Calzona 2007.

Giocondo's intention is to close the yawning gap between *opus* and *verba* by mediating between the ancient text and the bridge construction itself; theory and practice are disparate from one another and must be reconciled. The treatise focuses the passage describing the bridge's construction entirely through the question of its practical feasibility, ignoring its context. Giocondo decontextualizes and transforms the passage by adapting it to the cultural and intellectual concerns of his own reception area. In this regard his illustration of the bridge is consistent with the version of the text he proposes. The discourse of theory and practice is embodied in this image, which gives concrete form to the contents of the reference text and to Giocondo's emendation.

In the last section of his brief treatise Giocondo discusses the function of the *fibulae* (braces), which is to keep the *trabes* (transoms) separate:

> Quid autem sit fibula, & quomodo discludatur & reuinciatur non omnibus pervia est notitia, quamuis eius sit quotidianus usus. Vtuntur autem ea viri ac mulieres ad capita cingulorum, quibus circum se fluentes contineant vestes, traiecto per annulum altero cinguli capite, fibulaque reuincto, ut quanto plus trahitur, tanto fortius firmetur. Eiusmodi autem sunt et sellae multis Italiae urbibus communes quae clausae seruantur, et ad sedendi usum, quum discluduntur, & in contrariam partem reuinciuntur, eo fortius compressae firmantur.

> Not all people know readily what a *fibula* is or how it is held apart and fastened, even though it is used daily. Men and women use it at the ends of their belts to hold the garments together that flow about them. By pulling one end of the belt through a ring and fastening it with a *fibula* (belt buckle), the more one pulls the more stable the thing is made. Also of the same type are chairs which are known in many Italian cities and which are kept closed. When they are to be used for someone to sit on they are unfolded and fastened to the opposite side, become increasingly tighter the more they are pressed upon. (Giocondo 1544b, a 8^r)

Giocondo clearly addresses practical questions to the ancient text. He uses related, everyday objects like belt buckles and folding chairs to clarify the mechanical function of the braces employed by Caesar to stabilize the bridge. In addition, he adds three other technical examples from Vitruvius' *De architectura* that are mainly related to the construction of fortifications and thus belong to military discourse.[57] Furthermore – and probably based

---

[57] See Giocondo 1544b, a 8^r. He refers to Vitruvius, *De arch.* X, 2, 1: *Tigna duo ad onerum magnitudinem ratione expediuntur. A capite a fibula coniuncta et in imo divaricata eriguntur.* 'Two beams are prepared, in accordance with the size of the loads. They are fixed upright so that at the head they are joined together by a clasp and at the bottom they are spread apart' (trans. taken from Rowland in Vitruvius 1999, 120). Vitruvius, *De arch.* X, 2, 3: *Sin autem maioribus oneribus erunt machinae*

on these examples from Vitruvius – he supplements this list of examples by pointing out that even today (*hodie*) *fibulae* are used in the construction of walls and defences.[58]

These examples are meant to illustrate the mechanical principles by means of which certain structures are stabilized. Giocondo increasingly departs from Caesar's text and directs his attention solely to the question of its practical functionality and thus also to its facticity. On the one hand, Giocondo decontextualizes passages quoted from Vitruvius. He is not interested in their immediate concerns or contexts but rather in what they have to say about the mechanical principles by which *fibulae* function and the purposes to which they are put. He reads Vitruvius' work as an historical treasury of facts describing the workings of *fibulae* and explaining the stability they lend to various structures. On the other hand, he supplements and complements the intertextual references with examples from his own time. These derive from his own reception area, namely from everyday life, as well as from the field of military architecture. Thus the transformation of Caesar's text, which began with its being read in terms of theory and practice, is carried forward by means of intertextual and real-life examples.

The examples and quotations from Vitruvius are meant to clarify that the braces mentioned by Caesar have a specific mechanical use: they and they alone endow Caesar's Rhine bridge with the necessary stability. By explaining their function, Giocondo also implies that Caesar's passage on its own could lead to a misunderstanding of how the braces actually

---

comparandae, amplioribus tignorum longitudinibus et crassitudinibus erit utendum; eadem ratione in summo fibulationibus, in imo sucularum versationibus expediendum. 'Now if the machines are to be set up for greater loads, one will need to use greater lengths and thicknesses for the beams, and by the same principle for the clasp at the top and the rotation of the windlasses at the bottom' (trans. taken from Rowland in Vitruvius 1999, 121). Vitruvius, *De arch.* I, 5, 3: *Crassitudinem autem muri ita faciendam censeo, uti armati homines supra obviam venientes alius alium sine inpeditione praeterire possint, dum in crassitudine perpetuae tabulae oleagineae ustilatae quam creberrime instruantur, uti utraeque muri frontes inter se, quemadmodum fibulis his teleis conligatae aeternam habeant firmitatem.* 'I think that the thickness of the wall should be made in this manner: walking along its top, two armed men coming toward each other should be able to pass each other without difficulty; moreover, within its fabric, rods of scorched olive wood should be installed at as frequent intervals as possible, so that each of the faces of the wall, linked together by these rods (which act as clamps), will maintain an everlasting fixity' (trans. taken from Rowland in Vitruvius 1999, 28).

[58] See Giocondo 1544b, a 8ᵛ: *Huiusmodi autem fibulis, quibus tunc, & in colligandis muris, et in munitionum uallis utebantur, hodie quoque et nos utimur, transuersis in latum longuriis fibulatim dispositis, ut ictibus glandium, non uno loco tantum, sed tota uallis mole resistamus iuuantibus fibulis contineter in contrariam partem reuinctis.* 'Today we still use braces of the kind which were used then both for connecting walls and for constructing the ramparts of fortifications, as when we arrange the battens lengthwise by means of braces in order to resist the impacts of cannonballs not only at a single point but also along the entire foundations by using the ramparts. Here the braces help by being attached contiguously to the opposite side.'

worked.[59] Giocondo clearly wants to eliminate these uncertainties in the reference text. Nevertheless, he does not tackle this problem theoretically but rather explains it with real-life examples and intertextual references, thereby addressing the question of the bridge's feasibility as described by Caesar and closing the gap between theory and practice. Giocondo is concerned with how Caesar's bridge actually worked. Although Caesar does not intend for his description to be applied practically, Giocondo's treatment of it recasts it in terms of feasibility and functionality.

Thus the treatise is the product of a genre transformation. Knowledge is extracted from the ancient text, remodelled and subjected to technical questions in line with the requirements of a contemporary genre, the treatise, which did not even exist in antiquity. The treatise facilitates the explication of ancient knowledge in a closed thematic framework. In this respect it can be understood as a short commentary focusing on a very specific question and no others. Such a transformation strategy entails obscuring other issues raised in the ancient text and therefore has allelopoietical consequences. The ancient text is historicized, and its contents and their facticity are confirmed. Giocondo approaches the text scientifically, so to speak, inquiring about the theoretical feasibility of the bridge as described by Caesar. As a commentator he displays his humanistic skills by claiming the authority of his interpretation of the ancient text. Furthermore, he determines the meaning of the knowledge it contains by systematizing and explaining it. Therefore, the transformation of antiquity on display here is a function of the self-fashioning of the commentator, the explication of specific ancient knowledge and the concern to ascertain the facticity of this ancient knowledge.

It should also be noted, however, that the transformation of Caesar's text on the Rhine bridge results equally from Giocondo's biographical background and personal interests. Similarly, the short descriptions of Avaricum, Alesia, Uxellodunum and Massilia, with their related images, can probably be understood from this perspective. Giocondo's efforts to make Caesar's construction of the Rhine bridge comprehensible, to verify the chapter as historical-factual testimony and hence to establish the

---

[59] This is also pointed out by Edwards in Caesar 1994, 627:

> It is not clear what is meant by the *fibulae* mentioned by Caesar. Some authorities have thought that these are the diagonal ties, which are certainly required for the stability of the bridge, and which are probably implied by the phrase. The triangle formed by balks, transom, and tie resembles the shape, and performs the function, of a brooch. Other authorities hold that the *fibulae* were iron dogs, driven in to clamp each transom to its two pairs of balks at the points of juncture.

historicity of the bridge itself stem from both his humanistic ambitions and his interests as an architect. Thus the treatise itself reflects the notion that Giocondo's architectual work and his humanist efforts are difficult to separate from one another. Nevertheless, Giocondo's treatise must be interpreted specifically in the context of the field of humanist scholarship, too. It seems typical of the humanistic *Zeitgeist* to think about the feasibility and applicability of technical literary texts from classical antiquity and to write short treatises on contemporary and ancient architecture.[60] Even if the texts the humanists approach in a technical way are not necessarily technical in nature, the act of reading them in this way turns them, in the eyes of humanists, into technical texts.

Conclusion

With his decidedly technical reading of the chapter on the Rhine bridge's construction Giocondo brings to light a fundamental problem that has engaged scholars time and again, namely the practical feasibility of the bridge's description. At the same time he raises essential questions that still concern scholars today. For what audience are the *Commentarii* intended? Are their technical, military and strategic contents precise enough to be read as instructions for action or as material for imitation? If not, what is the purpose of such technical descriptions? Ultimately, Giocondo touches on the question of what functions technical writing had in ancient culture, which developed its own, different strategies for navigating successfully between theory and practice.

Through the appropriation strategies adopted by Giocondo – visualization of the bridge's construction according to Caesar's description; the systematic resolution of the bridge's elements by means of upper-case (in the image) and lower-case (in the treatise) letters; the paratextual function of the treatise within the edition; philological considerations; intertextual references to technical literature and other authorities; real-life examples; humanistic self-fashioning – the ancient passage is transferred into the semantic-discursive differential of theory and practice. This is especially the case with Giocondo's illustration of the Rhine bridge. Only through the relationship between the image and the text, and their respective involvement in the edition of Caesar's *Opera*, does the illustration receive its imaginative power and facilitate the understanding of the chapter as well as of the treatise.

---

[60] E.g. see Fiore 1998; Morresi 1998.

In addition, the perspective established by Giocondo entails the decontextualization of the bridge chapter, regardless of the short references to Caesar's deeds in Gaul at the beginning of the treatise. To be sure, these references place the bridge constructions in their historical context, but the chapter's other contents and larger context are treated as irrelevant.

In his treatise Giocondo attempts to restore the language of the reference text. In general he views the text as a theoretical foundation for an actual physical construction, but without claiming that its description should be put into practice. Rather, his intention seems to be to demonstrate that an understandable and logical 'plan' of the bridge's construction inheres in the reference text. The comprehensibility and theoretical feasibility of this 'plan' is articulated through the image as well as through the treatise.[61] As a paratext to the edition of Caesar, the *Expositio pontis* brings into relief certain aspects of the reference text that are of obvious interest to its humanist recipients. Thus it acts as a complementary textual space for the discussion of text-critical and practical-theoretical issues. In this sense the *Expositio pontis* can also be read as a small commentary.

These observations have obvious import for the relationship between words and deeds. With his reinterpretation of the ancient reference text, Giocondo inverts this relationship. Caesar's passage, as mentioned at the outset, proceeds rather from deeds to words. It is not meant to have a practical application but rather to demonstrate certain virtues of the Romans and of Caesar in particular. The new reading encouraged by Giocondo's treatise turns the humanist reader's attention instead towards the bridge itself and the method of its construction. By emending the supposedly corrupt manuscript tradition of the text, Giocondo aims to aid the text's linguistic and practical understanding – not in order, as has been said, for its instructions to be carried out, but rather to restore credibility to its theoretical feasibility as well as to the actual existence of the bridge.

Hence the ancient knowledge contained in Caesar's *De bello Gallico* is shored up, confirmed by displaying the feasibility and functionality of the bridge's construction. The gap between theory and practice identified by Giocondo is closed, *ut utraque constent, sensus scilicet, & constructio, tam ingeniosis, quam grammaticis, & operi verba sint conformia, & opus verbis.*

---

[61] See Giocondo 1544b, a 8ᵛ: *Ex dictis satis constare poterit & sensus, & constructio verborum Caesaris, & pontis forma, secundum figuram a nobis traditam.* 'From the aforesaid, both the sense and the construction of Caesar's words will be sufficiently known, and also the shape of the bridge according to the form passed down by us.'

REFERENCES

Primary Sources

Leon Battista Alberti 1485, *Leonis Baptistae Alberti Florentini viri clarissimi De re aedificatoria opus elegantissimum et quam maxime utile* (Florence: Nicolas Laurentio Alamano)

Gaius Iulius Caesar 1994, *The Gallic War*. With an English translation by H. J. Edwards (Cambridge, MA, London, repr.)

Giovanni Giocondo 1508, *Plinii Secundi Novocomensis epistolarum libri decem* (Venice: Aldus Manutius)

1509, *C. Crispi Sallusti de coniuratione Catilinae, eiusdem de bello Iugurthino, eiusdem oratio contra M. T. Ciceronem, M. T. Ciceronis oratio contra C. Crispum Sallustium; eiusdem orationes quatuor contra Lucium Catilinam, etc.* (Venice: Aldus Manutius)

1511, *M. Vitruvius per Iocundum solito castigatior factus cum figuris et tabula ut iam legi et intellegi possit* (Venice: Giovanni da Tridino)

1513a, *Vitruvius iterum et Frontinus a Iocundo revisi repurgatique quantum ex collatione licuit* (Florence: Filipo Giunta)

1513b, *Commentariorum de bello Gallico libri VIII, De bello civili Pompeiano libri III, De bello Alexandrino liber I, De bello Africano liber I, De bello Hispaniensi liber I* (Venice: Aldus Manutius)

1514, *Libri de rustica: M. Catonis lib. I, M. Terrentij Varronis lib. III, L. Iunii Moderati Columellae lib. XII ... Palladij lib. XIIII* (Venice: Aldus Manutius)

1544a, *IOANNES IVCVNDVS VERONENSIS IVLIANO MEDICO S.P.D.* in Glarean 1544a, a 2^r–a 4^r.

1544b, *IOANNES IVCVNDVS VERONENSIS Lib. IIII, EXPOSITIO PONTIS* in Glarean 1544a, a 7^r–a 8^v.

Heinrich Glarean 1544a, *C. IVLII CAESARIS COMMENTARIORVM LIBRI VIII. QVIBVS ADIECIMVS SVIS* in locis D. Henrici Glareani doctißimas annotationes (Basle: Nicolaus Bryling (=VD16 C 36))

1544b, *IN C. IVLII CAESARIS COMMENTARIORVM DE BELLO GALLICO LIBRUM IIII. Glareani annotationes* in Glarean 1544a, i8^v/144–k^r/145.

Aldus Manutius 1544, *ALDVS LECTORI S[ALUTEM DICIT]* in Glarean 1544a, b 8^{r-v}.

Marcus Vitruvius Pollio 1999, *Ten Books on Architecture*, trans. I. D. Rowland, commentary and illustrations by Th. Noble Howe, with additional commentary by I. D. Rowland and M. J. Dewar (Cambridge University Press)

Secondary Sources

L. Bergemann and M. Dönike, et al. 2011, 'Transformation: Ein Konzept zur Erforschung kulturellen Wandels' in H. Böhme (ed.), *Transformation: Ein Konzept zur Erforschung kulturellen Wandels* (Munich), 39–56

V. Biermann 1997, *Ornamentum: Studien zum Traktat De re aedificatoria des Leon Battista Alberti* (Hildesheim)

H. Böhme 2011, 'Einladung zur Transformation' in H. Böhme (ed.), *Transformation: Ein Konzept zur Erforschung kulturellen Wandels* (Munich), 7–37

R. Brenzoni 1960, *Fra Giovani Giocondo Veronese, Verona 1435–Roma 1515: Figura genialissima e tipica della versitilità rinascimentale italiana alla luce delle fonti coeve e dei documenti esposti cronologiamente* (Florence)

H. Burns 1999, 'Antike Monumente als Muster und als Lehrstücke: Zur Bedeutung von Antikenzitat und Antikenstudium für Albertis architektonische Entwurfspraxis' in K. W. Forster and H. Locher (eds.), *Theorie der Praxis: Leon Battista Alberti als Humanist und Theoretiker der bildenden Künste* (Berlin), 129–55

Arturo Calzona (ed.) 2007, *Leon Battista Alberti: Teorico delle arti e gli impegni civili del De re aedificatoria*. Atti dei convegni internazionali del Comitato Nazionale VI Centenario della Nascita di Leon Battista Alberti. Mantova, 17–19 ottobre 2002, Mantova, 23–25 ottobre 2003 (Florence)

F. Choay 1999, '*De re aedificatoria* als Metapher einer Disziplin' in K. W. Forster and H. Locher (eds.), *Theorie der Praxis: Leon Battista Alberti als Humanist und Theoretiker der bildenden Künste* (Berlin), 217–31

M. Deissmann (trans. and ed.) 1991, *Gaius Iulius Caesar: De bello Gallico – Der Gallische Krieg*, Latein/Deutsch (Stuttgart)

K. A. E. Enenkel (ed.) 2014, *Transformations of the Classics via Early Modern Commentaries* (Leiden)

K. A. E. Enenkel and H. Nellen (eds.) 2013, *Neo-Latin Commentaries and the Management of Knowledge in the Late Middle Ages and the Early Modern Period (1400–1700)*. Interdisciplinary Conference (Leuven)

F. P. Fiore 1998, 'The Trattati on Architecture by Francesco di Giorgio' in V. Hart and P. Hicks (eds.), *Paper Palaces: The Rise of the Renaissance Architectural Treatise* (Hong Kong), 66–85

A. Grafton 2001, *Leon Battista Alberti: Master Builder of the Italian Renaissance* (London)

R. Hafner and M. Völkel (eds.) 2006, *Der Kommentar in der frühen Neuzeit* (Tübingen)

T. R. Holmes 1913, *Cäsars Feldzüge in Gallien und Britannien*, trans. and rev. Wilhelm Schott, ed. Felix Rosenberg (Leipzig, Berlin)

S. Holtz, A. Schirrmeister and St Schlelein (eds.) 2012, *Humanisten edieren: Gelehrte Praxis im Südwesten in Renaissance und Gegenwart*. Veröffentlichungen der Kommission für geschichtliche Landeskunde in Baden-Württemberg, Reihe B: Forschungen (Stuttgart)

R. Kaiser 2013, 'Sola historia negligitur: Historiographisches Erzählen in Andreas Althamers Scholia zur *Germania* des Tacitus (1529)' in A. Heinze, A. Schirrmeister and J. Weitbrecht (eds.), *Antikes erzählen: Narrative Transformationen der Antike in Mittelalter und Früher Neuzeit [= Transformationen der Antike, 27]* (Berlin), 91–116

2014a, 'Understanding National Antiquity: Transformations of Tacitus's *Germania* in Beatus Rhenanus's *Commentariolus* (1519)' in K. Enenkel (ed.), *Transformations of the Classics via Early Modern Commentaries. Intersections 29* (Leiden), 261–78

2014b, 'Kanonisierung und neue Deutungsräume: Die Grenzen der Antike in Andreas Althamers *Commentaria* zur *Germania* des Tacitus (1536)' in W. Röcke, S. Möckel and A. Heinze (eds.), *Grenzen der Antike: Die Produktivität von Grenzen in Transformationsprozessen* (Berlin), 353–72

D. E. Koutroubas 1972, *Die Darstellung der Gegner in Caesars 'Bellum Gallicum'* (Heidelberg)

H. Locher 1999, 'Anmerkungen zur Aktualität des Theoretikers Leon Battista Alberti' in K. W. Forster and H. Locher (eds.), *Theorie der Praxis: Leon Battista Alberti als Humanist und Theoretiker der bildenden Künste* (Berlin), 1–7

U. Maier 1978, *Caesars Feldzüge in Gallien (58–51 v. Chr.) in ihrem Zusammenhang mit der stadtrömischen Politik* (Bonn)

G. Maurach 2003, *Caesar: Der Geschichtsschreiber. Kommentar für Schule und Studium* (Münster)

E. Mensching 1975, 'Caesars Interesse an Galliern und Germanen', *GGA* 227, 1: 9–21

M. Mislin 2006, 'The Planning and Building Process of Two Paris Bridges in the Sixteenth and Seventeenth Century' in Construction History Society (ed.), *Proceedings of the Second International Congress on Construction History*, vol. II (Cambridge), 2223–39

M. Morresi 1998, 'The Treatises and the Architecture of Venice in the Fifteenth and Sixteenth Centuries' in V. Hart and P. Hicks (eds.), *Hong Kong Paper Palaces: The Rise of the Renaissance Architectural Treatise* (Yale), 263–80

G. Most (ed.) 1999, *Commentaries – Kommentare* (Göttingen)

H. Oppermann 1968, *Julius Caesar in Selbstzeugnissen und Bilddokumenten* (Hamburg)

M. Pade (ed.) 2005, *On Renaissance Commentaries: Twelfth International Congress for Neo-Latin Studies at the University of Bonn, August 3–9, 2003*. International Congress of Neo-Latin Studies (Hildesheim)

D. Parker 1993, *Commentary and Ideology: Dante in the Renaissance* (Durham, NC, London)

J. Rüpke 1992, 'Wer las Caesars bella als commentarii?', *Gymnasium*, 99: 201–26

J. Scharloth 2005, 'Die Semantik der Kulturen: Diskurssemantische Grundfiguren als Kategorien einer linguistischen Kulturanalyse' in D. Busse, Th. Niehr and M. Wengeler (eds.), *Brisante Semantik: Neuere Konzepte und Forschungsergebnisse einer kulturwissenschaftlichen Linguistik* (Tübingen), 119–35

T. Schmitt 2000, 'Oikumene', *Der Neue Pauly*, vol. VIII (Stuttgart), 1138–40

Ch. Trzaska-Richter 1991, *Furor teutonicus: Das römische Germanenbild in Politik und Propaganda von den Anfängen bis zum 2. Jahrhundert n. Chr.* (Trier)

R. Urban 1989, 'Die Treverer in Caesars *Bellum Gallicum*' in Heinz E. Herzig and Regula Frei-Stolba (eds.), *Labor omnibus unus. FS Gerold Walser*. Historia: Einzelzeitschriften 60, 244–56

B. Vickers 1999, 'Humanismus und Kunsttheorie in der Renaissance' in K. W. Forster and H. Locher (eds.), *Theorie der Praxis: Leon Battista Alberti als Humanist und Theoretiker der bildenden Künste* (Berlin), 9–74

M. Weichenhan 2011a, 'Der Kommentar als Transformationsmedium des Textes' in Th. Wabe and M. Weichenhan (eds.), *Kommentare: Interdisziplinäre Perspektiven auf eine wissenschaftliche Praxis*. Apeliotes: Studien zur Kulturgeschichte der Theologie 10 (Frankfurt am Main), 9–25

2011b, 'Gassendis Kommentierung von Diogenes' Laertius Vitae philosophorum X – ein Beispiel für die Verwissenschaftlichung der Antike?' in Th. Wabe and M. Weichenhan (eds.), *Kommentare: Interdisziplinäre Perspektiven auf eine wissenschaftliche Praxis*. Apeliotes: Studien zur Kulturgeschichte der Theologie 10 (Frankfurt am Main), 91–125

H. Willich 1992, 'Giocondo, Fra Giovanni' in Ulrich Thieme and Fred C. Willis (eds.), *Allgemeines Lexikon der bildenden Künstler von der Antike bis zur Gegenwart, begründet von Ulrich Thieme und Felix Becker*, vol. XIV, 64–8

H. Wulfram 2001, *Literarische Vitruvrezeption in Leon Battista Albertis De re aedificatoria* (Munich)

CHAPTER 6

From Words to Acts?
On the Applicability of Hippocratic Therapy

Pilar Pérez Cañizares

The question 'From words to acts?', which provided the title for the conference from which this volume has arisen, immediately raises issues related to the applicability of medical treatment as transmitted in ancient texts, and it is to this topic that I shall turn my attention in this contribution.* More concretely, I shall focus on the transition from words to acts in the domain of therapeutics by examining the relationship between text and extra-textual applicability in therapeutic instructions. In other words, I shall try to establish to what extent the therapeutic instructions included in some of the Hippocratic treatises 'speak for themselves', that is, whether they could be interpreted and followed by readers, and whether there is evidence that hints at specialized knowledge shared by writers and targeted readers.

There has been much scholarly discussion on the prejudices regarding readers' abilities to understand particular types of text because of their grade of specialization;[1] on the one hand, the capability of laymen to understand and ultimately make use of medical texts has been questioned;[2] on the other, their major role as participants in medical debates, guardians of their own

---

* I thank Prof. Elisabeth Craik for her comments on an earlier draft of this chapter.
[1] The topic has been addressed by van der Eijk 1997, especially 86–9, and more recently in van der Eijk 2005, 29 ff.
[2] For instance, regarding the Hippocratic treatise *Aff.* (*De affectionibus*) some scholars consider that, due to its high degree of technical details, it cannot really be meant for laymen. See, for instance, Potter 1988, 4–5 and Totelin 2009, 15. See also Langholf 1996, 121, where it is suggested that the conflict is caused by the 'technical' content of the writing, as opposed to the capacities of laymen: 'Es könnte sein, daß der Verfasser traditionelle medizinische Texte mit einem an Laien gerichteten Proömium versah, sie nur wenig umarbeitete und im übrigen darauf vertrauen konnte, daß der Laie sie verstand.'

health, and ultimately patients with a deep interest in and understanding of medicine has been repeatedly shown.³

It is in the domain of therapy that medical texts better show their degree of applicability. The ultimate role of the physician lies in treating the patient for his or her condition so as to recover health. To express the general idea of applying a medical treatment, a wide range of Greek words is used in the Hippocratic treatises.⁴ Their different nuances and origins eventually show up in the general conception of the medical treatment shared by the Hippocratic doctors and above all in how this treatment was in practice carried out.

Erotian divided the Hippocratic writings into five different categories based on content and purpose, and it is under therapeutics that the examples listed are the most numerous; it is noteworthy that half of the forty or so books that Erotian attributes to Hippocrates fall under this heading. Further, within therapeutics Erotian distinguished between surgery and dietetics, under which, among others, the books on diseases and on gynaecology are listed.⁵

Whether dietetics, administration of drugs, manual therapy, surgery or even psychotherapy, the ways in which instructions are given range from very detailed ones, aiming at exhaustiveness, to ones characterized by vagueness or imprecision, including those making reference to therapies explained in other books.⁶ The heterogeneity of the *Corpus Hippocraticum* has given rise to different assessments regarding the style⁷ of the medical

---

³ See Lloyd 1979, 79, Nutton 2004, 52 and 70, Schiefsky 2005, 36 and Pérez Cañizares 2010.
⁴ See van Brock 1961, who studies the use, among others, of: ἰῆσθαι 'cure', ἰητρεύειν 'treat medically', θεραπεύειν 'treat', ἀκεῖσθαι 'heal', μελετᾶν 'care', ὠφελεῖν 'help', βοηθεῖν 'aid', μελεδαίνειν 'care for', μεταχειρίζεσθαι 'handle, treat', ἐπιχειρεῖν 'put one's hand to, attempt', ἐγχειρεῖν 'put hand to, attempt', φυλάσσειν 'guard, protect'.
⁵ Erotian's classification distinguishes the following categories: σημειωτικά, works on signs, where the books covering prognostics are included; αἰτιολογικά καὶ φυσικά, works on aetiology and on nature; θεραπευτικά, on treatment; ἐπίμικτα, mixed works, where the books of *Epid.* and *Aph.* are included; and τῶν δ' εἰς τὸν περὶ τέχνης τεινόντων λόγον, works devoted to the medical art. The works that Erotian listed under the heading 'therapy' are these: <θεραπευτικὰ> δέ· τῶν μὲν εἰς χειρουργίαν ἀνηκόντων· Περὶ ἀγμῶν, Περὶ ἄρθρων, Περὶ ἑλκῶν, Περὶ τραυμάτων καὶ βελῶν, Περὶ τῶν ἐν κεφαλῇ τραυμάτων, Κατὰ ἰητρεῖον, Μοχλικόν, Περὶ αἱμορροΐδων καὶ συρίγγων. εἰς δίαιταν· Περὶ νούσων α' β', Περὶ πτισάνης, Περὶ τόπων τῶν κατὰ ἄνθρωπον, Γυναικείων α' β', Περὶ τροφῆς, Περὶ ἀφόρων, Περὶ ὑδάτων. See Nachmanson 1918, 36, 12.
⁶ The treatise *Aff.* refers on ten occasions to what has been thought to be a collection of recipes called either Φαρμακῖτις or Τὰ φάρμακα instead of detailing the composition of the remedies prescribed. On this see Totelin 2009, 98 and Pérez Cañizares 2010.
⁷ See Thesleff 1966, 107: 'They [Hippocratic writings] cover a stylistic range extending from strictly technical texts to rhetorical logoi with only a core of technical matter, through a great variety of intermediate types.'

works attributed to Hippocrates,[8] and frequently these have been linked to questions of authenticity and date.[9]

If we follow the much-quoted aphorism VII 87,[10] it was commonly accepted that the therapeutic methods used by the Hippocratic physician differed in efficiency, the gradation starting with medicines (the least effective), continuing with surgery and finishing with cautery as the last instance. What fire could not cure was a hopeless case. It is interesting to note that the Latin tradition of this aphorism added at the beginning some words missing in the Greek, which included diet as the fourth element and the most moderate therapeutic measure.[11] It is very likely that this late introduction of a fourth element occurred centuries later, when the boundaries between the use of food and drink, either as components of a particular dietary plan or as simple drugs, were much more clearly traced than in the fifth and fourth centuries BCE.

The efficiency of these four therapeutic expedients corresponds to the difficulty of administering them, fire being the most effective of the resources, even if in the Hippocratic treatises cautery and incision are frequently mentioned as alternative methods of draining excess fluid.[12] That the performance of both surgical interventions required a certain degree of deftness is shown by the definition of manual dexterity (εὐχειρίη) given by the author of *Diseases* I, who mentions cutting or cauterizing without damaging sinews or veins as an example of the skills desirable in a practitioner, as well as hitting the pus when cauterizing or cutting in case of internal suppuration.[13]

---

[8] It is remarkable that, even centuries after scholars have mainly abandoned the so-called question of authenticity, the *Hippocratic Collection* is still commonly regarded as a corpus, as if it were the work of one author. On criticism of this view, see Craik 2009 and van der Eijk 2015. Also, in this first part of the chapter, the evidence will be taken from a specific group of the Hippocratic treatises, those covering nosology, as they bear a relationship among them, though to different degrees: *Morb.* I, *Morb.* II, *Morb.* III, *Aff.*, *Int.* (*De internis affectionibus*) and *Loc. Hom.* (*De locis in homine*).

[9] See for instance Littré 1839–61, I 473, who qualifies Galen's view on Hippocrates' conciseness, saying that excessive brevity is a feature of the style of writings that are in fact personal notes not prepared for publication, whereas treatises commonly accepted as genuine, such as *Progn.*, *Epid.* I, *Epid.* III and *Aer.* present an extensive exposition and a style comparable to that of Thucydides.

[10] *Aph.* VII 87 (L. IV 608.1 = Magdelaine 476.11= Jones IV 216.13) ὅσα φάρμακα οὐκ ἰῆται, σίδηρος ἰῆται· ὅσα σίδηρος οὐκ ἰῆται, πῦρ ἰῆται· ὅσα πῦρ οὐκ ἰῆται, ταῦτα χρὴ νομίζειν ἀνίατα.

[11] See Magdelaine 1994, 304 for the apparatus criticus, where it is noted: ante pr. ὅσα add. *quotquot dieta non curantur, medicaminibus curantur* lat. The words are included in the ancient Latin translation, dated to the fifth or sixth centuries CE.

[12] Above all the author of *Morb.* III refers frequently to the two methods, giving the practitioner the possibility of choosing between them. See, for instance, *Morb.* III 16 (L. VII 154 = Potter CMG 94 = Potter Loeb 54), *Int.* 9 (L. VII 212 = Potter VI 132), *Morb.* II 57 (L. VI 90 = Potter V, 300 = Jouanna X 2, 197), *Morb.* II 60 (L. VI 94 = Potter V 304 = Jouanna X 2, 199).

[13] *Morb.* I 10 (L. VI 158 = Wittern 26 = Potter V 120).

Depending on how serious the condition is, the therapeutic recommendations provided in the nosological treatises normally start with special dietetic measures and drugs. If these fail, then a more aggressive treatment has to be employed. The Hippocratic treatise *Internal Affections*, well known because of its accounts of several types of the same disease, is arguably the best example of a medical text, with long and detailed therapeutic accounts. One example taken from this text, the account of one of the five different types of dropsy,[14] will help to illustrate this. As with the whole treatise, the author of *Internal Affections* presents a very detailed exposition of therapy. Here the account refers to the third type of dropsy he describes, the one arising from the spleen and often caused by the ingestion of such fruits as apples, green figs or grapes. The treatment proposed consists of a series of dietetic prescriptions described with a considerable degree of detail, including recommendations of foods and drinks, avoidance of sweet things and exercising;[15] they are presented at the end of the account, as they are a fixed part of the treatment and should be carried out independent of whether the administration of drugs suffices or the case requires cautery:[16]

> Τοῦτον, ὅταν οὕτως ἔχῃ, κατ' ἀρχὰς μελετᾶν, ἄνω μὲν ἐλλέβορον διδούς, κάτω δὲ κνέωρον ἢ ἱππόφεω ὀπὸν ἢ κόκκον Κνίδιον· διδόναι δὲ καὶ ὄνειον γάλα ἑφθὸν ὀκτὼ κοτύλας μέλι παραχέας. καὶ ἢν μὲν ὑπὸ τούτων καθίστηται, ἅλις·

> When the case is such, treat him from the beginning by giving him hellebore [to clean] upwards and spurge-flax, juice of spurge or Cnidian berry [to clean] downwards; also give him eight cotyles of boiled ass-milk, adding honey to it. And if the disease is removed with these measures, it is enough. (*Int.* 25 (L. VII 230.13–17 = Potter VI 156.15–19))

---

[14] In the *Hippocratic Corpus* dropsy is a term used to designate a variety of diseases in which a gathering of fluids existed. *Int.* 22 to 26 are devoted to dropsy. Other examples with descriptions of symptoms and aetiology can be found in *Aff.* 20 and 22, *Loc. Hom.* 24–5, *Morb.* IV 57. For a general discussion of dropsy in these books, see Lonie 1981, 290 and Craik 1998, 172.

[15] *Int.* 25 (L. VII 230 = Potter VI 156):

> ταῦτα δὲ κατ' ἀρχὰς ποιέειν τοῦ νοσήματος, καὶ δίαιταν τήνδε προσφέρεσθαι, πυρετοῦ μὴ ἔχοντος· ἄρτῳ μὲν χρήσθω πυρίνῳ· ὄψον δ' ἐχέτω τάριχον Γαδειρικὸν ἢ σαπέρδην, καὶ κρέας τετρυμμένον οἰός, καὶ τὰ ὀξέα καὶ τὰ ἁλμυρὰ πάντα ἐσθιέτω, καὶ πινέτω οἶνον Κῷον αὐστηρὸν ὡς μελάντατον· τῶν δὲ γλυκέων ἀπεχέσθω· ἢν δ' ἐξανίστηται καὶ δυνατὸς ᾖ, παλαιέτω ἀπ' ἄκρων τῶν ὤμων, καὶ ταλαιπωρείτω περιόδοισι πολλῇσι δι' ἡμέρης, καὶ εὐωχείσθω ἃ προείρηται μάλιστα.

[16] For a general account of cautery in the Hippocratic treatises, see Craik 2009, 112, who mentions that cautery and cutting were alternative therapies in the classical period, but that the use of incisions to eliminate unwanted fluids prevailed later on.

On the Applicability of Hippocratic Therapy   97

The body should be cleansed by administering an emetic (hellebore) and other substances to purge the cavity downwards.[17] In case the drugs do not have the desired effect, then cautery is prescribed:

> ἢν δὲ μή, ὅταν μέγιστος ᾖ ὁ σπλὴν καὶ οἰδέῃ μάλιστα, καῦσαι μύκησι, τὰς κεφαλὰς πολλὰς ἀπολαβών, ἢ σιδηρίοισι, φυλασσόμενος ὅπως μὴ πέρην διακαύσῃς. (2 πολλάς θ: deest M)
>
> If not, when the spleen is at its biggest size and most swollen, cauterize it with fungi by grasping the ribs, or with irons, being careful not to burn through. (*Int.* 25 (L. VII 230.17–19 = Potter VI 156.19–23))

Similar medical material, containing accounts of excess fluid in the spleen and bearing similarities in symptoms and in part also treatment, appears in other Hippocratic treatises, in some of which the use of cautery to eliminate the excess of fluid is also mentioned.[18] The normal indication to carry out cauterization is usually just the verb 'to burn'; sometimes also an accusative, either expressing the part of the body to cauterize or the quantity of scars, or both. In this example there is an indication not to apply the cautery excessively, which could be harmful.[19] Anyway, these precautions seem in general to be taken for granted, as the instructions normally just mention the verb 'to burn', and details on how to proceed are mostly lacking. The success of the operation totally depended on manual dexterity; this must have been acquired at a first stage through the direct observation of a practising master and later by acquiring practical skills when acting repeatedly as a surgeon. That cautery could be either the wrong therapy or its application could be wrong is stressed by the author of *Diseases* I, who considers this one of the typical errors concerning the praxis of medicine.[20]

In some cases the choice between cautery and incision is left to the physician. Likewise, in this particular passage two different types of cautery are mentioned as suitable for treatment, the standard one being to apply

---

[17] On these procedures as a means of bringing about *katharsis* in the Hippocratic texts, see von Staden 2007.

[18] *Loc. Hom.* 24 (L. VI 314 = Craik 64), *Aff.* 20 (L. VI 230 = Potter V 36).

[19] Cf. also *Int.* 32 (L. VII 250 = Potter VI 184) καὶ καῦσαι, ὁκόταν παχύτατος καὶ μέγιστος ὁ σπλὴν γένηται· καὶ ἢν τύχῃς καύσας τοῦ καιροῦ, ὑγιέα ποιήσεις· Disease of the spleen is to be treated by cauterization, and 'if you hit the due measure in burning, you will make the patient healthy'. It is therefore stressed that too much heat could indeed be applied to the region treated with cauterization.

[20] Chapter 6 of *Morb.* I is devoted to correctness and incorrectness in medicine. After discussing various errors regarding the medical art, a list of examples of incorrect surgical treatments is presented; the last errors to be mentioned are falling short in depth or width when cutting or cauterizing, and cutting or cauterizing where it is not necessary. *Morb.* I 6 (L. VI 152 = Wittern 16 = Potter V 112) καὶ τάμνοντα ἢ καίοντα ἐλλείπειν ἢ τοῦ βάθεος ἢ τοῦ μήκεος· ἢ καίειν τε καὶ τάμνειν ἃ οὐ χρή.

red-hot irons (σιδηρίοισι) or to use fungi (μύκησι). Various are the problems of interpretation that this text poses, the main one being how exactly this type of cauterization was carried out.[21] This technique does not come up in any of the other Hippocratic texts; as is the case with some other details on therapy or diagnosis, the author of *Internal Affections* offers information that is lacking in the rest of the extant works of this time. In the whole book he recommends seven times to carry out cauterization by using μύκησι. The treatment is indicated in a variety of diseases,[22] and on two occasions the author recommends the use of irons for fleshy parts and the use of fungi for parts with bones and sinews.[23]

Any more details on how this cauterization using μύκησι was done are completely lacking in the treatise. Furthermore, the absence of evidence for this particular type of cautery in the rest of the *Corpus Hippocraticum* and in later Greek medical authors has contributed to divergent opinions on how to interpret καῦσαι μύκησι: apart from taking the word μύκης as a generic designation for fungus, the possibility that it may refer to a lamp-wick that was in fact used to burn the tissue was already considered by the

---

[21] The passage raises numerous interpretative difficulties. See, for instance, the English translation of this passage by Potter in the Loeb series: 'If not, when the spleen is at its greatest size and most swollen, cauterize it with fungi, holding their many heads away from the spleen, or with irons.' This interpretation coincides in general with that of the early modern editors such as Cornarius 1546, 271 *fungis urito, capita eius intercipiens, aut ferramentis*. Littré translates: 'on fera avec des champignons ou le fer, des cautérisations qui comprendront les extremités de la rate'. I shall just briefly refer to some of the issues, as they do not directly affect the argument I am concerned with in this chapter. First, the occurrence of ἀπολαβών with the rare meaning of 'grasp' is also attested in *Art.* 11 (L. IV 106 = Kühlewein II 128 = Withington 224) χρὴ δὲ ὧδε καίειν ταῦτα· ἀπολαβόντα τοῖσι δακτύλοισι κατὰ τὴν μασχάλην τὸ δέρμα, where the context is similar – the description of how to proceed with cauterization – and the use of τοῖσι δακτύλοισι serves to delimit the meaning of the verb. Second, τὰς κεφαλάς (that is, σπληνός) is an anatomical term referring to the upper parts of the spleen; cf. Kühn and Fleischer 1986-89 s.v. κεφαλή II. *pars extrema rerum*, and Irmer 1980, 273. More difficult to interpret is the occurrence of πολλάς, which, according to the *Index Hippocraticus*, is lacking in one of the two ancient manuscripts (M Marcianus gr. 269, tenth century). It could hint at a loss of text in the archetype, for instance πολλάς (ἐσχάρας) followed by a transposition. Cf. *Int.* 19 (L. VII 214 = Potter 134) μύκησι καῦσαι ὀκτὼ ἐσχάρας, τὰς κεφαλὰς ἀπολαβών τοῦ σπληνός, but it is within a critical edition that this should be considered. All these issues highlight the need for further scholarly research, especially a critical edition and commentary of *Int*.
[22] *Int.* 18 (L. VII 210 = Potter VI 130) disease that changes from nephritis and involves the vessels from head to foot; *Int.* 19 (L. VII 214 = Potter VI 134) the same disease, but involving the left vessel; *Int.* 24 (L. VII 226 = Potter VI 152) dropsy arising from the liver; *Int.* 28 (L. VII 240 = Potter VI 170) disease of the liver; *Int.* 30 (L. VII 244 = Potter VI 174) first disease of the spleen; *Int.* 51 (L. VII 292 = Potter VI 242) sciatica caused by phlegm.
[23] *Int.* 18 (L. VII 212 = Potter VI 132) καίειν δὲ χρὴ τὰ μὲν σαρκώδεα σιδηρίοισι, τὰ δὲ ὀστώδεα καὶ νευρώδεα μύκησι. Treatment for sciatica caused by phlegm: *Int.* 51 (L. VII 296 = Potter VI 248) ἢν δὲ μή, καῦσαι αὐτόν, τὰ μὲν ὀστώδεα μύκησι, τὰ δὲ σαρκώδεα σιδηρίοισι πολλὰς ἐσχάρας καὶ βαθείας.

eighteenth-century editor Stephan Mack;[24] this is also the interpretation given to the word in the *Index Hippocraticus*.[25]

Caelius Aurelianus, when discussing how to treat sciatica, gives detailed instructions about how to cauterize, and subsequently reviews other physicians' methods of cauterization. It is in this context that cauterization with fungi is mentioned, but nevertheless without identifying who used to carry out this method.[26] It is interesting to note that later on Caelius Aurelianus mentions the treatment of sciatica described in Hippocrates' *Places in Man*,[27] but it is impossible to know whether Caelius Aurelianus was referring precisely to *Internal Affections* when mentioning cautery with fungi. He may have had further medical texts that also contained this therapeutic advice.

The text of Caelius Aurelianus makes a couple of important details clear: first, tree fungi were used for cauterization and the procedure was similar to the traditional Chinese moxibustion, where moxa sticks are placed directly on the skin and burnt. Recent studies have shown that burning *fomes fomentarius*, a fungus known as tinder fungus, directly on the skin is a common practice of indigenous medicine shared by many ethnic groups in Europe and Asia.[28]

Returning to the Hippocratic passage, the diversity of interpretations on how the cauterization with fungi was carried out, due to the absence of detailed instructions, strongly contrasts with the tenor of the treatise, where the short mentions of this procedure and the variety of diseases for which it was useful indicate that it must have been a common practice. Similarly, in other Hippocratic texts it is rare to find extensive instructions on how to perform basic therapeutic treatments such as draining out fluids by cutting, or indeed cauterizing. A similar case is the application of cupping vessels, documented throughout the *Hippocratic Collection* but with no extant

---

[24] Littré in his apparatus criticus refers to Mack's interpretation by writing 'D'après Mack μύκης signifie ici, non le champignon de terre, mais celui qui se forme aux lampes.'

[25] See Kühn and Fleischer 1986-9 s.v. μύκης: II ellychnium. All the occurrences of this word in *Int.* are included under this meaning.

[26] The physicians are referred to as 'alii'. Cf. Caelius Aurelianus *Tard. Pass.* V 1, 20 (Bendz and Pape, CML VI 1, 2, 866) *alii ligneos fungos inserius ac superius angustos formantes patientibus apponunt locis, quos summitate accensos sinunt concremari, donec cinerescant atque sponte decidant. est enim haec ustio leni penetratione moderata.* 'Andere legen auf die erkrankten Stellen holzige Baumschwämme, denen sie unten und oben eine schmale Form geben und die sie, an ihrer höchsten Stelle angezündet, verbrennen lassen bis sie zu Asche werden und von selbst abfallen. Diese Form des Brennens erweist sich nämlich wegen des gelinden Durchdringens als maßvoll.'

[27] Caelius Aurelianus *Tard. Pass.* V 1, 24 (Bendz and Pape CML VI 1, 2, 868).

[28] Renaut 2004, 233–6.

specification on how to use them.[29] This shows that the Hippocratics were familiar from their medical practice with all these procedures and with the instruments used, and that most of the time the authors, as in the passage from *Internal Affections* before, drew on this implicit knowledge. In the case of the application of therapeutic measures that were very frequently used to treat a wide variety of conditions – for instance, the different diseases for which cauterization with fungi was advised – physicians would not need a written guide to action. The same applies to medical interventions such as cupping, draining fluids through incisions and in general applying manual therapy, as these can be considered basic techniques for a physician; also, patients were familiar with them, as they would probably either have experienced them themselves or seen how they were carried out.

My following example shows, I think, the very different circumstances for which a medical performance was described and communicated. The sophisticated removal of nasal polyps described in great detail in Chapter 33 of *Diseases* II shows how a specialized surgical intervention was explained step by step:

> Ἢν πώλυπος γένηται ἐν τῇ ῥινί, ἐκ μέσου τοῦ χόνδρου κατακρέμαται, οἷον γαργαρεών, καὶ ἐπὴν ὤσῃ τὴν πνοιήν, προέρχεται ἔξω, καὶ ἔστι μαλθακόν, καὶ ἐπὴν ἀναπνεύσῃ, οἴχεται ὀπίσω, καὶ φθέγγεται σομφόν, καὶ ἐπὴν καθεύδῃ, ῥέγχει. ὅταν οὕτως ἔχῃ, σπογγίον καταταμὼν στρογγύλον, καὶ ποιήσας οἷον σπεῖραν, κατειλίξαι λίνῳ Αἰγυπτίῳ καὶ ποιῆσαι σκληρόν· εἶναι δὲ μέγεθος ὥστ' ἐσαρτίζειν ἐς τὸν μυκτῆρα, καὶ δῆσαι τὸ σπογγίον λίνῳ τετραχόθι· μῆκος δ' ἔστω ὅσον πυγονιαῖον ἕκαστον· ἔπειτα ποιήσας αὐτῶν μίαν ἀρχήν, ῥάβδον λαβὼν κασσιτερίνην λεπτὴν ἐκ τοῦ ἑτέρου κύαρ ἔχουσαν, διείρειν ἐς τὸ στόμα τὴν ῥάβδον ἐπὶ τὸ ὀξύ, καὶ ἐπὴν λάβῃς, διέρσας διὰ τοῦ κύαρος τὸ λίνον, ἕλκειν ἔστ' ἂν λάβῃς τὴν ἀρχήν· ἔπειτα χηλὴν ὑποθεὶς ὑπὸ τὸν γαργαρεῶνα, ἀντερείδων, ἕλκειν ἔστ' ἂν ἐξειρύσῃς τὸν πώλυπον.

> If a polyp forms in the nose, it hangs down from the middle of the cartilage like a uvula. When the patient breathes out, the polyp moves outside, and

---

[29] The most thorough account of different types of cupping vessel can be found in *Medic.* 7 (L. IX 212 = Heiberg CMG I 22 = Potter VIII 306). In essence, there are two sorts: those with a small mouth and those with a large circle as a mouth. The former are used to attract fluids in a straight line, whereas the latter draw them from a larger and more dispersed area. As in the case of cautery, it is remarkable that the use of cupping vessels is recommended as a common therapeutic measure, and yet nowhere is it mentioned that the instruments had to be heated before being applied, nor do they refer to any other procedure to extract the air. Therefore their use must have been generally known. For the difference between a 'dry' or a 'wet' cupping depending on whether the fluids were just gathered together in the surface of the skin, or if they were actually extracted, see Craik 1998, 146. For recent discussion on cupping instruments as a representation of attraction in philosophical and medical literature, see Schiefsky 2005, 332. For their role as a distinctive sign for doctors in iconographic representations, see Jouanna 1999, 86.

it is soft; when the patient breathes in, the polyp moves back inside. The patient's voice is unresonant and when he sleeps, he snores.

When the case is such, cut down a sponge round and make it like a ball and wrap it with Egyptian linen so that it is hard; the size should fit into the nostril; bind it with threads in four places, the length of each being a cubit. Then make a single beginning of them, take a light tin rod with an eye at one end and draw the rod at an acute angle into the mouth; when you have gotten hold of it, draw (the thread) through the eye and pull (the rod) until you have the beginning (of the thread). Then, placing a forked probe under the uvula and exerting pressure, pull until you tear the polyp. (*Morb.*II 33 (L. VI 50 = Potter V 246 = Jouanna X 2, 167))

This passage describes a method of extracting nasal polyps with no parallel in the ancient medical texts.[30] It is a rather extraordinary example of a surgical technique that was still included in medical teaching books as late as 1888.[31] In fact, in *Diseases* II, 4 different methods of nose polyp removal are described: in addition to the above-mentioned method, the author also suggests cauterization with irons, extraction by tying it with a ligature or extraction with a scalpel.[32]

The difficulties involving the text have led to different interpretations,[33] mostly concerning details or particular steps of the surgery. However, the practical expertise that was needed in order to carry out such an intervention, together with the knowledge of the anatomy of the parts involved and the dexterity with the small surgical instruments, is apparent. That this surgery could not be learnt only from this text seems evident; nevertheless, such a detailed description would indeed be an adequate complement to practical training both as a reminder and to keep in mind the thought that such surgical intervention existed and was actually carried out by expert physicians. It is interesting to compare the account of the operation written by the author of *Diseases* II and the only other mention of this procedure extant in the *Hippocratic Collection*. In the treatise *Affections*

---

[30] For a parallel passage explaining this same technique in *Aff.*, see below. A similar procedure is described by Paul of Aegina, VI, 25 (Heiberg CMG IX 2, 64.28), who uses a knotted string instead of a sponge to push out the polyps and draw them into the oral cavity: λίνον οὖν παχὺ μετρίως, ὅσον σφήκωμα, ὡς ἀπὸ δυεῖν ἢ τριῶν δακτύλων κονδύλοις καταδήσαντες ἐνείρομεν διπυρήνου τρήματι καὶ τὸ ἕτερον πέρας τοῦ διπυρήνου διὰ τῆς ῥινὸς ἐμβαλοῦμεν ἄνω πρὸς τοὺς ἠθμοειδεῖς πόρους καὶ διὰ τῆς ὑπερῴας αὐτὸ καὶ τοῦ στόματος διεκβάλλοντες διασύρομεν ταῖς δύο χερσὶν ὥσπερ διαπρίζοντες τοῖς κονδύλοις τὰ σαρκώματα.

[31] See Vancil 1969, where, drawing on this and the following chapters of *Morb.* II, the author celebrates Hippocrates, calling him not only 'Father of Medicine', but also 'Father of Rhinology'.

[32] *Morb.* II 34–7 (L. VI 50 = Potter V 248 = Jouanna X 2, 168).

[33] See the edition by Potter, where part of the text is marked as corrupt. For a commentary on the passage, see Jouanna 1983, 167, 238 and Milne 1907, 77.

there is a parallel passage where this surgery to extract polyps is summarized in just one sentence:

> Ἢν δὲ ἐν τῇ ῥινὶ πώλυπος γένηται... ἐξαιρεῖται δὲ βρόχῳ διελκόμενος ἐς τὸ στόμα ἐκ τῆς ῥινός· οἱ δὲ καὶ φαρμάκοισιν ἐκσήπονται·
>
> If a polyp forms in the nose... it is extracted with a snare by pulling it through into the mouth from the nose. Others are made putrid with drugs. (*Aff.* 5 [[7,7]] (L. VI 214 = Potter V 246))

In contrast to the thorough description of the surgical intervention made by the author of *Diseases* II, in *Affections* the reader is confronted with information that is relevant not for performing this act, but for being aware of the procedure purely as interesting information or even as a warning in case the reader were indeed suffering from polyps. Note that *Affections* was intended to be read by laymen whose interest in medicine was the one natural and suitable for intelligent people,[34] but who would never practise medicine, never mind carry out surgery.

None of the oldest manuscripts[35] transmitting *Diseases* II contains any images at all. Much has been speculated about how widespread illustrated books were in antiquity. Just as the copies of the texts preserved in mediaeval manuscripts descend ultimately from the author's versions, the images eventually included in them must have been transmitted from antiquity too, although text and image do not necessarily belong to a single tradition.[36] This necessarily raises the question of whether images actually played a significant role in the communication of medical knowledge, in particular of therapy. That images were included in books in antiquity is clear from the different passages in the texts where authors themselves make reference to illustrations;[37] however, the purpose and function of the images and whether they constitute a non-verbal component actually pertinent to the transmission of particular medical knowledge are controversial.[38] On the one hand, the use of images in manuscripts is rather limited,[39] while on

---

[34] On this, see Pérez Cañizares 2010.
[35] Marcianus gr. 269, s. X (M) and Vindobonensis med. gr. 4, s. XI (θ).
[36] See Stückelberger 1994, Jones 1998 and Nickel 2005. For an overview of the scholarship on the origin of medical illustrations in mediaeval manuscripts and their role, see Lazaris 2010. I thank Professor Christian Brockmann for providing me with the essential bibliography on this topic.
[37] For a list of these passages, see Stückelberger 1994, 125–33.
[38] See for instance Jones 2006, who criticizes the anachronistic position of considering the images included in medical mediaeval manuscripts as 'medical illustrations' in a modern sense. Lazaris 2010 rejects the idea of images as a visual aid to help understand the texts better.
[39] An overview of extant illuminated manuscripts can be obtained by consulting the *Mackinney Collection of Mediaeval Medical illustrations*, available online at www.lib.unc.edu/dc/mackinney/index.html

the other the diversity of functions that these medical books could take – depending on for whom they were copied and for which purpose they were actually used – multiplies the possible roles that the images inserted in the books played and therefore leaves out the simplistic assumption of generalizing about their role as visual explanations of the text they refer to.[40] In the rare cases in which texts appear together with illustrations, a multiplicity of factors have to be taken into account to establish the relationship between them, as well as the role of the image.[41]

That physicians, probably with teaching purposes in mind, might have actually used images in Galen's time is suggested by a passage of *The Method of Medicine*, where Galen directs a barrage of criticism towards Julian, a physician belonging to the Methodist school of medicine. Among other acidic comments, Galen mentions that Julian, in his books, raises the question of whether painting is useful to the doctors.[42] Nevertheless, the lack of more evidence of a possible use of images made by doctors rather hints at its being a limited practice.[43]

Apollonius of Citium, who in the first century B.C. wrote a commentary on selected parts of the Hippocratic book *On Joints*, programmatically included images (ζωγραφική, ὑπόδειγμα) within his work with the purpose of providing a visual impression of the different techniques of orthopaedic reduction.[44] In this case the images were used to make possible a better understanding of the topic, as the author himself expressly

---

[40] For the difficulties in using illustrations as a guide to action in mediaeval surgical texts, see McVaugh 1998.

[41] Nickel 2005 presents examples of works in which text and image match and transmit the very same information and others in which there is an obvious disagreement between text and image. See also Lazaris 2010.

[42] Galen, *Meth. Med.* I 7, (Johnston and Horsley, I 84 = K. X 53.17) κατ' οὐδεμίαν αὐτῶν ἐτόλμησεν εἰπεῖν ὅ τί ποτ' ἐστὶ νόσος, καίτοι γε μηδὲν πρὸς ἔπος ἐν αὐταῖς διεξέρχεται μέχρι τοῦ καὶ τὰ τοιαῦτα ζητεῖν, εἰ ζωγραφία χρήσιμος ἰατροῖς ἐστιν· 'In none of them [the books], however, has he [Julian] ventured to say what a disease is. Indeed, he details nothing to the purpose in them, even going so far as to look into such things as whether painting is useful to doctors.' Like Marganne 2004, 159, I also interpret this passage as referring to the existence of illustrated medical books, related to the use of images with didactic or presentation purposes. Nevertheless, see Hankinson 1991, 145, who considers it just as an ironic remark: 'it is not clear what the point of the remark about painting is supposed to be'. In Tecusan 2004 there is no reference to any sort of use of illustrations by the Methodist physicians.

[43] Just as in the case of mediaeval manuscripts, the evidence from the papyri of Greco-Roman Egypt shows that illustrations were indeed used. However, the numbers indicate that they were far from being frequent and hint rather at a very marginal use. Marganne 2004 mentions 2 illustrated papyri from a total of about 250 papyri of medical content.

[44] Cf. Apollonius Citiensis, *In Hipp. De artic. comm.* I, 1 (Kollesch-Kudlien CMG XI 1,1, 14.7): ὁ δὲ κατὰ μέρος χειρισμός, ὃν τρόπον ὑποτετάχαμεν, οὕτως ἂν ἐπιτελοῖτο· τοὺς δὲ ἑξῆς τρόπους τῶν ἐμβολέων δι' ὑπομνημάτων, ζωγραφικῆς δὲ σκιαγραφίας τῶν κατὰ μέρος ἐξαρθρήσεων παραγωγῆς τε τῶν ἄρθρων ὀφθαλμοφανῶς τὴν θέαν αὐτῶν παρασχησόμεθά σοι.

declares a few times.[45] However, Apollonius also mentions the limitations of his visual resources for the purpose of being translated into actions, although he undoubtedly considered that they contributed to the comprehension of the surgical procedures;[46] the fact that the dedicatee was one of the Ptolemies excludes the possibility that these visual representations were meant, together with the descriptions, as a guide to action.[47]

The oldest surviving manuscript of Apollonius' work is a precious illuminated exemplar dated to around 900 CE, which contains an extensive collection of ancient surgical texts.[48] It is also the oldest testimony of Hippocratic treatises such as *On Fractures / On Joints*,[49] which present numerous examples of detailed medical interventions, such as the following, where it is explained how to fix a dislocated shoulder:

> χρὴ δὲ τὸν μὲν ἄνθρωπον χαμαὶ κατακλῖναι ὕπτιον, τὸν δὲ ἐμβάλλοντα χαμαὶ ἵζεσθαι ἐφ᾽ ὁπότερα ἂν τὸ ἄρθρον ἐκπεπτώκῃ. ἔπειτα λαβόμενον τῇσι χερσὶ τῇσι ἑωυτοῦ τῆς χειρὸς τῆς σιναρῆς, κατατείνειν αὐτήν, τὴν δὲ πτέρνην ἐς τὴν μασχάλην ἐμβάλλοντα ἀντωθεῖν, τῇ μὲν δεξιῇ ἐς τὴν δεξιήν, τῇ δὲ ἀριστερῇ ἐς τὴν ἀριστερήν. δεῖ δὲ ἐς τὸ κοῖλον τῆς μασχάλης ἐνθεῖναι στρογγύλον τι ἐναρμόσσον· ἐπιτηδειότα ται δὲ αἱ πάνυ σμικραὶ σφαῖραι καὶ σκληρα, οἷαι ἐκ τῶν πολλῶν σκυτέων ῥάπτονται· ἢν γὰρ μή τι τοιοῦτον ἐγκέηται, οὐ δύναται ἡ πτέρνη ἐξικνεῖσθαι πρὸς τὴν κεφαλὴν τοῦ βραχίονος· κατατεινομένης γὰρ τῆς χειρὸς, κοιλαίνεται ἡ μασχάλη· οἱ γὰρ τένοντες οἱ ἔνθεν καὶ ἔνθεν τῆς μασχάλης, ἀντισφίγγοντες, ἐναντίοι εἰσίν. χρὴ δέ τινα ἐπὶ θάτερα τοῦ κατατεινομένου καθήμενον κατέχειν κατὰ τὸν ὑγιᾶ ὦμον, ὡς μὴ περιέλκηται τὸ σῶμα, τῆς χειρὸς τῆς σιναρῆς ἐπὶ θάτερα τεινομένης· ἔπειτα ἱμάντος μαλθακοῦ πλάτος ἔχοντος ἱκανόν, ὅταν ἡ σφαῖρα ἐντεθῇ ἐς τὴν μασχάλην, περὶ τὴν σφαῖραν περιβεβλημένου τοῦ ἱμάντος, καὶ κατέχοντος, λαβόμενον ἀμφοτέρων τῶν ἀρχέων τοῦ ἱμάντος, ἀντικατατείνειν τινά, ὑπὲρ κεφαλῆς τοῦ κατατεινομένου καθήμενον, τῷ ποδὶ προσβάντα πρὸς τοῦ ἀκρωμίου τὸ ὀστέον. ἡ δὲ σφαῖρα ὡς ἐσωτάτω

---

[45] Cf. for instance Apol. Cit., *In Hipp. De artic. comm.* II, 10 (Kollesch-Kudlien CMG XI 1,1, 38.7) πρότερον δὲ τὰς λέξεις αὐτοῦ καταχωριῶ, εἶτ᾽ εἰρομένως τὸν τῶν ἐμβολῶν τρόπον δι᾽ αὐτῶν τῶν ὑποδειγμάτων ὑποτάξω πρὸς τό, καθάπερ καὶ ἐν τοῖς πρότερον, καὶ ἐν τούτοις εὐπαρακολούθητά σοι γενέσθαι τήν τε περὶ ἄρθρων θεωρίαν.

[46] Apol. Cit., *In Hipp. De artic. comm.* II, 10 (Kollesch-Kudlien CMG XI 1,1, 38.14) τὸ μὲν γὰρ ὑποδείγματος <οὗ> (add. Kollesch) ἔχει τρόπον ἐπὶ τὴν χρείαν μεταγόμενον, τὸ δ᾽ ἀπ᾽ αὐτῶν τῶν συμβαινόντων ἐναργῆ τὴν κατάληψιν ἐγχειρίζει.

[47] Marganne 2004, 158 mentions that illustrations included in medical books in the first century BC became works of art worthy enough to be dedicated to kings.

[48] Laurentianus Pl. 74.7, also known as the Nicetas codex, is digitally available via the website of the *Biblioteca Medicea Laurentiana*: http://teca.bmlonline.it/TecaRicerca/index.jsp

[49] Abbreviated B for the transmission of Hippocratic texts, the Laurentianus Pl. 74.7 contained originally the books *Fract.* (*De fracturis*), *Art.* (*De articulis*), *Off.* (*De officina medici*), *V.C.* (*De capitis vulneribus*), *Mochl.* (*Vectiarius*) and *Oss.* (*De natura ossium*), but the folios containing the end of *V.C.*, *Mochl.* and *Oss.* have been lost. See Hanson 1999, 15.

καὶ ὡς μάλιστα πρὸς τῶν πλευρέων κείσθω, καὶ μὴ ὑπὸ τῇ κεφαλῇ τοῦ βραχίονος.

The patient should lie on his back on the ground, and the operator should sit on the ground on whichever side the joint is dislocated. Then, grasping the injured arm with both hands he should make extension and exert counter-pressure by putting the heel in the armpit, using the right heel for the right armpit, and the left for the left. In the hollow of the armpit one should put something round fitted to it – the very small and hard balls such as are commonly sewn up from bits of leather are most suitable. For, unless something of the kind is inserted, the heel cannot reach the head of the humerus, for when extension is made on the arm the axilla becomes hollow and the tendons on either side of it form an obstacle by their contraction. Someone should be seated on the other side of the patient undergoing extension to fix the sound shoulder so that his body is not drawn round when the injured arm is pulled the other way. Take, besides, a fairly broad strap of soft leather, and after the ball is put into the armpit, the strap being put round and fixing it, someone, seated at the head of the patient undergoing traction, should make counter-extension by holding the ends of the strap, and pressing his foot against the top of the shoulder-blade. The ball should be put as far into the armpit and as near the ribs as possible, not under the head of the humerus. (*Art.* 4 (L. IV 82 = Kühlewein II 114 = Withington III 205))

The apparent difficulty in following instructions so as to carry out surgery has been stressed before with the example of nasal polyposis and appears again in the directions described in this book to treat dislocations and fractures. However, the authors' perspective is also necessary in order to shed light on the actual effectiveness of the whole communication process. This raises the questions of how medical knowledge gained from experience was communicated in written form,[50] how the authors experienced the activity of transmitting their technical procedures in an accomplished way and to what extent these written accounts were considered effective. The author of *On Fractures / On Joints* explicitly writes that this was not an easy task:

Ἀλλὰ γὰρ οὐ ῥηίδιον χειρουργίην πᾶσαν ἐν γραφῇ διηγεῖσθαι, ἀλλὰ καὶ αὐτὸν ὑποτοπεῖσθαι χρὴ ἐκ τῶν γεγραμμένων.

For the rest, it is not easy to give exact and complete details of an operation in writing; but the reader should form an outline of it from the description. (*Art.* 33 (L. IV 148 = Kühlewein II 151 = Withington III 258))

---

[50] See the general introduction in Kullmann and Althoff 1993.

However, difficult as it might have been, the author of *On Fractures / On Joints* systematically gathered a large variety of existing treatments for orthopaedic surgery. His critical attitude towards other physicians' practices,[51] his mention of cases in which a particular treatment he applied was not successful,[52] or his preference for a particular therapy among the existing ones[53] stress his desire to produce an extensive compendium of contemporary surgery.

The process of putting down in writing, of expressing in words, a concrete medical experience is intrinsically linked with the particular purpose of each author and involves various choices, including that of a particular genre. Thus this author decided to put down a series of therapeutic techniques he had previously learned and later on put them into practice himself; he also collected other procedures that he had seen done in public performances,[54] introducing a reasoned opinion about their efficiency. This stresses the importance of oral transmission for disseminating and acquiring medical knowledge, as it is the main source of the information this author introduces into his works; it also shows how written texts

---

[51] See, for instance, *Fract.* 1–2 (L. IV 414 = Kühlewein II 46 = Withington III 94), where the author attacks what he calls οἱ δὲ ἰητροὶ σοφιζόμενοι, the know-it-all physicians, as they usually make mistakes when bandaging a dislocated arm. οἶδα ἰητροὺς σοφοὺς δόξαντας εἶναι ἀπὸ σχημάτων χειρὸς ἐν ἐπιδέσει, ἀφ' ὧν ἀμαθέας αὐτοὺς ἐχρῆν δοκεῖν εἶναι. Also *Fract.* 30 (L. IV 518 = Kühlewein II 90 = Withington III 165), where the author disapproves of those physicians who in all cases of leg fracture fasten the foot to the bed.

[52] The author describes how he had once tried to reduce humpback by inflating a wineskin located under the patient's back while he was lying down. This attempt failed, but the author considers it worth telling, as it is possible to learn from failures too. *Art.* 47 (L. IV 212 = Kühlewein II 182 = Withington III 302) ἔγραψα δὲ ἐπίτηδες τοῦτο· καλὰ γὰρ καὶ ταῦτα τὰ μαθήματά ἐστιν, ἃ, πειρηθέντα, ἀπορηθέντα ἐφάνη, καὶ δι' ἄσσα ἠπορήθη. For a commentary on this passage, see Lloyd 1987, 125.

[53] See *Art.* 48 (L. IV 212 = Kühlewein II 182 = Withington III 302), where the author claims to know more methods for treating a dislocation of the vertebrae than the ones just presented, but says he will not present them because he has no faith in them: τρόπους δὲ ἄλλους κατατασίων, ἢ οἷοι πρόσθεν εἴρηνται, ἔχοιμ' ἂν εἰπεῖν, ἁρμόζειν ἂν δοκέοντας μᾶλλον τῷ παθήματι· ἀλλ' οὐ κάρτα πιστεύω αὐτοῖσι· διὰ τοῦτο οὐ γράφω.

[54] His experience is based on cases he practised himself and those in which he saw others apply a treatment. See, for instance, *Fract.* 4 (L. IV 430 = Kühlewein II 52 = Withington III 102), where it is stressed that a very strong extension is needed should both the forearm bones be broken. As a rule, most patients do not get enough extension, but the author also mentions that he once saw the bones of a child extended more than necessary. Illustrative of how these public performances must have been is a passage where he says practitioners who try to treat a humpback with succussions on a ladder do it just for show: *Art.* 42 (L. IV 182 = Kühlewein II 167 = Withington III 282) χρέωνται δὲ οἱ ἰητροὶ μάλιστα αὐτῇ οὕτως ἐπιθυμέοντες ἐκχαυνοῦν τὸν πολὺν ὄχλον· τοῖσι γὰρ τοιούτοισιν ταῦτα θαυμάσιά ἐστιν, ἢν ἢ κρεμάμενον ἴδωσιν, ἢ ῥιπτεόμενον ἢ ὅσα τοῖσι τοιούτοισιν ἔοικεν· καὶ ταῦτα κληΐζουσιν αἰεὶ καὶ οὐκέτι αὐτοῖσι μέλει, ὁποῖόν τι ἀπέβη ἀπὸ τοῦ χειρίσματος, εἴτε κακὸν εἴτε ἀγαθόν.

and oral discourse were interrelated, and how necessarily connected they were.⁵⁵

Evidence of the transition from oral to written culture in ancient Greece has been traced by scholarship in many domains.⁵⁶ In the field of medicine, literacy has been linked to changes such as the development of a scientific method,⁵⁷ and to the necessity to preserve, systematize and transmit the medical knowledge that had been acquired empirically.⁵⁸ Some of the Hippocratic treatises, as we read them today, were originally intended for oral delivery.⁵⁹ They are texts that present a particular series of stylistic and rhetorical devices, and were composed at a time when medical debates were crucial.⁶⁰

The author of *On Fractures / On Joints* is very conscious of the advantages of writing, among which the correction of mistakes in treatment⁶¹ and the standardization of surgical procedures are especially important.⁶² These bespeak the didactic setting of the treatise, which is described in the very first chapter as a δίδαγμα, a lesson.⁶³

---

⁵⁵ See Demont 1993, 195, who mentions that speeches meant for oral delivery eventually show characteristics of written texts. He also states that some intertextual features present in medical discourses such as *V.M.* (*De vetere medicina*) and *Acut.* (*De victu in acutis*), or *Morb.Sacr.* (*De morbo sacro*) and *Flat.* (*De ventis*), can only be explained by the existence of common written patterns.

⁵⁶ On literacy in ancient Greece, see Thomas 1992.

⁵⁷ Lonie 1983 and Miller 1990.

⁵⁸ Van der Eijk 1997, 98 and 2005, 38. Dean-Jones 2003 mentions the dissemination of previously restricted medical knowledge among a wider community – charlatans included – as a consequence of widespread literacy. Demont 1993 considers this transition as a good chance to expand the audience.

⁵⁹ Regarding works composed to be delivered orally, Jouanna 1999, 80 distinguishes between public speeches, intended for a broader audience, and didactic presentations, addressed to students or specialists. The treatises *Flat.* and *De arte* are examples of public rhetorical discourses, whereas *Gen.* (*De genitura*) / *Nat.Puer.* (*De natura pueri*) and *Aer.* (*De aere, aquis, locis*) are examples of courses in particular aspects of medicine. For *V.M.* as a work intended for oral delivery, see Schiefsky 2005, 36.

⁶⁰ The most illustrative example is probably the beginning of *Morb.* I, where the author mentions his target audience: people who want to participate successfully in a medical debate. See *Morb.* I 1 (L. VI 140= Potter V 98 = Wittern 2) ὃς ἄν περὶ ἰήσιος ἐθέλῃ ἐρωτᾶν τε ὀρθῶς, καὶ ἐρωτώμενος ἀποκρίνεσθαι, καὶ ἀντιλέγειν ὀρθῶς, ἐνθυμεῖσθαι χρὴ τάδε.

⁶¹ For a reference to the need for physicians to unlearn bad practice, see *Fract.* 25 (L. IV 500 = Kühlewein II 83 = Withington III 152) οὐ μέντοι γε ἂν ἔγραφον περὶ τούτου τοσαῦτα, εἰ μὴ εὖ μὲν ᾔδειν ἀσύμφορον ἐοῦσαν τὴν ἐπίδεσιν, συχνοὺς δὲ οὕτως ἰητρεύοντας· ἐπίκαιρον δὲ τὸ ἀπομάθημα· μαρτύριον δὲ τοῦ ὀρθῶς γεγράφθαι τὰ πρόσθεν γεγραμμένα, εἴτε μάλιστα πιεστέα τὰ κατήγματα, εἴτε ἥκιστα. Also *Fract.* 1 (L. IV 414 = Kühlewein II 47 = Withington III 96) ἐθέλω τῶν ἁμαρτάδων τῶν ἰητρῶν, τὰς μὲν ἀποδιδάξαι.

⁶² See Lonie 1983, who mentions the relationship between the increase in literacy associated with the sophist movement and the innovations introduced in a traditional manual craft such as the treatment of fractures.

⁶³ *Fract.* 1 (L. IV 414 = Kühlewein II 47 = Withington III 96).

That these treatises originated within a didactic context in which literacy was essential is also highlighted by the numerous references to a whole cluster of works this author planned to write.[64] Such references to further medical works are by no means uncommon in the *Corpus Hippocraticum*.[65] That said, the book *On Joints* contains the highest number of such allusions in the whole of the *Corpus*.[66]

Nevertheless, as was the case before with the example of nasal polyposis, these medical handbooks must necessarily have been conceived as a complement to practical instruction;[67] in this particular case, this is explicitly stated by the author of *On Fractures / On Joints* when he explains how healthy members should be used as a reference to be compared with dislocated or broken ones:

> Καὶ τοῦτο εἴρηται μὲν ὀρθῶς, παρασύνεσιν δὲ ἔχει πάνυ πολλὴν διὰ τὰ τοιαῦτα. καὶ οὐκ ἀρκεῖ μοῦνον λόγῳ εἰδέναι ταύτην τὴν τέχνην, ἀλλὰ καὶ ὁμιλίῃ ὁμιλεῖν.
>
> And this is correct, but due to such reasons there is certainly much misunderstanding. And it is not enough to know this art only theoretically, but also [it is necessary to] be acquainted with the practice. (*Art.* 10 (L. IV 102 = Kühlewein II, 126 = Withington III 220))

To conclude, the evidence shows that even if the texts sometimes 'speak for themselves', it is not surprising that the quantity of previous information shared by author and intended reader on a particular topic is decisive and that authors omit what they consider obvious for the audience they have in mind. Communication greatly depends on shared knowledge – empirical knowledge or knowledge from more formal training. Regarding therapy, everyday measures such as administering particular food can be easily followed as written instructions, whereas accounts of how surgery was carried out imply a prior theoretical understanding of anatomy and

---

[64] On this, see Craik 2002, Roselli 2006 and Pérez Cañizares 2015.

[65] Littré, in the first volume of his edition of Hippocrates, collected numerous references of this type, while searching for Hippocrates' lost works. See Littré 1839, I 54.

[66] The tense used is always the future, and the diversity of expressions used highlights the varied style of this author. The topics he planned to deal with are as follows: the rules of massage *Art.* 9 (L. IV 100 = Kühlewein II 125 =Withington III 220); the structure and function of glands *Art.* 11 (L. IV 108 = Kühlewein II 129= Withington III 226); the danger of wounds in temporal muscles *Art.* 30 (L. IV 142 = Kühlewein II 146 = Withington III 252); the parts of the body filled with water or mucus *Art.* 40 (L. IV 174 = Kühlewein II 163= Withington III 278); chronic diseases of the lung *Art.* 41 (L. IV 182 = Kühlewein II 167= Withington III 282); veins and arteries *Art.* 45 (L. IV 190 = Kühlewein II 171= Withington III 288); the movements and contractions of the uterus *Art.* 57 (L. IV 246 = Kühlewein II 202 = Withington III 332).

[67] On the surgical texts of the *Hippocratic Collection* (*Art. Fract. V.C.*) as a complement to practical instruction, see Craik 2010.

practical skills as to how to proceed and use the medical instruments needed. These surgical accounts cannot be interpreted as mere instruction manuals, but rather as adjuncts to practical learning. On the other hand, the abstraction needed to describe complex surgical interventions and to put them in written form bears considerable difficulties too, but represents an effective way of systematizing medical knowledge.

REFERENCES

G. Bendz and I. Pape 1993, *Caelius Aurelianus Teil II, Chronische Krankheiten III–V Indizes*, CML VI 1 (Berlin)

E. M. Craik 1998, *Hippocrates: Places in Man* (Oxford)
  2002, 'Phlegmone, Normal and Abnormal' in A. Thivel and A. Zucker (eds.), *Le Normal et le pathologique dans la collection hippocratique*. Actes du Xème colloque international hippocratique, Nice, 6–8 October 1999 (Nice), 285–301
  2009, 'Hippocratic Bodily "Channels" and Oriental Parallels', *Medical History*, 53: 105–16
  2010, 'The Teaching of Surgery' in H. F. J. Horstmannshoff (ed.), *Hippocrates and Medical Education*. Selected Papers Read at the XIIth International Hippocrates Colloquium, Universiteit Leiden, 24–6 August 2005 (Leiden), 223–33

L. Dean-Jones 2003, 'Literacy and the Charlatan in Ancient Greek Medicine' in Y. Harvey (ed.), *Written Texts and the Rise of Literate Culture in Ancient Greece* (Cambridge), 97–121

P. Demont 1993, 'Die Epideixis über die Technē; im V. und IV. Jh.' in W. Kullmann and J. Althoff (eds.), *Vermittlung und Tradierung von Wissen in der griechischen Literatur* (Tübingen), 181–210

R. J. Hankinson 1991, *Galen: On the Therapeutic Method Books I and II* (Oxford)

M. Hanson 1999, *Hippocrates: On Head Wounds*, CMG I 4,1 (Berlin)

J. L. Heiberg 1924, *Paulus Aegineta, Libri V–VII*, CMG IX 2 (Leipzig, Berlin)

D. Irmer 1980, 'Die Bezeichnung der Knochen in Fract. und Art.' in M. D. Grmek (ed.), *Hippocratica*. Actes du Colloque hippocratique de Paris, 4–9 September 1978 (Paris), 265–83

I. Johnston and G. H. Horsley 2011, *Galen: Method of Medicine, Books 1–4* (London, Cambridge, MA)

P. M. Jones 1998, *Medieval Medicine in Illuminated Manuscripts* (London, Milan)
  2006, 'Image, Word and Medicine in the Middle Ages' in J. A. Givens, K. Reeds and A. Touwaide (eds.), *Visualizing Medieval Medicine and Natural History, 1200–1550*, 1–24

J. Jouanna 1983, *Hippocrate X: Maladies II* (Paris)

J. Kollesch, F. Kudlien 1965, *Apollonii Citiensis In Hippocratis De articulis commentarius* (in linguam Germanicam transtulerunt J. Kollesch et D. Nickel), CMG XI 1,1 (Berlin)

H. Kühlewein 1902, *Hippocratis opera quae feruntur omnia*, CMG II (Berlin)

J. H. Kühn and U. Fleischer 1986–9, *Index Hippocraticus* (Göttingen)
W. Kullmann and J. Althoff 1993, *Vermittlung und Tradierung von Wissen in der griechischen Literatur* (Tübingen)
V. Langholf 1996, 'Nachrichten bei Platon über die Kommunikation zwischen Ärzten und Patienten' in R. Wittern and P. Pellegrin (eds.), *Hippokratische Medizin und antike Philosophie* (Hildesheim), 113–42
S. Lazaris 2010, 'L'Illustration des disciplines médicales dans l'antiquité: Hypothèses, enjeux, nouvelles interprétations' in M. Bernabò (ed.), *La collezione di testi chirurgici di Niceta*. Florence, Biblioteca Medicea Laurenziana, Plut. 74.7 (Rom e), 99–109
E. Littré 1839–61, *Oeuvres complètes d'Hippocrate*, 10 vols. (Paris)
G. E. R. Lloyd 1979, *Magic, Reason and Experience: Studies in the Origin and Development of Greek Science* (Cambridge University Press)
  1987, *The Revolutions of Wisdom: Studies in the Claims and Practice of Ancient Greek Science* (Berkeley)
I. M. Lonie 1981, *The Hippocratic Treatises 'On Generation', 'On the Nature of the Child', 'Diseases IV'* (Berlin, New York)
  1983, 'Literacy and the Development of Hippocratic Medicine' in F. Lasserre and Ph. Mudry (eds.), *Formes de pensée dans la Collection Hippocratique*. Actes du IVe Colloque International Hippocratique, Lausanne, 21–6 September 1981 (Geneva), 145–61
C. Magdelaine 1994, 'Histoire du texte et édition critique, traduite et commentée des Aphorismes d'Hippocrate.' Unpublished PhD thesis, Université Sorbonne-Paris IV
M. H. Marganne 2004, 'Aport de la Papyrologie à l'histoire de la médecine', *Histoire des Sciences Médicales*, 38, 2: 157–64
M. McVaugh 1998, 'Therapeutic Strategies: Surgery' in M. D. Grmek (ed.), *Western Medical Thought from Antiquity to the Middle Ages* (Cambridge, MA), 273–90
G. Miller 1990, 'Literacy and the Hippocratic Art: Reading, Writing and Epistemology in Ancient Greek Medicine', *Journal of the History of Medicine*, 45: 11–40
J. Milne 1907, *Surgical Instruments in Greek and Roman Times* (Oxford)
E. Nachmannson 1918, *Erotiani vocum Hippocraticarum collectio cum fragmentis* (Gothenburg)
D. Nickel 2005, 'Text und Bild im antiken medizinischen Schrifttum', *Akademie Journal*, 1: 16–20
V. Nutton 2004, *Ancient Medicine* (London)
P. Perez Cañizares 2010, 'The Importance of Having Medical Knowledge as a Layman: The Hippocratic Treatise Affections in the Context of the Hippocratic Collection' in H. F. J. Horstmannshoff (ed.), *Hippocrates and Medical Education*. Selected Papers Read at the XIIth International Hippocrates Colloquium, Universiteit Leiden, 24–6 August 2005 (Leiden), 87–99
  2015, 'The Hippocratic Treatise Affections in the Context of the Hippocratic Corpus' in L. Dean-Jones and R. J. Hankinson (eds.), *What is Hippocratic about the Hippocratic Corpus?* (Leiden), 83–98

P. Potter 1980, *Hippokrates: De Morbis III*. CMG I 2,3 (Berlin)
1988, *Hippocrates Vols.* V *and* VI. (London, Cambridge, MA)
L. Renaut 2004, 'Marquage corporel et signation religieuse dans l'antiquité'. Thèse de doctorat, École Pratique des Hautes Études (Paris)
A. Roselli 2006, 'Strategie spositive nei trattatti ippocratici: Presenza autoriale e piano espositivo in Malattie IV e in *Fratture e Articolazioni'* in M. M. Sassi (ed.), *La costruzione del discorso filosofico nell'età dei presocratici* (Pisa), 259–83
M. Schiefsky 2005, *Hippocrates on Ancient Medicine: Translated with Introduction and Commentary* (Leiden)
A. Stückelberger 1994, *Bild und Wort: Das illustrierte Fachbuch in der antiken Naturwissenschaft, Medizin und Technik* (Mainz)
M. Tecusan, *The Fragments of the Methodists: Methodism outside Soranus. Vol.* I: *Text and Translation* (Leiden 2004)
H. Thesleff 1966, 'Scientific and Technical Style in Early Greek Prose', *Arctos* 4: 89–127.
R. Thomas 1992, *Literacy and Orality in Ancient Greece* (Cambridge University Press)
L. M. V. Totelin 2009, *Hippocratic Recipes: Oral and Written Transmission of Pharmacological Knowledge in Fifth- and Fourth-Century Greece* (Leiden)
K. Usener 1990, '"Schreiben" im *Corpus Hippocraticum*' in W. Kullmann, and M. Reichel (eds.), *Der Übergang von der Mündlichkeit zur Literatur bei den Griechen* (Tübingen), 291–9
N. van Brock 1961, *Recherches sur le vocabulaire médical du grec ancien: Soins et guérison* (Paris)
M. E. Vancil 1969, 'A Historical Survey of Treatments for Nasal Polyposis', *The Laryngoscope*, 79, 3: 435–45
P. J. van der Eijk 1997, 'Towards a Rhetoric of Ancient Scientific Discourse: Some Formal Characteristics of Greek Medical and Philosophical Texts (Hippocratic Corpus, Aristotle)' in E. J. Bakker (ed.), *Grammar as Interpretation: Greek Literature in Its Linguistic Contexts* (Leiden), 77–129
2005, *Medicine and Philosophy in Classical Antiquity* (Cambridge)
2015,'On "Hippocratic" and "Non-Hippocratic" Medical Writings' in L. Dean-Jones and R. Rosen (eds.), *Ancient Concepts of the Hippocratic* (Leiden), 17–47
H. von Staden 2007, 'Purity, Purification and Katharsis in Hippocratic Medicine' in M. Vöhler and B. Seidensticker (eds.), *Katharsiskonzeptionen vor Aristoteles: Zum kulturellen Hintergrund des Tragödiensatzes* (Berlin) 21–52
E. T. Withington 1928, *Hippocrates Vol.* III (London, Cambridge, MA)
R. Wittern 1928, *Die Hippokratische Schrift De morbis I: Ausgabe, Übersetzung und Erläuterungen* (Hildesheim, New York)
1998, 'Gattungen im Corpus Hippocraticum 'in W. Kullmann, J. Althoff and M. Asper (eds.), *Gattungen wissenschaftlicher Literatur in der Antike* (Tübingen), 17–36

CHAPTER 7

Naso magister erat – sed cui bono?
On Not Taking the Poet's Teaching Seriously

Alison Sharrock

The major difficulty for Ovid in making claims about the utility of his *Ars Amatoria*, a difficulty which marks his work out as different from all the other texts considered in this volume, is that what he is teaching was illegal when he wrote it. It is not the case, of course, that seeking to have sex with women is against either the letter or the spirit of the Augustan moral legislation, since slaves and prostitutes are readily available, and married procreation is positively encouraged, but seeking to have illicit sex with women whom you persuade to want to give in to you, rather than whom you marry, pay, own or force – that is certainly sailing close to the winds of the law.[1] The only way to claim that the subject matter of the *Ars Amatoria* is perfectly legal requires the pretence that the intended object of seduction is some sort of independent woman who might accept a man out of curiosity, personal pleasure or affection, but who is not subject to the constraints of the law or to the economic constraints that would make such a situation likely – and all that is distinctly implausible.[2] The other possibility is to claim that the work has absolutely nothing to do with Real Life and is just a literary game. Many scholars since Ovid's time have taken the latter option as almost self-evident.[3]

Ovid himself is in no doubt, in his erotodidactic cycle, of the utility of his work; nor does he feel the need to make any apology for its aesthetics. He opens the *Ars Amatoria* with the well-known and confident claim that this poem has the answers for everyone:

> SI QUIS in hoc artem populo non novit amandi,
> hoc legat et lecto carmine doctus amet.

---

[1] Gibson 2003, 25–32. If the argument of McGinn 1991 is correct regarding the provisions of the extremely unclear *Lex Iulia*, even the relationships with other people's concubines apparently recommended by the poem as legal and safe would be on dangerous ground.
[2] James 2003 argues forcefully for appreciation of the economic constraints under which the elegiac beloved operates.
[3] Heath 1985 describes the *Ars Amatoria* (*Ars am.*) as didactic in form but not didactic in purpose.

If anyone in this people does not know the art of loving, let him read this, and having read the poem, let him love learnedly.[4] (*Ars am.* 1.1–2)

In the last poem in the cycle, *Remedia Amoris*, Ovid's claims for utility are even stronger, since this poem has the power to stop the reader committing suicide (the poet is addressing Cupid here):

> cur aliquis laqueo collum nodatus amator
>   a trabe sublimi triste pependit onus?
> cur aliquis rigido fodit sua pectora ferro?
>   invidiam caedis pacis amator habes.
> qui, nisi desierit, misero periturus amore est,
>   desinat, et nulli funeris auctor eris.

Why has any lover tied his neck in a noose and hung, a sad burden, from a high roof beam? Why has anyone dug into his own breast with a hard knife? Though a lover of peace, you [i.e. Cupid] are blamed for slaughter. Let him desist who, unless it stops, is going to perish from miserable love, and you will be the instigator of no funeral. (*Rem. am.* 17–22)

Although the *Remedia* is directed first at men, the poet is at pains to point out that its life-saving utility works for women also:

> e quibus ad vestros si quid non pertinet usus,
>   at tamen exemplo multa docere potest.
> utile propositum est saevas extinguere flammas
>   nec servum vitii pectus habere sui.

If there is anything in this advice which doesn't pertain to your needs, nonetheless it can teach a great deal by example. It is a useful proposal to extinguish the savage flames and to have a heart which is not a slave of its own vice. (*Rem. am.* 51–4)

In the Ovidian way, both these programmatic claims for utility are subject to ironic subversion. The opening invitation to 'anyone who doesn't know' will later be undermined (or perhaps confirmed?) by the claim in the poet's defence of the *Ars Amatoria* that he is teaching *quod nemo nescit, amare* ('what no one doesn't know – loving'),[5] that is, that there is nobody who comes into the category of the reader for whom the *Ars* is useful, because they know it already (implication: their behaviour is not Ovid's fault!). In the case of utility to women, which begins with protection from suicide,

---

[4] All quotations from the *Ars Amatoria* and *Remedia Amoris* (*Rem. am.*) are from Kenney 1994. All translations are my own. Quotations from other Latin works are from relevant OCT editions, unless otherwise stated.
[5] *Tristia* (*Tr.*) 1.1.112.

the proof texts which show that if various famous women had made use of Ovid's poem they would not have killed themselves also show that such use of the poet's advice would cut swathes through the canon of Greco-Roman literature. At *Rem. am.* 55–68, a series of mythological lovers would not have suffered and so made poetry if they had followed Ovid's advice: Phyllis, a character from the Trojan cycle but more immediately from Ovid's *Heroides*, would not have committed suicide in despair of the return of Demophoon;[6] Dido would not have supplied the material for the peerless *Aen.* 4 (or Ovid's *Heroides* 7); Medea would have fallen out of the tragic corpus (and *Heroides* 12); Tereus and his family would not have become birds, thus depriving literature from Aristophanes' *Birds* to Ovid's own *Metamorphoses*; Pasiphae would not have produced the Minotaur (with consequences for the *Ars Amatoria* itself); Helen would have stayed with Menelaus, thus destroying the entire Homeric cycle (and the first pair of the double *Heroides*); and Scylla would not have stolen her father's purple lock (to the loss of neoteric epyllion and the *Metamorphoses*). It is worth noting that, in common with many Ovidian *exempla*, this catalogue does not quite do what it claims to do. The first two women would have been saved from suicide, but after that their salvation, valuable though it might be, would not so much have saved their lives as have condemned them to insignificance and oblivion. No reader of poetry, moreover, could possibly celebrate such a loss.

Nonetheless, for the suffering woman herself such a proposal might indeed seem 'useful'. The language of utility within the erotodidactic cycle is given an extra frisson which is surely not active in most didactic and technical works. *utilis* is a slightly suggestive word, implying sexual potency and effectiveness, such that its obvious didactic purpose is overlaid with a knowing wink. That the word has such connotations can be seen from the poem in the middle of the *Amores*, when Ovid claims that his Callimachean delicacy does not come at the cost of erotic strength:

> saepe ego lascive consumpsi tempora noctis,
> <u>utilis</u> et forti corpore mane fui.

> Often I have used up the whole night in sexual activity, and been <u>useful</u> [use for something] and of strong body in the morning. (*Am.* 2.10.27–8)

The double-entendre is very clear in the explicitly sexual passage at the end of *Ars am.* 2. In describing the benefits of foreplay:

[6] See Kennedy 2006.

On Not Taking the Poet's Teaching Seriously

> fecit in Andromache prius hoc fortissimus Hector
> nec solum bellis utilis ille fuit;

> This is what the hero Hector did first to Andromache and he was not useful only for war. (*Ars am.* 2.709–10)

After having recommended slow delay, Ovid changes tack and indicates that sometimes it can be beneficial to be in a hurry:

> cum mora non tuta est, totis incumbere remis
> utile et admisso subdere calcar equo.

> When delay is not safe, it is useful to lean forward with all your oars and to apply the whip to the running horse. (*Ars am.* 2.731–2)

A more delicate play on the eroticism of utility comes in the *Remedia*, when Ovid is giving the advice that one might get over an affair by getting involved in another affair, and suggests that his own *Ars Amatoria* is the place to look to put that precept into action:

> quaeris ubi invenias? artes tu perlege nostras:
> plena puellarum iam tibi navis erit.

> You ask where you may find one [i.e. a new girl]? Read my *Arts*: your ship will soon be full of girls. (*Rem. am.* 487–8)

Immediately following this precept, a celebration of the poet's erotodidactic power introduces the related advice that our rejected lover should feign lack of concern.

> quod siquid praecepta valent mea, siquid Apollo
> utile mortales perdocet ore meo . . .

> If my precepts have any power, if Apollo teaches mortals anything useful through my mouth . . . (*Rem. am.* 489–90)

Utility, then, for Ovid, is an erotic matter – and erotic advice is useful.

As Horace says explicitly, so Ovid implicitly but clearly has the goal to mix *utile dulci* ('the useful with the sweet', Horace, *Ars poet.* 343), *lectorem delectando pariterque monendo* ('by delighting the reader and equally advising him', 344). This topos, perhaps, is the defining mark of didactic literature, with its twofold teleology as literature and as teaching. In Cicero's rhetorical works, especially those like *De Oratore* which use the dialogue form, the two purposes are particularly closely entwined, since much of the teaching takes place in the form of the thing taught (speeches). The topos occurs at *Brut.* 185–200, and at *De or.* 1.130, 2.144, 2.121. One might expect that such a topos would differentiate didactic literature from those technical discourses which present themselves as having only the simple purpose of teaching. It is, however, a trope which extends quite far into the more technical end of the didactic spectrum. Vitruvius ends the opening chapter of his first book (*De architectura* 1.1.18) with a self-deprecating *captatio benevolentiae* to Augustus, in which he begs the reader (Augustus and *qui ea volumina sunt lecturi*, 'those who are going to read these books') to forgive his failings in expression, on the grounds that, while he has made extravagant claims for the wide range of knowledge required by an architect, such an architect does not need to be expert in all those things, only cognisant of them, and that he, in particular, is primarily an architect, and neither a rhetorician nor a philosopher. Of his authority as an architect, by contrast, he is confident. Even he, however, hopes that the work may be of some interest to those other than the builders who will directly learn from it:

> Namque non uti summus philosophus nec rhetor disertus nec grammaticus summis rationibus artis exercitatus, sed ut architectus his litteris inbutus haec nisus sum scribere. De artis vero potestate quaeque insunt in ea ratiocinationes polliceor uti spero, his voluminibus non modo aedificantibus sed etiam omnibus sapientibus cum maxima auctoritate me sine dubio praestaturum.

> For it is not as a great philosopher nor an eloquent speaker nor a teacher practised in the highest elements of the art, but as an architect who has dipped a toe into these letters that I have strived to write these things. With regard to the power of the art, indeed, and the principles which are in it, I promise, as I hope, that in these volumes I shall excel without doubt and with the greatest authority, not only for builders but indeed for all the wise.[7]

Fögen's study (2009) of the linguistic characteristics of Roman technical literature emphasises the similarities, in this and related regards, between the architect and rhetoricians such as Cicero and Quintilian, and also other technical writers such as Galen and Isidore. The requirement of interdisciplinarity, for example, expressed in this first prologue, is widely shared among technical authors as well as those, like Cicero and Ovid, whose

---

[7] Does Vitruvius mean a contrast here between those who read the work in order to learn how to build and those who read out of interest in the science and/or literature (if such a distinction is meaningful in the ancient world)? The Loeb translator takes him to intend a contrast between inexperienced builders and those who already have some knowledge, but the context of the dedication to Augustus and the reader makes the former interpretation more likely, as is the interpretation also based on the lines by Fögen 2009, 113. Callebat 1982, 716 assumes a mixed audience, including practitioners and lay people, reading for pleasure.

work could more obviously be expected to address a broadly educated audience looking for educated and educative entertainment as well as technical information. It seems, then, that it is almost impossible to impose clear divisions along the spectrum of didactic writing from Ovid's sophisticated games, through Cicero's retirement activities, to Vitruvius' account of architecture.

If the presence of the *dulcis* element in Ovid's erotodidactic cycle goes without saying, the contribution of utility is explicit. There is a great deal in the cycle about what is useful to the lover, not surprisingly, since such is the purpose of the work. What is relevant for our considerations here is the discourse of utility rather than its actuality. I shall mention three examples where the teacher draws particular attention to the utility of his teaching: benefit through pain, every little helps and the failings of rivals.

A common topos of modern, especially popular, medicine and education (and fitness regimes) is the claim that something has to hurt to help. Although this is not particularly common in Ovid's work, it does arise at *Rem. am*:[8]

> dura aliquis praecepta vocet mea; dura fatemur
> esse, sed ut valeas multa dolenda feres.
>
> Someone may call my precepts harsh; I admit that they are harsh, but in order to be well you must bear many painful things. (*Rem. am.* 225–6)

The difficult and painful treatment here, which will be compared in the next couplet with drinking bitter medicine and being refused food when ill, was that you should leave Rome, although that is where your heart is. For any Roman, a lover or not, such an instruction might indeed seem like bitter medicine. The primary sphere of such language must be medicine, where technical writers regularly refer to the problems of patients refusing unpleasant medicine, to the extent that it is said to be easier to treat a slave (whom one can force) than a rich man![9] Ovid's engagement

---

[8] Henderson 1979 on this passage develops a nice double metaphor of a journey and building (a journey through a building) for this section of the poem, where 'the patient is exposed to an increasingly difficult and distressing series of cures up to roughly the geographical centre of the Tractatio... thereafter the going becomes progressively easier (cf. 495–6), for he has been through the fire and merits encouragement' (70–1). Apart from a general reference to Lucretius, however, he does not say much about parallels for 'this is going to hurt'.

[9] I am grateful to David Langslow for this point, and for directing me to Cels. 3.21.2: *facilius in servis quam in liberis tollitur* [i.e. dropsy] *quia, cum desideret famem, sitim, mille alia taedia longamque patientiam, promptius iis succurritur qui facile coguntur quam quibus inutilis libertas est.* He also mentioned a passage of Alexander of Tralles (2.375.11–15), in which the author discusses the difficulty

with the topos, however, may well owe more to the use of quasi-medical language in philosophical writing. Lucretius' honeyed cup, which hides the bitter medicine for children, is only the most famous example of such a metaphor.[10]

The somewhat related idea of *per ardua ad astra* is old.[11] In the *Ars*, the difficult task of putting up with a rival is placed in the context of such virtuous effort:

> ardua molimur, sed nulla nisi ardua virtus;
> difficilis nostra poscitur arte labor.
> rivalem patienter habe: victoria tecum
> stabit, eris magni victor in Arce Iovis.

We strive to the heights, but there is no virtue which is not an uphill struggle; a difficult task is demanded by my Art. Put up with a rival patiently: victory will stand with you and you will be victor in the citadel of great Jupiter. (*Ars am.* 2.537–40)

The application of positive moral value to the effort and suffering involved in the pursuit of one's goals, whatever the absolute moral standing of those goals, resonates strongly with the discourse of the modern health and fitness lobby. As regards the utility of the didactic enterprise, it must be remembered that there is a strand in ancient thought which would take the most noble, and therefore the most useful, intellectual activity to be philosophy.[12]

Elsewhere in the cycle, in a manner foreshadowing the contemporary Tesco slogan 'Every little helps', Ovid makes a series of points which he admits are minor, but which add up to be some real help to the trainee lover.

> tu tantum numero pugna praeceptaque in unum
> contrahe: de multis grandis acervus erit.

You just fight with lots of weapons [the previous lines have shown that in nature the little can bring down the large] and gather my precepts into one: there will be a great pile from the many bits. (*Rem. am.* 423–4)

---

of dealing with patients, especially rich ones, who are unwilling to take medicines or undergo enemas for gastric problems, despite their efficacy. While this text is of course much later than Ovid, it is likely to reflect earlier traditions, especially given the conservative tendencies of medical writing.

[10] Seneca makes regular use of medical metaphors, e.g. *Ep.* 1.8.1–2 *salutares admonitiones, velut medicamentorum utilium compositiones, litteris mando, esse illas efficaces in meis ulceribus expertus, quae etiam si persanata non sunt, serpere desierunt.*

[11] See Janka 1997, ad loc., who compares Hes. *Erg.* 289–90.   [12] Callebat 1982, 709.

The sentiment that a large heap can be made from little parts is proverbial, as is suited to didactic discourse.[13] Specifically here, the teacher is drawing attention to the value of even apparently little bits of advice, little bits of his poem, when they are put together. Further examples occur at *Ars am.* 2.285, 2.293, 2.331, 3.353, *Rem. am.* 715.

A teacher is always more or less in competition with other teachers for his pupils' attention, competition which encourages denigration of the opposition, to such an extent that 'arguing with rivals' can be an effective vehicle for exposition of one's own programme. The medics and the philosophers seem to be particularly inclined to criticise their colleagues.[14] Although compared with some other technical writers, Ovid has relatively little to say about the failings of his intellectual rivals, what he does say is carefully pitched. The most extensive and consistent rejection of alternative methods comes in the references to magic as a way of procuring or curing love, an idea that occurs twice in *Ars am.* 2 (2.99–108, 415–25) and again in *Rem. am.* 249–90, including an account of the ineffectual witch Circe.

> fallitur, Haemonias si quis decurrit ad artes,

Anyone who rushes off to Haemonian arts is deceived... (*Ars am.* 2.99)

> sunt quae praecipiant herbas, satureia, nocentes
> sumere; iudiciis ista venena meis.

There are women who teach you to take up harmful herbs, savory; those are poisons in my judgement. (*Ars am.* 2.415–17)

> viderit, Haemoniae siquis mala pabula terrae
> et magicas artes posse iuvare putat.

He is mistaken, whoever thinks that the baneful crops of the Haemonian land and magic arts can help. (*Rem. am.* 249–50)

All three examples just quoted are heavily marked with didactic language. For all the ironic complexities involved in the refusal of magic, powerful

---

[13] Henderson 1979, ad loc.
[14] From the most technical end of the didactic spectrum come two examples supplied to me by Aimee Schofield: Philo, *Belopoeica* 76.21–77.6, where Philo describes Dionysius of Alexandria's repeat shooting catapult as basically useless, and Heron, *Belopoeica* 73.6–74.4, where Heron criticises the way other writers set out their artillery treatises. Cf. also Vitruvius 6.6:

> Cum autem animadverto ab indoctis et inperitis tantae disciplinae magnitudinem iactari et ab is, qui non modo architecturae sed omnino ne fabricae quidem notitiam habent, non possum non laudare patres familiarum eos, qui litteraturae fiducia confirmati per se aedificantes ita iudicant: si inperitis sit committendum, ipsos potius digniores esse ad suam voluntatem quam ad alienam pecuniae consumere summam.

metaphor as it is for both love and poetry,[15] the pose of rejection functions to throw emphasis on the teacher both as rational scientist and – especially – as more effective than the opposition. Perhaps surprising is that otherwise so little is said about alternative erotic advisers. We do not have to go as far as the sex manuals attributed to Philaenis and others[16] to find possible alternatives: Priapus in Tibullus 1.4 is the most explicit nearby example of a teacher of love, but there is also Ovid's *alter(a) ego* Dipsas in his own *Am.* 1.8, not to mention hers in the Acanthis of Prop. 4.5 (both, it should be noted, represented as witches).[17] Moreover, as Gibson reminds us,[18] Ovid has a fine tradition of rivals in erotodidaxis, ranging from Plato to Propertius, but Ovid does not take his own advice about suffering rivals gladly. He is not going to offer any free advertising to the real opposition, but rather picks an easy target in the witch of myth and the social underworld. At the same time, he uses such interactions as he does have with the 'arguing with rivals' to position himself on the side of rationality within the traditions of intellectual debate. Ovid presents himself as the Lucretius of love, taking the high road of scientific superiority over the *falsa ratio* of the opposition. Like many philosophical works, *De rerum natura* is thick with denigration of rival philosophical schools.[19] Ovid wins both ways.

Despite Ovid's presentation of magic as the despicable inferior alternative to artistic loving, he does offer advice in the *Remedia* on the avoidance of aphrodisiac foods (801),[20] avoidance that he describes as being *utilis* to the recovering lover. We may assume from this that such foods would come under medicine/lifestyle rather than magic – unless, of course, it's a bluff, which would be entirely possible in this poem, which may be entirely bluff. Teaching us about what foods to avoid in order to depress erotic interest may have been a feature of the poem that justified its popularity in times and places when eroticism of this nature was suppressed, but in its own context, in a poem where so much of the generic and imagistic effects work contrary to the proposed anti-erotic purpose, there is just as much risk that the pupil might learn what to seek out when being told what to avoid.[21]

Ovid ends the lesson for both men and women with the boast that each side can truly say *Naso magister erat*. That Ovid was his pupils'

---

[15] Sharrock 1994a, Chapter 2.   [16] Parker 1992.
[17] See Gibson 2003, 14–21 on the erotodidactic tradition, which he describes as, particularly for the third book's address to women, a more important influence on Ovid's poem than conventional didactic literature.
[18] Gibson 2003 13–19.   [19] See Sharrock 1994a, 51–2.
[20] Faulkner 2011, 183, on Callim. *Epigr.* 46 and the fudging of personas between poet and doctor: 'dietetics had in the fourth century come to hold an important place in medical therapy'.
[21] Rosati 2006.

teacher/master in their erotic exploits might seem like an unsurprising claim, which we expect to see duplicated in a wide range of didactic and technical literature. The question arises, however, whether the right to claim magisterial status depends on expertise in the area ('subject knowledge', in the terminology of Her Majesty's Inspectorate), or on teaching skills. Ovid's claim for success in the epilogue to Book 2 is precisely as a practitioner of his art rather than as a teacher of it.

> quantus apud Danaos Podalirius arte medendi,
>   Aeacides dextra, pectore Nestor erat,
> quantus erat Calchas extis, Telamonius armis,
>   Automedon curru, tantus amator ego.
>
> As great as Podalirius was among the Greeks in the art of healing, as Achilles was in fighting [with his right hand] and Nestor in giving advice [in his heart], as great as Calchas was with entrails, and Ajax with weapons, Automedon with his chariot – so great a lover am I. (*Ars am.* 2.735–8)

Several of the heroes and skills taking a bow at this point appeared at the very opening of the *Ars Amatoria* as proof texts for the poet's skill as a teacher (*Ars am.* 1.3–8), and a teacher of Cupid himself. Automedon's chiasmatic appearance, in particular, links the passages together, displaying Ovid's use of two possible directions in which the teacher might go to claim efficacy for his lessons: that, like Isocrates, he may not be great at doing but is brilliant at teaching,[22] or, like Vitruvius, that his right to teach depends on his personal expertise in the subject matter.

But who are the lucky beneficiaries of Ovid's advice? He told us that the men were *si quis . . . in hoc populo*, i.e. any Roman[23] man who does not already know – everyone or no one. The women 'readers', apparently, are to be considered as not respectable and (Ovid hopes) not covered by the Augustan moral legislation. But this is a fudge. Even the distinction between courtship directed towards marriage and courtship directed towards seduction/prostitution is not always kept firmly distinguished in the corpus. The celebratory first poem of *Amores* 2 muddies the water most explicitly:

> me legat in sponsi facie non frigida virgo
> et rudis ignoto tactus amore puer.

---

[22] Isocrates famously lacked the physical stamina for rhetorical performance, but was an outstanding teacher of rhetoric and theoretician of education. See Clark 1957, 6–10. Cicero, by contrast, might reasonably make the claim to base his right to pontificate on rhetorical education precisely on his own position as Rome's greatest orator.
[23] See Volk 2006 on the Romanness of Ovid's erotodidaxis.

> Let the virgin who is not cold in front of her betrothed read me, and the
> inexperienced boy touched by unknown love. (*Am.* 2.1.5–6)

These young pupil-readers have the requisite lack of previous knowledge, thus clearly requiring a level 1 course, but they also look remarkably like the respectable Roman youth whom Ovid will later deny having corrupted. The notorious disclaimers of the *Ars Amatoria* pretend that they can place the implied woman of the relationship in a social category which makes the poem's teaching unproblematic, because such women lack the long dress of the Roman matron (*Ars am.* 1.31–4) etc., while explicit statements distinguish lovers from a married couple (but say nothing about whether either of them might be married to someone else):

> hoc [i.e. arguing] decet uxores, dos est uxoria lites;
>   audiat optatos semper amica sonos.
> non legis iussu lectum venistis in unum;
>   fungitur in vobis munere legis Amor.

> This suits wives. Arguing is a wife's dowry. Let a girlfriend always hear what
> she wants to hear. Not on the order of the law have you come into one bed;
> Love performs the role of law for you. (*Ars am.* 2.155–8)

Contrary to what Ovid may try to claim later, those readers, including Augustus, have a good case who see the potential for adulterous advice even in the instruction nominally addressed to supposedly unproblematic women, since there is a very fine line between advice on how to cheat on a paramour and how to cheat on a husband:

> qua vafer eludi possit ratione maritus
>   quaque vigil custos, praeteriturus eram.
> nupta virum timeat, rata sit custodia nuptae:
>   hoc decet, hoc leges duxque pudorque iubent.
> te quoque servari, modo quam vindicta redemit,
>   quis ferat? ut fallas, ad mea sacra veni.

> I was going to pass over the question of how to elude a cunning husband
> and a wakeful guard. Let a bride fear a man/her husband, let the guarding
> of a bride be firm. This is decent. The laws, the Emperor, and Shame order
> this. But who could put up with you also being guarded, you who have only
> just been freed? In order to cheat, come to my rites. (*Ars am.* 3.611–16)

The 'official' meaning of this passage is to set up a clear opposition between wives and the female recipients of Ovid's advice, presented here firmly in the guise of freedwomen.[24] The trouble is, it doesn't work. Not only could wives learn from the lesson directed to freedwomen, but also the justification for the advice is that his readers should attempt to undermine their guards because it is not right for them to be guarded. This implies that they are being guarded, just like their more respectable counterparts. Gibson (1998) suggests the likelihood that even if it is clear (as clear as mud, as they say) that the women addressed are freedwomen concubines rather than freeborn wives, Ovid's suggestion is still illegal, and that tricking their guards is by no means fine in accordance with the Augustan legislation. In the battle for the drawing up of lines between categories, Ovid no more has absolute control than does the Emperor.

Moreover, an inherent flaw in Ovid's attempted defence is the problem that, so often in the *Ars Amatoria* and indeed the *Remedia*, women are supposed to be doing this for love. Girls are supposedly excessively driven by desire (for example, *Ars am.* 1.347–50, with the stories of female excessive desire described there); the most successful student of Ovid is the one who can make a girl really unhappy because of her love for him (*Ars am.* 2.448). The girl whose social position makes Ovid's advice legal is not likely to be pursuing a relationship for love. Either Ovid and his readers are very naive about how prostitutes function, or, more likely, the implied woman of the relationship is not clearly defined[25] in social or marital status, but may be as varied in this way as she is varied in physical appearance requiring different kinds of *cultus* (and even sexual positions).

The problem is, why should women be interested in Ovid's advice? What use is it for them? If it were straightforwardly advice on how to get the best financial deal out of a lover (and there are indeed snippets along the way),[26] then it would have an obvious utility for women. If it were advice on how to contract an advantageous marriage, then likewise. But in fact it is advice on how to manage extramarital but sentimental affairs, in which the primary motivation is emotional rather than material – in a world in which extramarital sex brings her dangers and potential damage to which men are immune. The first challenge for Ovid is to persuade girls to be involved at all. The result is that Ovid's persuasion to his female audience

---

[24] See Gibson 2003, 336–74 for Ovid's conflation of two aspects of the manumission process. As he says in 337, 'how many *libertinae* would thank Ovid for highlighting this socially sensitive event?'
[25] See Gibson 1998 on there not being much distinction between prostitutes and respectable women in the poem.
[26] James 2003 reads Roman love elegy as coming into being against a background of financial need on the part of the *puella*, for whom advice on how to make a killing would be undoubtedly useful. It is not clear to me, however, that Ovid is always entirely honest about which side he is on.

has to look much more like an act of seduction than a rational argument about utility. Why should the reader be interested in Ovid's advice:

> nostra sine auxilio fugiunt bona: carpite florem,
> qui, nisi carptus erit, turpiter ipse cadet.

> Without help, our blessings flee: pluck the boom, which, unless it is plucked, will of its own accord fall foully. (*Ars am.* 3.79–80)

What has she got to lose?

> ut iam decipiant, quid perditis? omnia constant;
> mille licet sumant, deperit inde nihil.
> conteritur ferrum, silices tenuantur ab usu;
> sufficit et damni pars caret illa metu.

> What do you lose, even if they are deceiving you? Everything is settled; even if a thousand take from there, nothing perishes. Iron is worn away, stones grow thin from use; that place is sufficient and has no fear of loss. (*Ars am.* 3.89–92)

All this, of course, is a lie.[27] Both Ovid and his readers would know that the answer to the first question is, in fact, a great deal.

We might say, then, that the drive of *Ars am.* 3 is erotic persuasion. For all that the book delineates at length exact rules of self-prettification to be followed by girls of different appearance, and makes suggestions about how to manipulate lovers in their several kinds, the mask slips sufficiently often for us to suspect that the coach might still be helping the other team. The force of *Remedia*, beloved (with whatever degree of conscious irony) of the mediaeval monastic curriculum for its anti-erotic properties, also regularly creates an atmosphere more likely to encourage an emotional and erotic mindset than to free the reader from the discourse of love. Even the relatively straightforward first two books of the *Ars Amatoria* are as much concerned with seduction of the reader into the erotic world as with the details of instruction.

Is the nature of the erotodidactic cycle primarily a matter of training to a specific skill, or is it more like the contemporary literate education, which is very theoretical and has a strong sense of itself as 'kingmaker',[28] where the goal is to create a rounded and fully socialised person who can make appropriate contributions to society irrespective of specific knowledge? I would suggest that it is formally the former but finally the latter. What Ovid was doing was, we might say, not so much teaching a set of techniques

---

[27] See Gibson 2003, 122–4, for the mixture of euphemistic delicacy and outrageous suggestion here.
[28] Morgan 1998, Chapter 7, on the nature of learning is important.

but inculcating a mindset, education in its fullest sense by persuasion and attraction as much as by training. I have argued (in Sharrock 1994a) that behind the teacherly relationship of instruction in the *Ars Amatoria* is a seductive attempt to mould the reader, as trainee lover, into the mindset of elegiac erotic discourse, while in 1994b I suggested that we see in the relationship between this poem and Virgil's *Georgics* the development of a kind of political education: just as the *Georgics* does not really teach the nitty-gritty of farming, but creates the mindset of a Roman citizen, so the *Ars Amatoria* is not so much a manual of courtship techniques but a work designed to shape the reader into the antithesis of a certain kind of conservative Roman ideal. If this is correct, then we can perhaps see an interaction, albeit ironic, between what Ovid is doing and what is at the heart of Roman liberal education. Even though this education may well teach certain facts, its true goal is to inculcate values. Ultimately, the product of such an education is not just someone who knows things but, as Cato famously said, *vir bonus dicendi peritus*.[29]

It is a topos of training in oratory that what is required is a general liberal education.[30] The background to this claim may be in part directed towards the goal of aristocratic effortless superiority and in part a response to the old argument as to whether good speaking can be taught.[31] Such a question is at issue also in Ovid's erotodidaxis: the cycle's opening claim for the universal applicability of its teaching is followed by an argument for (or rather, rhetorical statement of) its very status as an art. In case the reader should be in any doubt that love is an art which can be taught, Ovid proves it with parallels:

> arte citae veloque rates remoque moventur,
>   arte leves currus: arte regendus Amor.
> curribus Automedon lentisque erat aptus habenis,
>   Tiphys in Haemonia puppe magister erat:
>
> By art swift ships move with sail and oar, by art light chariots: Love must be ruled by art. Automedon was skilled with chariots and pliant reins,[32] Tiphys was master in the Haemonian ship. (*Ars am.* 1.3–6)

---

[29] Sen. *Controv.* 1.9.10 about what Cato said: *Orator est, Marce fili, vir bonus, dicendi peritus*, 'an orator, Marcus my son, is a good man skilled in speaking.'
[30] See, for example, Cic. *De or.* 1.6. Such an idea leaks into less self-consciously upper-class discourse when Vitruvius delineates a range of subjects with which the architect should be acquainted, although not expert. See Fögen 2009, 114–19, including the contribution of the *vir bonus* ideal to the Vitruvian ideology.
[31] Clark 1957, Chapter 1.
[32] The eroticism of these terms will become clear to us when we have learned more about how imagery works: see above.

Then he sets himself up as the teacher of Love, i.e. with Cupid as his pupil, playing on the ambiguity of types of genitive. In these two aspects of Ovid's self-presentation, he may perhaps have an eye on more directly educational didactic writing. Cicero's rhetorical works, drawing on the old debate about the teachability of speaking mentioned above, itself part of the widespread head-to-head between nature and nurture in ancient thought, pretend that there can be some doubt as to whether rhetoric is an art. Grube (1962, 245): 'Such questions as whether rhetoric is or is not an *ars*, one of the perennial problems of ancient rhetoric, often lead to very dull discussions (as, for example, in the second book of Quintilian), but Cicero enlivens the argument by the clash of opinions and personalities.'[33]

Despite Grube's strictures on the first Roman professor of rhetoric, Quintilian's defence of the impact of his subject matter has relevance for Ovid's didactic. When Quintilian comes to codify his vast teaching experience into a major work of literary instruction, he too finds it necessary to argue the case for oratory (the goal of the liberal education with which he is concerned) as an art, in inevitably conventional terms. As it is for Ovid, although of an entirely different nature, so Quintilian's goal is as much the general nurturing of a mindset as it is detailed instruction: education as much as training. Most of Chapter 2.15 is taken up with an extensive discussion of the nature of rhetoric, pierced by worries over its sophistic potential. Variations on 'the power of speech to persuade' will not do as a definition of rhetoric for Quintilian, because that would also describe the activity of *meretrices, adulatores, corruptores* ('prostitutes, flatterers, seducers', 2.15.11) – though this might do only too well for Ovid! Summing up towards the end of the disquisition on the nature of rhetoric, Quintilian exposes his liberal (in the sense of *ingenuus*) credentials:

> nos autem ingressi formare perfectum oratorem, quem in primis esse virum bonum volumus, ad eos qui de hoc opere melius sentiunt revertamur.
>
> But having begun to form the perfect orator, whom most of all we want to be a good man, let us return to those who have the better opinions about this matter. (Quintilian, *Inst.* 2.15.33)

Thinking better about such things, for Quintilian, means adding a moral element to the power of persuasion. His definition *bene dicendi scientia* ('the science of speaking well', 2.15.38) requires the moral as well as the technical sense of 'well'. Next, Quintilian considers *an utilis rhetorice* ('whether rhetoric is useful', 2.16.1), and defends it from the accusation that there

---

[33] See Cic. *De or.* 1.92–3, 109–10, 145–6, 2.28–33, 232.

can be damaging rhetoric by pointing out that there can be damaging food but we still eat, roofs might fall down but we still have houses, etc., while although there may be bad orators there are also bad doctors and bad generals, but we don't (the implication is) reject medicine or military strategy as a result (2.16.5–6). A rhetorical *praeteritio* puts the case for the utility of rhetoric in a way that could not be denied except by the denial of Roman social identity:

> nam ut omittam, defendere amicos, regere consiliis senatum, populum, exercitum in quae velit ducere, quam sit utile conveniatque bono viro, nonne pulchrum vel hoc ipsum est, ex communi intellectu verbisque, quibus utuntur omnes, tantum adsequi laudis et gloriae, ut non loqui et orare sed, quod Pericli contigit, fulgurare ac tonare videaris?
>
> For to pass over how useful and appropriate it is for a good man to defend his friends, to control the senate by his counsels, to lead the people and the army in the direction he wants, is it not fine, from common understanding and words used by all, to pursue so much praise and glory that you seem not to speak and plead but, as in the case of Pericles, to flash thunder and lightning?

The subject that Quintilian is teaching, then, is how to be a good Roman. This, I suggest, is a good parallel for the inculcation of a mindset which is the goal of Ovid's erotodidactic poetry.[34] I do not suggest that Quintilian is in any sense directly influenced by the elliptical playfulness of Ovid here, but rather that Ovid could be playing with a tradition in educational writing.

Whether Ovid intended to instruct a lover, a citizen, or a reader, the claim that his didactic poetry had nothing to do with real life is undermined by one particular reader, Augustus. Unless we think that the poem is a smokescreen and the real reason for the exile is the unexplained error, then life and literature clash in Ovid's biography in a serious way. As is well known, the poet was sent into exile – or rather, as he likes to remind us, relegated – to Tomi on the shores of the Black Sea, in AD 8. He spends two substantial books of poetry, plus an all-out invective poem, arguing with Augustus and the world about his situation, the weather and the nature of poetry. His most intense poetic self-defence is the one-poem second book of the *Tristia*.[35] His first critical move is to pick up again the idea

---

[34] The part of Quintilian's argument (2.17.9–11) against those who say that rhetoric is not an art because people spoke persuasively before it became codified and taught has a parallel in the passage in *Ars am.* 2 in which Ovid acknowledges that there was sex before rhetorico-erotic sophistication developed (*Ars am.* 2.479–80).

[35] Important here is Barchiesi 2001, Chapter 4.

with which he played in *Am.* 3.12, which is to stress the fictional nature of poetry, the implication being that any readers who take him at his word are desperately naive.

> exit in immensum fecunda licentia vatum,
>   obligat historica nec sua verba fide:
> et mea debuerat falso laudata videri
>   femina; credulitas nunc mihi vestra nocet.
>
> The fruitful licence of poets goes out far and wide, and does not bind their words with historical accuracy: my woman also ought to have seemed to have been praised falsely; now your credulity harms me. (*Am.* 3.12.41–4)

The joke there, of course, is that the whole poem is a double bluff, in which the poet/lover warns other lovers/readers off his mistress with the claim that she is 'just' a literary invention.[36] In *Tristia* 2, Ovid goes through his poetic CV, ostensibly to show Augustus how good he has been at praising prince and principate. He introduces his epic *Metamorphoses* with the same kind of comment about the status of fiction.

> inspice maius opus, quod adhuc sine fine reliqui,
>   in non credendos corpora versa modos:
>
> Inspect my greater work, which I have up to now left unfinished, in which bodies are changed into modes that are not to be believed. (*Tr.* 2.63–4)

A considerable amount of space is taken up with suggesting to Augustus that he has not actually read the poem properly:

> non ea te moles Romani nominis urget,
>   inque tuis umeris tam leve fertur onus,
> lusibus ut possis advertere numen ineptis,
>   excutiasque oculis otia nostra tuis.
>
> That great mass of the Roman name does not weigh you down nor is the burden carried on your shoulders so light that you could turn your godhead to my silly games, and examine my leisure activities with your eyes. (*Tr.* 2.221–4)

There are all sorts of generic games going on here, with Augustus taking the role of Aeneas,[37] but while the emperor is busy in that role he is also constructed as a bad reader of literature. Had he not been, he would have realised that poetry is not serious, not about real life.

---

[36] It almost seems as though Ovid had foreknowledge of the creative critical line in feminist readings of Roman elegy that derives from the seminal Wyke 1987. Specifically on the *Amores*, see Keith 1994.

[37] *moles*: Verg. *Aen.* 1.32, *tantae molis erat Romanam condere gentem*. The shouldered burden clearly recalls the *pietas* image of Aeneas carrying Anchises from burning Troy, and the prophetic shield by means of which Aeneas picks up the fates of his descendants.

> at si, quod mallem, vacuum tibi forte fuisset,
> nullum legisses crimen in Arte mea.

> But if, as I would have preferred, you had by chance had a free moment, you would have read that there is no crime in my Art. (*Tr.* 2.239–40)

This couplet echoes one of the *Ars Amatoria*'s disclaimers:

> nos Venerem tutam concessaque furta canemus
> inque meo nullum carmine crimen erit.

> We sing safe Venus and allowed thefts and there is no crime in my song. (*Ars am.* 1.33–4)

The line is itself quoted a few lines later in *Tr.* 2.250. But, famously, there is indeed *crimen* in Ovid's *carmen*. So should we take him seriously, or not?

Throughout the exile poetry, Ovid is engaged with a problem of responsibility in literature that still resonates today. How far are the producers of cultural items in control of the reception of their work? To what extent are they responsible for bad uses to which it may be put? How far can they be blamed if the material falls into the wrong hands? Does literature make any difference to life anyway? Current debates about the effects of television and computer games on children (and not only children) show that the question is still a live one. Ovid tries to use his quotation of the disclaimer to prove that he had made it absolutely clear that his teaching was X-rated. But, as he (or, according to some editors, an interlocutor) admits, the trouble is it's hard to stop children getting hold of over-18 material – or indeed (and it is not so very different, from the point of view of the paternalistic persona presented here), to stop respectable women reading works designed for non-respectable women.

> ecquid ab hac omnes rigide submovimus Arte,
> quas stola contingi vittaque sumpta vetat?
> at matrona potest alienis artibus uti,
> quodque trahat, quamvis non doceatur, habet.

> Did I not strictly remove from this Art those women whom the long dress and the ribbons she wears forbid to be touched? But a matron can make use of the arts of another, and has something that she may extract from it, although she isn't the subject of the teaching. (*Tr.* 2.251–4)

Well, the answer to that is that there is plenty of other material which is just as bad and does not even carry a health warning:

> nil igitur matrona legat, quia carmine ab omni
> ad delinquendum doctior esse potest.

> Let a matron not read anything then, since from any poem she could learn something of delinquency. (*Tr.* 2.255–6)

He then picks up some examples of unimpeachably moral Roman literature, Ennius' *Annales* (*nihil est hirsutius illis*, 'nothing is hairier than that', *Tr.* 2.259) and Lucretius' *De rerum natura* (*requiret, Aeneadum genetrix unde sit alma Venus*, 'she will ask how kindly Venus became the mother of the people of Aeneas', 261–2). From this point, he moves to the argument that 'it isn't the gun's fault'.[38]

> non tamen idcirco crimen liber omnis habebit:
> nil prodest quod non laedere possit idem.
> igne quid utilius? siquis tamen urere tecta
> conparat, audaces instruit igne manus.
>
> But it isn't the case that for this reason every book will be charged: there is nothing useful which cannot also be harmful. What is more useful than fire? But if someone prepares to burn down a house, he equips his audacious hands with fire. (*Tr.* 2.265–8)

Ovid's argument here is interestingly similar to that which Quintilian uses to justify rhetoric despite the bad use to which it is put. Just because something can be misused this does not mean it is inherently bad.

Another line of attack, also active in the modern debate, is to claim that poetry has no effect on the reader. If the respectable can set eyes on the obscene without harm, then surely matrons can read the *Ars* without being incited to adultery.

> corpora Vestales oculi meretricia cernunt,
> nec domino poenae res ea causa fuit.
> at cur in nostra nimia est lascivia Musa,
> curve meus cuiquam suadet amare liber?
>
> Vestal eyes see prostitutes' bodies, and this matter has not been a cause of punishment for the master. But why is wantonness excessive in my Muse, or how come my book persuades anyone to love? (*Tr.* 2.311–14)

Not only is the final line exactly what the *Ars Amatoria* was designed to do, but also an astute reader of Ovid, which the poet tells us that Augustus is not, might remember that he said earlier that:

> in medio passimque coit pecus: hoc quoque viso
> avertit vultus nempe puella suos.

---

[38] In fairness, his point is not quite as irresponsible as my modern parallel would suggest: while the argument of the gun lobby, that it is people who shoot, not guns, is rendered specious by the number of accidents, Ovid's example is indeed of something with genuinely good uses.

> The herd copulates anywhere and everywhere: when she sees this, without doubt a girl turns aside her face. (*Ars am.* 2.615–16)

This would seem to suggest that seeing things can indeed sometimes have an effect on the seer. Indeed, Ovid's elliptical and, I'm sure, deliberately frustrating reference to his supposed error in *Tristia* 2 itself talks about the damage done by accidental seeing (*Tr.* 2.105–10).

A further strand of Ovid's self-defence is to attempt (or, more probably, to pose as attempting) to undo the connections between teaching and learning. It is a given of educational theory that one has to learn in order to teach. Ovid poses as regretting having ever learnt – but he claims that what he learnt and taught was literature, not life.

> ei mihi! quo didici? cur me docuere parentes,
>   litteraque est oculos ulla morata meos?
> haec tibi me invisum lascivas fecit ob Artes,
>   quas ratus es vetitos sollicitare toros.
> sed neque me nuptae didicerunt furta magistro,
>   quodque parum novit, nemo docere potest.
> sic ego delicias et mollia carmina feci,
>   strinxerit ut nomen fabula nulla meum.

> Alas! For what purpose did I learn? Why did my parents teach me, and why did any letter delay my eyes? This made me hateful to you because of my playful Arts, which you have thought trouble forbidden beds. But no brides have learnt thefts from me as teacher, and no one can teach what he hardly knows. So I have made delights and soft songs, in such a way that no story has grazed my name. (*Tr.* 2.343–50)

It was literature he learnt, then, and for which he regrets his parents' care (unlike Vitruvius, who thanks his parents for his education). Having learnt was what enabled him to teach – but, he claims, he did not teach, because one cannot teach what one has not learned, and he has not learned to be a lover ('honestly, Augustus, despite what I said about *tantus amator ego*'). This claim, however, is itself a topos of poetry's engagement with life, powerfully expressed by Catullus' violent abuse of anyone who suggests that just because his poems are erotic his life must be 'soft' (*Cat.* 16). Central to Ovid's version of this argument is a non-equation between literature and life, summed up as follows:

> crede mihi, distant mores a carmine nostri:
>   vita verecunda est, Musa iocosa mihi;
> magnaque pars operum mendax et ficta meorum
>   plus sibi permisit compositore suo.

> Believe me, my morals are quite different from my poem: my life is chaste, my Muse playful; and the great part of my works, deceitful and fictional, has allowed more to itself than to its composer. (*Tr.* 2.353–6)

But the very fact that Ovid is in a position of arguing this case shows its falsity. If literature has nothing to do with life, how come he is in exile, and what would be the point in trying to use literature to effect a return? Augustus was clearly not convinced by the argument that literature and life are unconnected.

The poem's afterlife, likewise, gives the lie to any suggestion of such a separation between literature and life.[39] Some critics take the view that mediaeval readers failed to see the irony of the *Ars Amatoria*, in part because they did not have enough information about the Augustan background. For example, Desmond (2006, 36):

> The irony and rhetorical excess of the *Ars Amatoria* are relatively transparent to the reader able to situate the *Ars* in relation to Augustan poetics and imperial politics: read in that context, the *Ars* suggests that heteroerotics is one effect of imperialism. The mediaeval reception of the *Ars*, however, illustrates the contingent nature of literary irony: lacking much specific knowledge or information regarding the Augustan age, mediaeval readers lacked the framework within which to appreciate the ironic texture of the poem. Instead, mediaeval readers treat the *Ars* as an ethical treatise on love and seduction.

This is an updated version of the classic C. S. Lewis view (1936) of mediaeval love poetry as 'Ovid misunderstood'. But other critics suggest that mediaeval readers understood the irony of the poem perfectly well, even if the manner in which such irony played out for their culture was different from that of the Augustan age, as indeed both it and the original reading are different from the modern manifestation of Ovidian irony.[40] Critical reading that acknowledges the culturally contingent nature of interpretation, therefore, needs to appreciate the apparently serious and direct mediaeval use of the poem on its own terms.

It may be difficult for modern readers to appreciate the high status of didactic poetry and prose within mediaeval education and literary culture.[41]

---

[39] On Ovid in the Middle Ages, see Dimmick 2002.
[40] An example of a sympathetic ironic reading of the mediaeval reading of the *Ars Amatoria* is P. L. Allen 1989.
[41] See Hexter 1986, 17:

> Whatever else it is, the *Ars Amatoria* is a didactic work. Now not only in the mediaeval school but among much wider circles of mediaeval readers and authors didactic poetry was prized and cultivated, as it was well into the 18th century, to a degree hard for current tastes to

Ovid's poem functioned both as a canonical text from the golden age of the classical heritage which, for all its paganism and immorality, maintained a high status among mediaeval schoolmen – a status conveniently enhanced, for Ovid's didactic cycle, by the ostensible rejection of sex offered by the *Remedia*, and also as an inspiration and model for mediaeval reinterpretations, texts that themselves were directed towards the personal formation of the reader.[42] A text like the *Ars Amatoria* is described by mediaeval readers as 'ethical', not because it is morally good, but because the sphere of activity with which it is concerned is behaviour.[43] A mediaeval introduction to a text of the *Ars Amatoria*, published in Huygens, shows clearly the interpretation of the poem in didactic terms.

> Intentio sua est in hoc opere iuvenes ad amorem instruere, quo modo debeant se in amore habere circa ipsas puellas, materia sua est ipsi iuvenes et puellae et ipsa precepta amoris, quae ipse iuvenibus intendit dare... Ethicae subponitur, quia de moribus puellarum loquitur, id est quos mores habeant, quibus modis retineri valeant.

> His intention in this work is to instruct young men towards love, how they ought to behave in love around the girls themselves; his material is the young men themselves and the girls and very precepts of love, which he himself intends to give to the young men... It is included under ethics, because it speaks about the behaviour of girls, that is what customs they have and by what means they may be retained. (Huygens 1970, 33)

This introduction may be doing little more than repeating what Ovid says himself. When it comes to the *Remedia*, however, the mediaeval commentator expatiates on the interactions of poetry and life as he reads Ovid.

> Ovidius iste amandi librum composuit, ubi iuvenes amicas acquirere, acquisitas benigne tractare docuit, et puellas id idem instruxerat. Quidam autem iuvenes voluptati nimium obedientes non solum virgines, verum et ipsas matronas et consanguineas minime vitabant, virgines coniugatis sicut non uxoratis se pariter subiungebant. Unde Ovidius ab amicis et ab aliis in maximo odio habebatur; postea penitens, quos offenderat sibi reconciliari

---

appreciate or even comprehend... Certainly the range of mediaeval testimonies on Ovid's erotic poetry gives one the impression that the two things mediaeval readers and writers liked most were to love and to learn.

[42] Scholars of mediaeval Latin poetry have rightly seen the mediaeval arts of love as, in addition to their direct didactic function, having a role in the teaching of writing poetry, i.e. arts of composition. See P. L. Allen 1989, Adams 2007, 55 n. 3.

[43] See Hexter 1986, 16, 47. On the sense of 'ethical' here, see J. B. Allen 1982 and Dimmick 2002, 268: 'poetic texts generally belong under moral philosophy: ethics provides a set of implicit norms of behaviour, which poetry dramatises'.

> desiderans vidensque hoc non melius posse fieri quam si dato amori medicinam adinveniret, hunc librum scribere aggressus est, in quo pariter iuvenibus et puellis irretitis <consulit>, qualiter contra illicitum amorem se armare debeant.

> That man Ovid composed a book of loving, in which he taught young men to acquire girlfriends, to treat them well once they had been acquired, and the same man instructed girls in the same matter. But certain young men, excessively devoted to voluptuousness, by no means avoided not only virgins, but even matrons themselves and their relatives, virgins joined themselves equally to the married as to those who did not have wives. For this reason Ovid became a source of great hatred to his friends and others. Later regretting it, wishing to reconcile himself with those whom he had offended and seeing that this could be done in no way better than if he should produce an antidote for the love which he had created, went on to write this book, in which he took thought equally for entangled young men and girls, how they ought to arm themselves against illicit love. (Huygens 1970, 34)

The author goes on to develop the connection with medicine in the *Remedia* and then to lay out the material and purpose of the poem. It is perhaps not surprising that the *Remedia* should be particularly popular in the mediaeval curriculum and among mediaeval readers, since it is ostensibly a rejection of love.[44] What is more significant is the extent to which mediaeval commentators interpret the poems as intimately connected with the lives of first-century BC Romans. I have been unable to see mediaeval *accessus* other than those published by Huygens, but I repeat some observations from Hexter's discussion of Hafn. 2015B.[45] The author, Hexter says, 'relates the story how Ovid saw his contemporaries committing suicide over unhappy love affairs and undertook the work to remedy this'. The mediaeval commentator also 'discusses what the benefit of the work would have been for Ovid's contemporaries'. At this point, according to Hexter, the mediaeval writer addresses himself directly to the commentary's readers, using the second person, for his discussion of the *utilitas* of the text he is introducing. His introduction ends with 'the value of the work for current and future readers'. In the commentary in the same manuscript, the author regularly reminds his students of the didactic purpose of the poem.[46] A critic hostile to the idea that the *Ars Amatoria* and *Remedia Amoris* can be expected to have any direct utility for readers, whether ancient, mediaeval or modern, might say that the authors of these *accessus* are driven by a cultural need to justify the time spent in scholarship on and reading of such texts by an appeal to (or an imaginative construction of) their potential utility, much in the way that the British government has currently decided that research

---

[44] Hexter 1986, 18. [45] Ibid. 47. [46] Ibid. 52.

in every field of academic endeavour should be able to display the 'impact' that it has had on the life of the nation. On such a reading, the *Ars* would not really be useful, but discourse would be attempting to create an illusion of usefulness for a particular ideological purpose. There may be some truth in such an approach, but discourse creates reality, and anyone who plays at learning from literature might well actually end up doing so. Perhaps, however, we can go further in receiving the utility of Ovid's erotodidaxis in the Middle Ages, not so much the widespread application of his detailed advice in this period, but through his contribution to the development of a mindset that is directed towards having an effect, for better for worse, on people's lives.

The argument of Adams (2007) is that the mediaeval arts of love were concerned, at least in part, to inspire love in the reader and thus to encourage marriage. The argument is this: Ovid taught his reader to act like a lover in order to feel like a lover; the mediaeval teachers, such as those depicted in the *Roman de la Rose*, did the same; this had a real resonance with actual lives and loves, particularly at this time (the twelfth and thirteenth centuries), because it was the time in which the church was beginning to stress the importance of personal consent in the construction of a marriage. Thus, ironically, Ovid's poem contributes to encouraging socially approved marriage. Adams bases this reading of the art of love in the context of a wider mediaeval discourse of 'arts', in which inner feeling and outer expression are intimately entwined and mutually reinforcing.[47]

Whether we want to go as far as Adams in seeing the love encouraged by Ovid's contribution to the mediaeval *artes amatoriae* as directed towards marriage, it seems extremely likely that Ovid's programme of acting like a lover in order to become one was taken seriously in the Middle Ages. We do not have to ditch all the wonderful political irony of Ovid's *Ars amatoria* in order to understand that the schoolmen and the troubadours might have had a point.

> est tibi agendus amans imitandaque vulnera verbis;
> haec tibi quaeratur qualibet arte fides.

---

[47] Adams 2007, 63:

> Although the mediaeval *artes amatoriae* aim to produce a different type of subject from the treatises on the gestures of prayer, they share important assumptions... Like their religious counterparts, the secular *artes amatoriae* offer their readers emotional templates, that is, programmes of gesture and words to manage powerful and potentially unruly inward emotion. Ovid's love writings were of course the principal model for the secular works. But one of the reasons clerics writing for lay audiences... responded to Ovid as an expert in love... is that in the words of the *magister amoris*, they discovered a physiology of emotion compatible with their own assumptions about the relationship between gesture and the inner state, along with a program for modulating the mind through physical movement.

You must act like a lover and imitate wounds with words; this credence should be sought by you through whatever art you like. (*Ars am.* 1.611–12)

The insight that mediaeval readers have well perceived, which Ovid, for all his denials, knew and taught well, is behaviour creates reality.

> saepe tamen vere coepit simulator amare;
> saepe, quod incipiens finxerat esse, fuit.
> quo magis, o, faciles imitantibus este, puellae:
> fiet amor verus, qui modo falsus erat.

But often a trickster has begun to love truly; often he has become what at the beginning he had feigned to be. All the more therefore, girls, be kind to imitators: a love will become true, which just now was false. (*Ars am.* 1.615–18)

REFERENCES

T. Adams 2007, 'Performing the Medieval Art of Love: Medieval Theories of the Emotions and the Social Logic of the *Roman de la Rose* of Guillaume de Lorris', *Viator*, 38, 2: 55–74

J. B. Allen 1982, *The Ethical Poetic of the Later Middle Ages: A Decorum of Convenient Distinction* (Toronto)

P. L. Allen 1989, 'Ars amandi, ars legendi: Love Poetry and Literary Theory in Ovid, Andreas Capellanus, and Jean de Muen', *Exemplaria*, 1: 181–205

A. Barchiesi 2001, *Speaking Volumes: Narrative and Intertext in Ovid and Other Latin Poets* (London)

L. Callebat 1982, 'La Prose du "De architectura" de Vitruve' in H. Temporini and W. Haase (eds.), *Aufstieg und Niedergang der römischen Welt* II 30.1: 696–722

D. L. Clark 1957, *Rhetoric in Greco-Roman Education* (New York)

M. Desmond 2006, *Ovid's Art and the Wife of Bath: The Ethics of Erotic Violence* (Ithaca, NY, London)

J. Dimmick 2002, 'Ovid in the Middle Ages: Authority and Poetry' in P. R. Hardie (ed.), *The Cambridge Companion to Ovid* (Cambridge University Press), 264–87

A. Faulkner 2011, 'Callimachus' Epigram 46 and Plato: The Literary Persona of the Doctor', *Classical Quarterly*, 61: 178–85

Th. Fögen 2009, *Wissen, Kommunikation, und Selbstdarstellung: Zur Struktur und Charakteristik römischer Fachtexte der frühen Kaiserzeit.* Zetemata 134 (Munich)

R. K. Gibson 1998, 'Meretrix or Matrona? Stereotypes in Ovid *Ars Amatoria* 3', *Papers of the Leeds Latin Seminar*, 10: 295–312

2003 *Ovid, Ars Amatoria Book 3* (Cambridge University Press)

G. M. A. Grube 1962, 'Educational, Rhetorical, and Literary Theory in Cicero', *Phoenix*, 16: 234–57

M. Heath 1985, 'Hesiod's didactic poetry', *Classical Quarterly*, 35: 245–63

A. A. R. Henderson 1979, *Ovid: Remedia Amoris* (Edinburgh)
R. J. Hexter 1986, *Ovid and Medieval Schooling: Studies in Medieval School Commentaries on Ovid's Ars Amatoria*, Epistulae ex Ponto, *and* Epistulae Heroidum (Munich)
R. B. C. Huygens 1970, *Accessus ad auctores. Bernard d'Utrecht: Conrad d'Hirsau, Dialogus super auctores* (Leiden)
S. L. James 2003, *Learned Girls and Male Persuasion: Gender and Reading in Roman Love Elegy* (Berkeley)
M. Janka 1997, *Ovid: Ars Amatoria, Buch 2, Kommentar* (Heidelberg)
A. M. Keith 1994, 'Corpus eroticum: Elegiac Poetics and Elegiac Puellae in Ovid's *Amores*', *Classical World*, 88: 27–40
D. F. Kennedy 2006, 'Vixisset Phyllis, si me foret usa magistro: Erotodidaxis and intertextuality' in R. K. Gibson, S. Green and A. R. Sharrock (eds), *The Art of Love: Bimillennial Essays on Ovid's Ars Amatoria and Remedia Amoris* (Oxford), 54–74
E. J. Kenney 1994, *Ovidi Nasonis Amores, Medicamina Faciei Femineae, Ars Amatoria, Remedia Amoris iteratis curis edidit E. J. Kenney*. 2nd edn. (Oxford)
C. S. Lewis 1936, *The Allegory of Love: A Study in Medieval Tradition* (Oxford)
T. A. J. McGinn 1991, 'Concubinage and the Lex Iulia on Adultery', *Transactions of the American Philological Association*, 121: 335–75
T. Morgan 1998, *Literate Education in the Hellenistic and Roman Worlds* (Cambridge University Press)
H. N. Parker 1992, 'Love's Body Anatomized: The Ancient Erotic Handbooks and the Rhetoric of Sexuality' in A. Richlin (ed.), *Pornography and Representation in Greece and Rome* (Oxford), 90–111
G. Rosati 2006, 'The Art of *Remedia Amoris*: Unlearning to Love?' in R. K. Gibson, S. Green and A. R. Sharrock (eds.), *The Art of Love: Bimillennial Essays on Ovid's Ars Amatoria and Remedia Amoris*(Oxford), 143–65
A. R. Sharrock 1994a, *Seduction and Repetition in Ovid's Ars Amatoria 2* (Oxford)
  1994b, 'Ovid and the Politics of Reading', *Materiali e Discussioni per L'analisi dei Testi Classici* 33: 97–122
K. Volk 2006, '*Ars Amatoria Romana*: Ovid on love as a cultural construct' in R. K. Gibson, S. Green and A. R. Sharrock (eds.), *The Art of Love: Bimillennial Essays on Ovid's Ars Amatoria and Remedia Amoris* (Oxford), 235–51
M. Wyke 1987, 'Written Women: Propertius' scripta puella', *The Journal of Roman Studies*, 77: 47

CHAPTER 8

From technē *to* kakotechnia
Use and Abuse of Ancient Cosmetic Texts

Laurence Totelin

Starting with Xenophon in the *Oeconomicus*, ancient male authors have expressed concern at women using cosmetics and adorning themselves.[1] A passage of Lucian's *Dialogue of the Courtesans* is only one text among many conveying an anti-cosmetic message:

> After all, one could perhaps put up with the conduct of the men. But the women –! That is another thing women are keen about – to have educated men living in their households on a salary and following their litters. They count it as <u>an embellishment</u> if they are said to be cultured, to have an interest in philosophy and to write songs that are hardly inferior to Sappho's. To that end they too trail hired rhetoricians and grammarians and philosophers along, and listen to their lectures – when? It is ludicrous! – either while <u>their toilets</u> are being made and their hair dressed, or at dinner; at other <u>times</u> they are too busy! And often while the philosopher is delivering a discourse, the maid comes in and hands her a note from her lover, so that the lecture on chastity is kept waiting until she has written a reply to the lover and hurries back to hear it.[2]

In addition to reminding us that all women are by definition voraciously sexual, this passage suggests, in ironic tones, that for these creatures, beauty

---

[1] On cosmetics in the ancient world, see, inter alia, Grillet 1975, Dayagi-Mendels 1989, Wyke 1994, Richlin 1995, Saiko 2005, Stewart 2007, Olson 2009.

[2] Lucian, *On Salaried Posts in Great Houses* (*Merc. Cond.*) 36 (trans. A. M. Harmon, with changes by E. Hemelrijk my emphasis):

> Καίτοι φορητὰ ἴσως τὰ τῶν ἀνδρῶν. αἱ δὲ οὖν γυναῖκες – καὶ γὰρ αὖ καὶ τόδε ὑπὸ τῶν γυναικῶν σπουδάζεται, τὸ εἶναί τινας αὐταῖς πεπαιδευμένους μισθοῦ ὑποτελεῖς συνόντας καὶ τῷ φορείῳ ἑπομένους· ἐν γάρ τι καὶ τοῦτο τῶν ἄλλων <u>καλλωπισμάτων</u> αὐταῖς δοκεῖ, ἢν λέγηται ὡς πεπαιδευμέναι τέ εἰσιν καὶ φιλόσοφοι καὶ ποιοῦσιν ᾄσματα οὐ πολὺ τῆς Σαπφοῦς ἀποδέοντα – διὰ δὴ ταῦτα μισθωτοὺς καὶ αὐταὶ περιάγονται ῥήτορας καὶ γραμματικοὺς καὶ φιλοσόφους, ἀκροῶνται δ' αὐτῶν – πηνίκα; γελοῖον γὰρ καὶ τοῦτο – ἤτοι μεταξὺ <u>κομμούμεναι</u> καὶ τὰς κόμας παραπλεκόμεναι ἢ παρὰ τὸ δεῖπνον· ἄλλοτε γὰρ οὐκ ἄγουσι σχολήν. πολλάκις δὲ καὶ μεταξὺ τοῦ φιλοσόφου τι διεξιόντος ἡ ἅβρα προσελθοῦσα ὤρεξε παρὰ τοῦ μοιχοῦ γραμμάτιον, οἱ δὲ περὶ σωφροσύνης ἐκεῖνοι λόγοι ἑστᾶσι περιμένοντες, ἔστ' ἂν ἐκείνη ἀντιγράψασα τῷ μοιχῷ ἐπαναδράμῃ πρὸς τὴν ἀκρόασιν.

On this passage, see Hemelrijk 1999, 37–8, Levick 2007, 112.

and philosophical knowledge were two types of embellishment. Lucian's woman does not perceive the difference between true and false beauty. And because she does not understand this, instead of writing verse or philosophical thoughts, she ends up writing to her lover, thus misusing any education she may have. Writing, knowledge, cosmetics and gender relationships are the themes explored in this chapter, which deals with the cosmetic recipes that have been preserved in Greek and Latin, the bulk of which are to be found in encyclopaedias and medical texts. Indeed, whilst only five recipes are preserved from Ovid's poem *Medicamina faciei femineae*, hundreds, probably even thousands, have come down to us through medical and scientific treatises.[3] Indeed, it is difficult to find a major medical compilation that does not contain cosmetic recipes. Thus, the Hippocratic treatise *Diseases of Women* (fifth to fourth centuries BC) includes a few recipes for breath-fresheners, face creams and remedies against freckles;[4] Scribonius Largus' *Composite Remedies* (first century AD) preserves three recipes for toothpaste;[5] Celsus (from the time of the Emperor Tiberius) in Book 6 of *On Medicine* transmits recipes against alopecia and spots, and remedies to whiten scars;[6] Pliny the Elder (AD 23–79) mentions recipes for hair-dyes, face creams and depilation creams;[7] Galen's (AD 129–?199/216) first book of *On the Composition of Medicines according to Places* deals mostly with cosmetic afflictions such as alopecia, dandruff and hair loss, and includes recipes for hair-dyes and depilatories (see below for more detail); and the second book of the pseudo-Galenic *On Remedies Easily Procured* contains recipes for hair-dyes and against hair loss.[8] Recipes for face creams, depilatories, hair products, etc. are also found in the medical compilations of Latin late antique authors such as Theodorus Priscianus (fourth or fifth century AD) and Marcellus of Bordeaux (turn of the fourth and fifth centuries AD);[9] and of Greek Byzantine compilers such as Oribasius (a contemporary of the Emperor Julian), Aetius (fl. c. AD 530–60) and Paul of Aegina (seventh century AD).[10]

[3] On Ovid's *Medicamina faciei femineae* (*Medic.*), see P. Green 1979, Saiko 2005, 186–215.
[4] Hipp. *Mul.* 2.185–91 (L. VIII 366–70). See Saiko 2005, 86–101 on dermatology in the *Hippocratic Corpus* (but with no reference to the gynaecological texts).
[5] Scribonius, *Compositiones* 59–60.
[6] Celsus, *De medicina* (*Med.*) 6.4 (alopecia); 6.5 (remedies against spots and other freckles); 6.6.25 (remedies to whiten scars). See Saiko 2005, 225–35.
[7] See e.g. 32.67–8 (hair-dyes); 32.84 (face cream); 32.135 (depilatory; see below for more detail on this recipe). On Pliny and cosmetics, see Richlin 1995, 198–9, Vons 2000, Saiko 2005, 234–53.
[8] [Galen], *De remediis parabilibus* (*De remed. parab.*) 2.1 and 2.2 (14.390–3 K.). See Meilhac-Léonnelli 2003, 178.
[9] For Theodorus, see Meilhac-Léonnelli 2003. Marcellus, *De medicamentis* 7 (CML V 1, 100–10 Liechtenhan): hair-dyes and remedies against hair loss.
[10] See below for references to recipes in Aetius and Paul. For Oribasius, see, for instance, *Eclogae medicamentorum* 5 (CMG VI 2, 2, 187 Reader).

It might be tempting to believe that the medical tradition is somewhat more 'scientific' and less biased against women than the elegiac, moralistic and religious traditions when it comes to cosmetics. Nothing could be further from the truth: several medical compilers convey strongly prejudiced anti-cosmetic messages. Thus Celsus writes in the sixth book of his *On Medicine*: 'To treat pimples, freckles and spots is almost ridiculous, yet women cannot be torn away from caring for their appearance.'[11] Here Celsus establishes a firm link between cosmetics and women's wishes. Similarly, Pliny the Elder apologises for the frivolity of one of the cosmetic recipes (to whiten the skin and smooth wrinkles) he transmits, but adds that 'because of the desires of women, I must not omit it'.[12] Galen too reiterates this message in an oft-mentioned passage of his *On the Composition of Medicines according to Places*, where he differentiates between medical and frivolous cosmetics:

> The cosmetic part of medicine [*kosmetikon*] differs from the art of embellishment [*kommōtikon*] in this way. The aim of the art of embellishment is to induce a gained beauty, whereas the aim of the cosmetic part of medicine is to keep everything in the body according to nature, whereby beauty according to nature also follows. In reality, to make the colour of the face whiter or redder with remedies, or make the hair on the head thick, or flame-coloured, or black, or augmented in length as women do, these and other similar [stratagems] are the vices of embellishment and not the work of medicine. But because of the overlap between these fields, royal women or even the kings [i.e. emperors] themselves sometimes assign to us [i.e. Galen] tasks to do with embellishment, and it is not possible to refuse them by explaining that embellishment differs from the cosmetic part of medicine.[13]

---

[11] Celsus, *Med.* 6.5 *Paene ineptiae sunt curare varos et lenticulas et ephelidas, sed eripi tamen feminis cura cultus sui non potest.*

[12] Pliny, *Historia Naturalis* (*HN*) 28.184 *Frivolum videatur, non tamen omittendum propter desideria mulierum, talum candidi iuvenci XL diebus noctibusque, donec resolvatur in liquorem, decoctum et inlitum linteolo candorem cutisque erugationem praestare.* 'The following may appear frivolous, but because of the wishes of women, I must not omit it: the knuckle-bone of a white bull-calf is boiled for forty days and forty nights, until it is reduced to jelly, and, smeared over a linen cloth, gives whiteness to the skin and clears away wrinkles.' On this passage, see Richlin 1995, 198, Vons 2000, 351.

[13] Galen, *De compositione medicamentorum secundum locos* (*De comp. med. sec. loc.*) 1.2 (12.434–5 K.):

> Τίνι διαφέρει τοῦ κομμωτικοῦ τὸ κοσμητικὸν τῆς ἰατρικῆς μέρος. Τῷ μὲν κομμωτικῷ σκοπός ἐστι κάλλος ἐπίκτητον ἐργάσασθαι, τῷ δὲ τῆς ἰατρικῆς μέρει τῷ κοσμητικῷ τὸ κατὰ φύσιν ἅπαν ἐν τῷ σώματι φυλάττειν, ᾧ καὶ τὸ κατὰ φύσιν ἕπεται κάλλος. ἀπρεπὴς γὰρ ὀφθῆναι κεφαλὴ πάθος ἀλωπεκίας ἔχουσα, καθάπερ γε κἂν ἐκ τῶν ὀφθαλμῶν αἱ βλεφαρίδες ἐκπέσωσι καὶ τῶν ὀφρύων αἱ τρίχες. οὐ μόνον δ᾽ εἰς κάλλος, ἀλλὰ καὶ πολὺ πρότερον εἰς αὐτὴν τὴν ὑγείαν τῶν μορίων αἱ τρίχες αὗται συντελοῦσιν, ὡς ἐν τοῖς περὶ χρείας μορίων ἐδείχθη. τί δεῖ λέγειν περὶ λειχήνων ἢ ψώρας ἢ λέπρας ὡς παρὰ φύσιν ταῦτα; τὸ μέντοι λευκότερον τὸ χρῶμα τοῦ προσώπου ποιεῖν ἐκ φαρμάκων ἢ

For the physician from Pergamum too, cosmetics are mainly the concern of women, at all levels of society. Since *On the Composition of Medicines according to Places* was probably written during the Severan period, the 'royal women' to whom Galen is referring have sometimes been identified with Julia Domna and Julia Maesa, two princesses known for their complicated coiffures.¹⁴ Whatever the identity of these ladies, one must note that Galen also talks about men here. Galen's 'kings' may include any of the following emperors the physician cared for: Marcus Aurelius, Commodus, Septimius Severus and Caracalla. Galen also admits that the boundary between his *kosmētikē technē* (positively connoted) and his *kommōtikē technē* (negatively connoted) is somewhat porous. A few pages later in the same treatise, the physician adds that he has accepted to include recipes for hair-dyes in his pharmacological works because he knows of women who harmed themselves, or even died, by using harsh products on their heads.¹⁵ Sometimes one has to list remedies of the fraudulent art (*kakotechnia*) of embellishment to prevent their harming people.¹⁶ Whether Galen is being genuine here or masking his true reasons for including recipes for embellishing products, one cannot tell.

The passages of Celsus, Pliny and Galen remind us that these authors writing on medical topics participated in the same anti-cosmetic culture as Xenophon, Lucian or Tertullian.¹⁷ They should act as a warning that all aspects of medical cosmetology must be analysed critically: authorship – is it significant that many recipes are attributed to women?; substances – did

> ἐρυθρότερον ἢ τὰς τρίχας τῆς κεφαλῆς οὔλας ἢ πυρρὰς ἢ μελαίνας ἢ καθάπερ αἱ γυναῖκες ἐπὶ μήκιστον αὐξανομένας, ταῦτα καὶ τὰ τοιαῦτα τῆς κομμωτικῆς κακίας ἐστίν, οὐ τῆς ἰατρικῆς τέχνης ἔργα. διὰ δὲ τὴν κοινωνίαν τούτων ἐνίοτε καὶ βασιλικαὶ γυναῖκες ἢ οἱ βασιλεῖς αὐτοὶ προστάττουσιν ἡμῖν καὶ τὰ τῆς κομμωτικῆς, οἷς οὐκ ἔνεστιν ἀρνεῖσθαι διδάσκοντας διαφέρειν τὴν κομμωτικὴν τοῦ κοσμητικοῦ μέρους τῆς ἰατρικῆς. διὰ τοῦτο οὖν ἔδοξέ μοι καὶ τὰ ὑπὸ Κρίτωνος γεγραμμένα φάρμακα διαφυλακτικὰ τριχῶν καὶ αὐξητικὰ κατὰ τὸ πρῶτον τῶν κοσμητικῶν ἐφεξῆς ὑπογράψαι.

On this passage, see, e.g., Swain 1996, 371, Meilhac-Léonelli 2003, 178–9, Saiko 2005, 122–4, Flemming 2007, 268.

¹⁴ For dates, see Fabricius 1972, 23–4. Identification with Julia Domna and Julia Maesa: Hemelrijk 1999, 124, Levick 2007, 113. Some scholars have argued that Galen was part of Julia Domna's circle: see Bowersock 1969, 106 for references and counter-arguments. On the Severan women's coiffures, see Bartman 2001, 14–17.

¹⁵ Galen, *De comp. med. sec. loc.* 1.3 (12.442 K.): οὐ μόνον γὰρ ἐν κινδύνῳ γενομένας οἶδα πολλάκις γυναῖκας, ἀλλὰ καὶ ἀποθανούσας ἐκ τοῦ καταψυχθῆναι τὴν κεφαλὴν ὑπὸ τῶν τοιούτων φαρμάκων. 'For I have often seen women not only put themselves in danger, but actually die, from over-cooling their heads with such drugs.' On haircare in the ancient world, see Bartman 2001.

¹⁶ Galen, *De comp. med. sec. loc.* 1.3 (12.445 K.): τούτοις ἐφεξῆς ὁ Ἀρχιγένης ἐκ τῆς κομμωτικῆς κακοτεχνίας ὄντα καὶ ταῦτα ἔγραψε. 'After those, Archigenes wrote [recipes] that belong to the fraudulent art of cosmetics.'

¹⁷ Tertullian's *On Woman's Apparel* (*De cultu feminarum*) is roughly contemporary with Galen's *On the Composition of Medicines according to Places* (*De comp. med. sec. loc.*).

women really put dung on their face?; and intended readership – are these recipes truly responding to the wishes of women? The gender question is important here: the ancient cosmetic tradition is very much centred on women – gender-neutral claims against the inclusion of cosmetics in ancient medical texts are extremely rare.[18] Yet it is clear that both men and women used cosmetics in the ancient world, and it is one of the aims of this chapter to place men right at the centre of research on cosmetics. Linked to the question of gender is the question of this volume – the question of practice and practical applicability. As with all ancient recipes (medical or otherwise), one has to ask whether the cosmetic recipes transmitted in medical texts were actually used, and, if so, in what way? Medical writers claimed loud and clear that their writings in general, and their lists of recipes in particular, were useful. Such claims are found, for instance, in Galen's preface to *On the Composition of Medicines according to Types*; and in Theodorus Priscianus' prefaces to the *Remedies Easily Procured*, just before recipes for hair-dyes.[19] But what were the cosmetic recipes included in medical writings useful for? Did physicians include them in order to protect women against dangerous products? Or did they have any ulterior motives? And were there dangers of abuse? Of course, the answer to these questions may vary from one text to the other, or even from one recipe to the other, and here I can only offer general observations.

In this chapter, a detailed analysis of some ingredients included in ancient cosmetics and of the Table of Contents to Crito's lost *Cosmetics* will show that there were large areas of overlap in antiquity between cosmetological, gynaecological and sex manuals. All had procreation as one of their main purposes. This reproductive aim did not preclude sexual pleasure, however, and I shall suggest that some readers might have (mis?-)used ancient cosmetic texts as pornography.

Ingredients

Reading ancient cosmetic recipes is not the most appetizing of experiences.[20] These remedies were meant to treat unpleasant afflictions

---

[18] A gender-neutral comment can be found in the *Medicina Plinii* (fourth century AD) 1.5 (CML 3, 12 Önnerfors): *Potest uideri supervacuum inter remedia corporis ponere ea quae ad decorum pertinent. Sed quosdam pudet aut ipsos rubeos esse, aut in tantum luxuriate indulgent ut deliciae eorum inter se dissentiant uolentium in pueris rufos capillos, in uiris recusantium.* On this passage, see Meillhac-Léonelli 2003, 177.

[19] Theodorus, *Euporiston* 1 (4 Rose) Galen, *De comp. med. per gen.* 1.1 (13.365 K.). See Formisano 2001, 29 on the trope of utility.

[20] Analyses of ingredients included in ancient cosmetics are found in P. Green 1979, Saiko 2005 and Olson 2009, 294–303.

and sometimes included less than savoury ingredients such as urine, horses' teeth and burnt mice. The following sample of recipes, written down by of Heracleides of Tarentum (third to second centuries BC), extracted by Statilius Crito (from the time of the Emperor Trajan – see below) in his treatise *Cosmetics* and transmitted in Galen's *On the Composition of Medicines according to Places*, lists mouse droppings, burnt sea-urchins, burnt hairs of bears, burnt frogs and various other burnt ingredients:[21]

> Concerning the [remedies] prescribed by Heracleides against chronic alopecia. Against alopecia, as [transmitted by] Heracleides of Tarentum.
> [Remedy] of Orestinus against chronic alopecia: <u>mouse droppings</u>, 2 drachmai; frankincense, 2 drachmai; dilute with sharp vinegar. Anoint, having shaved the places [of application] beforehand.
> Another: <u>sea-urchins burnt with their shells</u>, 2 drachmai; the same amount of Aleppo gall; the same amount of <u>burnt bitter almonds</u>; <u>mouse droppings</u>, 1 drachma. Dilute with vinegar; crush well and apply.
> Another according to Orestinus: <u>burnt hairs of bear</u>; <u>burnt maidenhair</u>; <u>burnt root of reed</u>; <u>burnt fig leaves</u>; <u>burnt Cilician cloth</u>; of each the same amount; take up with bear fat and cedar oil. When for use, foment, having cleansed the place beforehand with fig leaves and wiped clean with soda; anoint with the remedy during the day. When the surface of the skin is ulcerated, apply a cerate made of gruel, wheat- and barley-flour.
> Another prepared by Othon the Sicilian: <u>soot of very small frogs burnt</u> in a vase, one portion; <u>mouse droppings</u>; white hellebore; <u>burnt root of reed</u>; white pepper; of each the same amount. Take up with vinegar and use, having cleansed and shaved the place [of application] beforehand.[22]

Classicists will be familiar will the reference in Horace's twelfth epode to rouge made of crocodile dung, an ingredient also mentioned in a cosmetic

---

[21] This is a prime example of the complexity of recipe transmission in the ancient world, on which see Fabricius 1972, Hanson 1997. Galen makes it clear that he knew the cosmetic recipes of Heracleides second-hand at *De comp. med. sec. loc.* 1.3 (12.445–6 K.): ἀλλὰ καὶ τὰ τοῦ Ἡρακλείδου καὶ Κλεοπάτρας, ὅσοι τ' ἄλλοι μετ' αὐτοὺς ἐν τῷ μεταξὺ γεγόνασιν ἰατροὶ φάρμακα πάντα συνήθροισεν ὁ Κρίτων. 'Crito collected all the remedies of Heracleides, Cleopatra, and of all the other physicians active in the intermediary period.'

[22] Galen, *De comp. med. sec. loc.* 1.1 (12.402–3 K.):
[Περὶ τῶν ὑπὸ Ἡρακλείδου γεγραμμένων πρὸς ἀλωπεκίας χρονίας.] Πρὸς ἀλωπεκίας ὡς Ἡρακλείδης ὁ Ταραντῖνος. Ὀρεστίνου πρὸς τὰς κεχρονισμένας ἀλωπεκίας: <u>μυοχόδων</u> < β'. λιβανωτοῦ < β'. ὄξει δριμυτάτῳ διαλύσας ἐπίχριε, προξυρήσας τοὺς τόπους. ἄλλο. <u>ἐχίνων θαλασσίων σὺν τοῖς ὀστράκοις κεκαυμένων</u> < β'. κηκίδων ὀμφακιτίδων τὸ ἴσον, <u>ἀμυγδάλων πικρῶν κεκαυμένων</u> τὸ ἴσον. Μυοχόδων <α'. ὄξει διαλύσας καὶ τρίψας ἐπιμελῶς ἐπιτίθει. ἄλλο ὡς Ὀρεστῖνος. <u>ἀρκτείων τριχῶν κεκαυμένων</u>, <u>ἀδιάντου κεκαυμένου</u>, <u>καλάμου ῥιζῶν κεκαυμένων</u>, <u>συκῆς φύλλων κεκαυμένων</u>, <u>κιλικίου ῥάκους κεκαυμένου</u> ἑκάστου τὸ ἴσον ἀναλάμβανε στέατι ἀρκτείῳ καὶ κεδρίᾳ. ἐπὶ δὲ τῆς χρήσεως πυρία, ἀνατρίβων τὸν τόπον συκῆς φύλλοις καὶ νίτρῳ ἀποσμήχας, κατάχριε τῷ φαρμάκῳ καθ' ἡμέραν, ἑλκωθείσης δὲ τῆς ἐπιφανείας ἐπιτίθει κηρωτὴν διὰ λεκίθου καὶ πυρίνων καὶ κριθίνων ἀλεύρων, χυλῷ προσφάτῳ ἐπίχριε. ἄλλο Ὄθωνος Σικελοῦ ἐπιτετευγμένον. <u>βατράχων τῶν σμικροτάτων κεκαυμένων ἐν χυτριδίῳ</u>, τῆς σποδοῦ μέρος α'. <u>μυοχόδων</u>,

context by Pliny, Galen and Clement of Alexandria, and by Ovid in the *Ars Amatoria*.[23] While the process and product of defecation might not have been as hidden in antiquity as they are today, it is clear that the ancients thought of excrement as disgusting.[24] There is also ample evidence to prove that cosmetic products themselves were considered repulsive.[25]

Remedies including dung and burnt parts of animals seem to reflect the desperation of men and women suffering from debilitating alopecia or trying to achieve the perfect complexion.[26] Indeed, the woman who has recourse to crocodile dung in Horace is a desperate old, stinking hag. If ever such ingredients were used in ancient cosmetics, they must have created a situation whereby the prescriber exerted total control over the patient who obediently smeared excrement onto his or her face. Opportunities for abuse of power might have been lurking everywhere in the business of cosmetics, a business that has as its main aim the control of the body.[27]

However, one has to exert caution when faced with such ingredients for two reasons. First, they tend to distract the attention from other, more sensible-sounding ingredients.[28] Some ancient cosmetic recipes would still be considered effective today, as is argued by Peter Green (1979) in an article on Ovid's recipes in the *De medicamine*.[29] For instance, the 'costly and pleasant' ointment of Cleopatra transmitted by Aetius would have done exactly what is written on the tin: it would have smelt lovely, and would have indeed been rather costly with its numerous rare and imported ingredients.[30] It would have also invoked the pleasures – or vices, depending

ἐλλεβόρου λευκοῦ, καλάμου ῥίζης κεκαυμένης, πεπέρεως λευκοῦ, ἑκάστου τὸ ἴσον, ὄξει προσανάλαβε καὶ χρῶ προανατρίβων καὶ ἀποξυρῶν τὸν τόπον.

[23] Clement of Alexandria, *Paedagogus* 3.2.7.3; Galen, *De comp. med. sec. loc.* 1.1 (12.408, 416 K.); Galen, *De simplicium medicamentorum temperamentis ac facultatibus* (*De simpl. medicament. temp.*) 8.56 (12.47–8 K.); Horace, *Epod.* 12.11; Ovid, *Ars am.* 3.270 (but no mention of dung); Pliny, *HN* 28.108, 184. See Richlin 1983, 111–12; Hendry 1995, Olson 2009, 297.
[24] See e.g. Galen, *De simpl. medicament. temp.* 10.1 (12.249 K.): πολὺ δ' αὐτοῦ βδελυρώτερον ἡγοῦμαι τὴν κόπρον εἶναι. 'But I consider excrement to be much more loathsome than this [i.e. earwax].' See von Staden 1992, 8 on disgust for excrement in antiquity. See Toulalan 2007, 228–9 for a warning that defecation may have been considered differently in different periods. On Dreckapotheke, see von Staden 1991, von Staden 1992, Hanson 1998, 89, Totelin 2009, 212–14.
[25] See Richlin 1995, 195, Olson 2009, 303–4 for references.
[26] On desperation, see Olson 2009, 297.   [27] See Richlin 1995, 186.
[28] See Douglas 1966, 119, Totelin 2009, 212–13.   [29] See also Saiko 2005, 202–15.
[30] Aetius, *Tetrabiblon* 8.6 (CMG VIII 2, 408 Olivieri). Ἄλλο σμῆγμα Κλεοπάτρας βασιλίσσης πολυτελὲς εὐῶδες. Κόστου σμύρνης τρωγλίτιδος ἴρεως ναρδοστάχυος ἀμώμου φύλλου κασσίας σχοίνου ἄνθους ἀνὰ α μυροβαλάνου λίτρας δ νίτρου ἀφροῦ λίτρας β, κόψας σήσας χρῶ· ποιεῖ εἰς ὅλον τὸ σῶμα. 'Other ointment of the Queen Cleopatra, costly and pleasant. *Kostos*, troglodytic myrrh, iris, spikenard, amomum, leave of cassia, flower of reed, of each one ounce, *myrobalanon*, 4 pounds, sodium carbonate, two pounds: crush, sieve, and use. It works on the entire body.'

on one's angle of approach – of the Orient. Other cosmetics would have been merely ineffective. Here we are lucky to have some material evidence. A tin canister (such tins are mentioned in numerous medical texts) dating to the second century AD was found in a recent excavation at the site of the Roman London temple, with its cream content intact. Chemical analysis revealed the product was made of animal fat and contained tin. While tin is not known to have any effect on the skin, it is harmless. The researchers found the cream pleasant on application.[31]

Second, code words might sometimes have been used in cosmetic recipes. Thus the substitution list in the Corpus of Magical Papyri gives 'crocodile dung' as the code word for 'Ethiopian soil'.[32] That there were links between cosmetology and magic is made clear by the words used to refer to cosmetics both in Latin (*medicamina*) and in Greek (*pharmaka*), which have associations with poisons and spells.[33] At a more concrete level, the Justinian digest stipulates that unguent-dealers could be prosecuted for handling lethal poisons, and Pliny indicates that cosmetic afflictions were sometimes the result of magic spells: 'Hair lost by mange is restored by the ash of mice, their heads and tails, or their whole bodies, especially when this affliction is the result of sorcery.'[34]

Whilst it is possible that references to dung and other dirty ingredients are sometimes code words, I do not think it applies in all cases. Galen himself certainly testifies to having used dung, even though he found it repulsive.[35] In fact, the use of dung in cosmetic recipes makes some sort of sense. As suggested by Ann Hanson, such ingredients are probably used, like excrement in farmer's fields, as a means to fertilise bald heads or sickly skins.[36] The use of burnt products in skin remedies can be explained in a similar fashion: ash does have a fertilising effect on fields, so why not postulate the same effect on the skin? Using ash resulting from the burning

---

[31] Evershed et al. 2004.
[32] PGM 12.414. On this list, see Scarborough 1991, 159–60, Laskaris 2005.
[33] See Richlin 1995, 186, Olson 2009, 304–9. On the meaning of the word *pharmakon*, see Artelt 1968.
[34] *Iustiani digesta* 48.8.3.3.: *Alio senatus consulto effectum est, ut pigmentarii, si cui temere cicutam salamandram aconitum pituocampas aut bubrostim mandragoram et id, quod lustramenti causa dederit cantharidas, poena teneantur huius legis.* See Olson 2009, 305–6. On people active in the drug trade in Rome, see Korpella 1995. Pliny the Elder, *Historia Naturalis* 29.107 (trans. W. H. S. Jones): *Alopecias cinis et e murium capitibus caudisque et totius muris emendat, praecipue si veneficio acciderit haec iniuria, item irenacei cinis cum melle aut corium combustum cum pice liquida.*
[35] Galen, *De simpl. medicament. temp.* 10.19 (12.291 K.).
[36] Hanson 1998, 89. See also W. H. S. Jones 1957, 462. On the fertilising power of dung, see, for instance, Varro, *De re rustica* 1.38: 'Cassius states that next to pigeon dung human excrement is the best [fertiliser], and in the third place goat, sheep, and ass dung; that horse dung is least valuable, but good on grain land; for on meadows it is the most valuable of all, as is that of all draught animals which feed on barley, because it produces a quantity of grass.'

of ingredients that were particularly associated with fertility (such as the horn of a deer) might have been thought even to increase this fertilising effect on the skin.[37]

Thus 'fertilising' seems to have been one of the important functions of ancient cosmetics. We are brought back to the link between cosmetics and sexuality postulated in the ancient literature, although here it has more positive, or at least more ambivalent, connotations. It is worth noting that the other medical context in which dung is also regularly used is gynaecology, where it appears in remedies against sterility.[38] There too excrements are highly ambivalent ingredients. They have the power to fertilise, but they also have degrading connotations – as is argued by Heinrich von Staden (1991, 1992), they can be seen as a means to cleanse dirt (impure women) with dirt.

In sum, whilst one should not exaggerate the importance of dirty ingredients in ancient cosmetic recipes, which make use of a great variety of products, it is fair to say that they add to the sexual character of medical cosmetology, especially if these products can be interpreted as 'fertilisers' of the skin or the hair. The link between cosmetology and sexuality becomes even more apparent when looking at the types of ailment treated in ancient cosmetic treatises and the names of the authorities to whom cosmetic recipes and texts were attributed.

Content and Authorship of Ancient Cosmetic Treatises

Research on ancient cosmetics has focused on perfume and make-up. Thus, in his *Sex in the Ancient World from A to Z*, John G. Younger defines cosmetics as 'agents to color the skin and to hide blemishes and perfumes to counteract body odor'.[39] It seems, however, that ancient cosmetic treatises encompassed much more than face packs and products for curling the hair, as is testified by the contents of Statilius Criton's treatise *Cosmetics*. This text is unfortunately lost, but Galen transmits extensive extracts and the table of contents to its four books (see Appendix 1 below).[40] Besides chapters devoted to hair-dyes, skin remedies (both of the medicinal and the make-up types), toothpastes, deodorants, depilatories, and remedies against the effects of childbearing (stretch marks and other marks), one finds more

---

[37] On deer horn, see Totelin 2007.   [38] See, for instance, Hipp., *Mul.* 1.89 (L. VIII 214.8–13).
[39] Younger 2005, 38. Modern definitions of cosmetics also encompass much more than products to be applied to the face – see Vons 2000, 346.
[40] On Crito, see Scarborough 1985, Scarborough and Touwaide 2008. On Tables of Contents in the ancient world, see Riggsby 2007.

unexpected headings. Book 2 included recipes to preserve the virginity of both boys and girls, and recipes against excessive wetness and coldness in women (2.10–12); Book 3 had remedies for removing tattoos (3.10); and Book 4 (4.14–16) included recipes against prolapse of the intestines and the anus, hydroceles (water in the scrotum) and leipodermos (an affliction which affects the prepuce).[41] Nor were all products in Crito's work to be applied to the human body: Book 2 contained recipes for perfumes for clothes and bedchambers (2.19–21).

Two points need to be stressed. First, behind all the outcries against women's frivolity and depravity, there is much that pertains to boys and men in the ancient cosmetic tradition. In fact, complaints against women's vanity and disgusting habits might have served the purpose to conceal, to mask, a much more complex reality, where men consumed as many cosmetics as women.[42] Both sexes may have worried about the unsightly appearance of alopecia; both sexes may have suffered from anal prolapse, and in general may have taken care of their anal region's appearance; but remedies against first-beard rashes (3.13) and afflictions of the scrotum are most definitively for men.[43] One may retort that the men interested in such products were effeminates, *cinaedi*. Certainly Paris the Dancer, who created a depilatory transmitted by Crito and preserved by Galen, might have fallen into that category.[44] Overindulging in depilation was seen as a mark of the effeminate in Roman literature, as were wearing too much perfume, curling one's hair and generally taking too keen an interest in one's looks.[45] Men in all sections of society could be accused of excessive interest in their physical appearance, but emperors were particularly vulnerable to such charges. We have already mentioned Galen's 'kings', to whom we should add Julius Caesar, who depilated and was obsessed with his hair; Domitian, who was very sensitive about his baldness and allegedly authored a work on the care of the hair; and Otho, whom Juvenal compares to Semiramis

---

[41] On tattoos in the ancient world, see C. P. Jones 1987. On coldness in women, see Carson 1990. On virginity, see Sissa 1987.
[42] Wyke 1994 and Richlin 1995 both insist on the fact that men should not use make-up.
[43] On care of the anus' appearance in the ancient world, see Richlin 1983, 38 and 41. On care of the beard in antiquity, see Bartman 2001.
[44] Galen, *De comp. med. sec. loc.* 1.4 (12.454 K.): Πάριδος τοῦ ὀρχηστοῦ παραχρῆμα τὰς τρίχας αἴρει. σανδαράχης γο α′. ἀσβέστου < α′. ἕψε μεθ' ὕδατος ἐπιμελῶς καὶ ὅταν ἄρξηται ἀναζεῖν, ἀνελόμενος χρῶ. '[Depilatory] of Paris the dancer; it removes hair on the spot: realgar, 1 ounce, unslaked lime, 1 drachma. Boil with water carefully and, when it starts to bubble up, remove and use.'
[45] On the characteristics of the effeminate, see Richlin 1983, 3 ff.; Williams 2010, 141 ff. Some of the best ancient descriptions of *cinaedi* are to be found in Seneca, *Controv.* 1, intro 8–9, and in the speech of Scipio Aemilianus (Gell. 6.12). On depilation in the ancient worlds, see Kilmer 1982, Younger 2005, 55 (s.v. hair).

and Cleopatra.⁴⁶ However, whilst too polished an appearance in a man was frowned upon, unkemptness was also considered unacceptable; and men who took care of their looks were thought to attract more women.⁴⁷ The Roman man had to strike the right balance between effeminacy and untidy masculinity, and a moderate use of cosmetics might have helped him to do just that. Thus, whilst I would agree with Wyke's statement whereby 'the Roman discourses of adornment... articulate the female as always unreal, as an image constructed in a mirror, and as the opposite and the inferior of the male', I would suggest that these same discourses articulate the male as unreal too.⁴⁸ And when one unreal gender (female) is constructed as the opposite of another unreal (male) gender, the whole edifice of gender construction starts to look like a game of smoke and mirrors.

The second point to note in relation to the Table of Contents to Crito's *Cosmetic* is that the work had two focal points: the face and the sexual/anal region. This same dual focus is found, to a certain extent, in modern cosmetology, which deals with intimate soaps and wipes alongside products to be applied to the face. Today as in the past, humans have recourse to cosmetics to attract each other and enhance sexual experiences – the face serves as 'advertisement' for the sex. In the ancient world, that link between face and sex was made even more explicit, in particular in the case of women. The same word (mouth: Latin *os*, Greek *stoma*) referred to both the mouth and the vagina; and medicaments to be applied to the mouth or nostrils were recommended in the treatment of uterine ailments, as well as in fertility tests.⁴⁹ Here it is interesting to note that some of the first cosmetic recipes preserved in Greek are to be found in one of the gynaecological treatises of the *Hippocratic Corpus: Diseases of Women II*, where recipes for toothpaste and freckle removers (Chapters 185 and 190) are interspersed with remedies for cracked nipples and ascarides (Chapters 186 to 188). This link between cosmetology and gynaecology endured throughout antiquity and well into the early modern period.⁵⁰ This connection is particularly visible in the curious 'Metrodora' compilation. This text, on whose date scholars have

---

⁴⁶ Suetonius; *Divus Julius* 45.2; *Domitianus* 18; Juvenal (on Otho): 2.99–109.
⁴⁷ See Richlin 1983, 136–7, Gleason 1995, Wyke 1994, 135.   ⁴⁸ Wyke 1994, 148.
⁴⁹ For remedies to be applied to the mouth and nostrils to treat the womb, see e.g. [Hippocrates], *Mul.* 2.127 (L. VIII 272–4). For fertility tests, see, for instance, [Hippocrates], *Steril.* 214 (L. VIII 416.2–5). On conceptions of the female body in antiquity, see e.g. Dean-Jones 1994, King 1998. Although they are not particularly clear, there are some indications that a similar link was postulated between the male anus and the mouth. See Dean-Jones 1994, 72–3.
⁵⁰ Some mediaeval treatises combining gynaecology and cosmetics have been studied recently. See M. H. Green 2001: 113–24 (on *De palliandis muliebribus*, a text that is part of the Salernitan Trotula ensemble), Cabré 2000, Caballero-Navas 2008. See below for more detail on the early modern period.

been unable to settle (dates vary from the first to sixth centuries AD),[51] starts with a collection of gynaecological treatments which are very similar to those found in the gynaecological treatises of the *Hippocratic Corpus*.[52] From Chapter 32, however, the material looks less 'Hippocratic', and one finds recipes to make deflowered girls appear to be virgins (32); methods to recognise whether a girl is a virgin or not (33); remedies to prevent a woman from being adulterous (34); concoctions to make a woman acknowledge her adultery (35); and aphrodisiacs (*hēdonika*) (36). From comparison with Crito's Table of Contents, we know that these remedies can fall under the label of 'cosmetics'. Chapters 37 to 50 are again more strictly gynaecological in nature (they deal with breast and lactation issues). However, Chapter 51 starts another cosmetic section: remedies to make the breast bright and beautiful; recipes to brighten the face, including a recipe used by 'Berenice, queen of Egypt, later called Cleopatra' (52–5); deodorants (56); and perfumes (57). Admittedly, the 'Metrodora' treatise may be the result of a slow process of accretion; but even so, the combination of cosmetic and sexually related products with strictly gynaecological remedies must be significant.

A similar connection between gynaecology and cosmetology can be observed in the recipes attributed to three women: Salpe, Elephantis and Cleopatra (see Appendix 2 for the recipes attributed to these women).[53] Whether male authors were talking through 'ventriloquised' female voices, or whether women were actually active in the field of ancient cosmetology, will not preoccupy us here.[54] Instead, I want to stress how sexual knowledge was attributed to these female authorities. Salpe, to whom Pliny attributed a depilatory recipe (to be used on boys, it should be stressed), is called an *obstetrix* (midwife): she apparently created various other drugs, including an aphrodisiac. Salpe – or whoever was hiding behind that mask –may have

---

[51] This text is found in a single manuscript of the Laurentian Library in Florence (MS Laur. Plut. 75.3) under the heading Ἐκ τῶν Μητροδώρας, περὶ τῶν γυναικείων παθῶν τῆς μήτρης. The edition of Metrodora remains Kousis 1945; see also Del Guerra 1994. For a date in the first century, see Deichgräber 1932, Nutton 1995, 49. For a date in the Byzantine period, see Kousis 1945, Congourdeau 1993, Rubio Gómez 1996, Touwaide 2000. Parker 2008 gives it a date between AD 50 and 400. It is interesting too that the 'Metrodora' compilation seems to be attributed to a woman, even though the female name Metrodora is not really attested. See Dean-Jones 1994, 33, Flemming 2007, 258.

[52] See Totelin 2009, 276–9 for similarities between the 'Metrodora' compilation and Hippocratic gynaecological texts.

[53] Other cosmetic recipes are attributed to women in ancient medical texts: Scribonius Largus, *Compositiones* 59, 60 (toothpastes used by Octavia, the sister of Augustus, Augusta and Messalina). Poppaea: Pliny, *HN* 11.238; 28.183; Juvenal 6.462–3.

[54] On 'ventriloquism', see Toulalan 2007, 137.

been involved in the sex market. Athenaeus mentions a woman from Lesbos, named or nicknamed Salpe, author of *paignia*, that is, 'frivolous works', which are believed to have been of a pornographic nature by scholars.⁵⁵ Elephantis too was known as a writer on cosmetics by Galen (remedy on alopecia), as an authority on abortives by Pliny, and as an author of erotic texts by several poets and other authors.⁵⁶ According to Suetonius, Elephantis was particularly appreciated by the Emperor Tiberius. Cleopatra was, according to Galen and other Greek medical authorities, the author of a treatise entitled *Cosmetics*, of which several recipes have been preserved. In the Latin medieval tradition, on the other hand, Cleopatra was known as an authority on gynaecology; and a book on aphrodisiacs by Cleopatra is mentioned by the Arabic author Qustā ibn Lūqā (AD 820–912). If we piece together these three different traditions, we get an image of a Cleopatra active in cosmetology, gynaecology and sexual advice, like Elephantis and Salpe.⁵⁷ We can only speculate, but it might be the case that the original Cleopatra's *Cosmetics* included recipes for aphrodisiacs and gynaecological treatments, as does the Metrodora compilation. It might also be the case that Elephantis' sex manuals or Salpe's *Paignia* included cosmetic recipes. In fact, the tradition of sex manuals (of which we know very little), that of cosmetic manuals (of which we know slightly more) and that of gynaecological treatises (of which we have many surviving texts) might have largely overlapped in the ancient world.⁵⁸ Gynaecological treatises in the ancient world are, for the most part, pronatalist: they aim at making women more fertile.⁵⁹ Treatises written in the Hippocratic tradition prescribe sexual intercourse as a cure for some gynaecological ailments. They recommend women initiate sexual relationships when they have become healthy.⁶⁰ It is then, and despite the claims to the contrary found in ancient literature, that the cosmetics included in gynaecological and cosmetic treatises might have come in handy, in making men and women more attractive to each

---

⁵⁵ Athenaeus 322a. On Salpe, see Davidson 1995, Bain 1997. The women mentioned here are also discussed by Gazzaniga 1997, Parker 1997, Flemming 2000, 39 ff., Flemming 2007, Plant 2004.
⁵⁶ Martial 12.43.4; Suetonius, *Tib.* 43.2; Tatianus, *Ad. Gr.* 34.3; Suda, s.v. Ἀστυάνασσα.
⁵⁷ The Latin tradition consists of the *gynaecia* (which exists in a long and a short version, and which covers various gynaecological ailments and their treatments) and the *pessaria* (a collection of pessary recipes). See M. H. Green 2000, 8–10. Neither text has ever been edited according to modern standards. Arabic tradition: Ullmann 1970, 127. The Jewish tradition also reports that the queen herself was involved in embryological studies pursued through the dissection of pregnant slaves (Tosefat Niddah 4.17 and BT Niddah 30b). On Cleopatra and the medical tradition, in addition to works mentioned in n. 21, see Sbordone 1930, Marasco 1995, Marasco 2003, Scarborough 2012.
⁵⁸ On sex manuals in the ancient world, see Parker 1992, King 1994.
⁵⁹ Soranus is unique in seeing virginity as a good in itself.
⁶⁰ See Totelin 2007, Totelin 2009, Chapter 5.

other, and in facilitating sexual intercourse. A beautiful face, a disease-free body and a sweet-scented bedchamber can be seen as aids for healthy sexual relationships that will lead to procreation. In a pronatalist context, ancient cosmetic recipes could be particularly useful.

Towards a Conclusion: Use and Abuse of Ancient Cosmetic Material

On the face of it, ancient cosmetics texts are practical texts, interminable lists of recipes, to be transformed into remedies to improve beauty. These recipes are sometimes accompanied by moral statements on women's vanity. Medical writers reluctantly deal with this topic, following the desires of frivolous females. However, cosmetics have as their main purpose dissimulation, and when one scratches the mask, under the layers of moral foundation, and female rouge, one finds a reality that is rather more complicated. First, there is quite a lot that concerns men – and men only – in the cosmetic tradition. It would be simplistic to label all men who had recourse to cosmetics in the ancient world as effeminate. Second, wherever we look in the field of ancient cosmetology, we find sexual content and sexual connotations. From the fertilising ingredients cosmetic recipes contain, to the sexual afflictions mentioned in cosmetic works, to the authorities to whom remedies are attributed, one is constantly brought back to sex. There might have been large areas of overlap between ancient gynaecology, cosmetology and sexual manuals. Sexual manuals in the ancient world, such as those of Elephantis, were read for sexual gratification. Could cosmetic texts and gynaecological texts that include cosmetics also be read in that way? Here it is worth looking at other more-documented periods. Sarah Toulalan, in her *Imagining Sex: Pornography and Bodies in Seventeenth-Century England* (2007), finds evidence of medical authorities (such as Culpeper) being accused of writing obscene works;[61] lists of books appropriate to a 'love academy' that include midwifery manuals;[62] and generally, a great overlap between midwifery books and pornography. Thus, for instance, the infamous Aristotle's *Masterpiece* (1684) and its predecessor *The Problems of Aristotle, with other Philosophers and Physicians* (1647) were half midwifery handbooks, half sex manuals.[63] Interestingly, the *Master Piece* was banned in Britain until the 1960s, as it was considered pornographic in nature. To the evidence collected by Toulalan, one could add the letters (in Latin) composed by a seventeenth-century forger allegedly exchanged between the

---

[61] Toulalan 2007, 16, 53.
[62] Toulalan 2007, 17. The list is found in the *Practical Part of Love*, 1660. See also King 1994.
[63] Toulalan 2007, 67. On the *Masterpiece*, see Fissell 2003.

physician Soranus of Ephesus, Marc Antony and Cleopatra regarding the queen's unbridled sex drive.[64] These letters take the form of a 'real' medical correspondence, including 'serious' gynaecological information, as well as recipes for medicaments. Some of these recipes, for instance those for an-aphrodisiacs, would not have been out of place in an ancient cosmetic treatise. But the playful tone of the letters seems to indicate that they were read not for medical reasons, but for pleasure. It is also worth mentioning that these letters were transmitted alongside Petronius' *Satyricon*, another work filled with sexual allusions.

Toulalan uses the category 'pornography' to refer to some midwifery books composed in the seventeenth century, acknowledging the difficulties in defining the word. 'One person's pornography is another person's sex education', and a large variety of works can create a sexual response in their readers.[65] Nor is it easy to guess the intention of an author. Thus the writer of a medical treatise might proclaim an educative purpose for his work but deliberately include words, descriptions or depictions that some readers might find titillating. It is also important not to take an ahistorical view of pornography. Seventeenth-century pornography is different from modern pornography and may be different from pornography in other periods. In the seventeenth century, according to Toulalan, pornography encompassed a variety of texts that dealt with sex that:

> could be used both for education and for stimulation; they might be informative but they were also arousing. Thus some texts can be seen as more 'pornographic' than others, but no text, whatever its nature or origin, was entirely 'pornographic' in the modern sense that it contained only explicit representations of the sexual body and acts of sexual intercourse.[66]

Toulalan also notes that, unlike modern pornography, much pornography in the seventeenth century deals with reproduction and generation (pregnancy, childbirth, lactation etc.) because 'sexual pleasure was understood as not complete pleasure if it did not have the possibility of conception'.[67]

---

[64] *Titii Petronii arbitri equitis Romani Satyricon, cum fragmento nuper Trangurii reperto. Accedunt diversorum poetarum lusus in priapum, pervigilium veneris, Ausonii canto nuptialis, cupido crucifixus, Epistolae de Cleopatra, & alia nonnulla. Omnia commentariis, & notis doctorum virorum illustrata concinnate Michaele Hadrianide, Amstelodami* 1669. The letters are discussed and translated in Hanson 2008. Although Soranus was active in the first century AD, Cleopatra and Soranus were sometimes presented as contemporaries. See Tzetzes, *Chiliades* 6.293–6: 'She [i.e. Cleopatra] used, on the one hand, Dexiphanes to [build] machines, on the other hand, the doctors from Ephesus, Soranus and Rufus, for all the ways of embellishing the forms of the face, for every feminine ailment, and other treatments.'

[65] Toulalan 2007, 2.
[66] Toulalan 2007, 271. On the definition of pornography, see also Parker 1992.
[67] Toulalan 2007, 64.

It is important to note that many of the seventeenth-century medical works studied by Toulalan have their origins in the ancient tradition. They claim to be authored by ancient writers (such as Aristotle) or build upon ancient works. One therefore wonders whether the seventeenth-century tradition of reading medical treatises for sexual gratification might also have its origin in the ancient world. Gynaecological treatises that include cosmetic recipes and cosmetic treatises, with their partly sexual contents, might have had as one of their purposes to help men and women in conceiving children. However, there is the possibility that, like Lucian's woman misusing philosophical knowledge, readers of cosmetic works could misused their content and used them as pornography. Male (and female) readers might have fantasised over the oft-mentioned desires of women to cover their face in disgusting, ill-smelling ingredients; they might have found sexual pleasure in reading lists of aphrodisiacs, virginity-restorers, sexual treatments; they might have found sexual advice in cosmetic treatises. Ancient cosmetic works may have appealed to some for their obscene content. *Technē* can sometimes turn into *kakotechnia*.

Appendix 1: Table of Contents to Crito's *Cosmetics*, apud Galen, *On the Composition of Medicines according to Places* 1.3 (12.444.6–9 K.)

Chapters of the first book of Crito's *Cosmetics*	1. Τοῦ πρώτου τῶν Κρίτωνος κοσμητικῶν κεφάλαια
[Remedies] to preserve the hair	1.1 διαφυλακτικὰ τριχῶν
[Remedies] to grow the hair	1.2 αὐξητικὰ τριχῶν
[Remedies] to protect the hair	1.3 προφυλακτικὰ τριχῶν
Dyes for grey hairs	1.4 βάμματα πολιῶν
Dyes that make the hair blond or golden	1.5 βάμματα ὥστε ξανθὰς καὶ χρυσιζούσας ποιεῖν
Washing products for the hair	1.6 σμήξεις τριχῶν
Protective unguents	1.7 ἐπίχριστα προφυλακτικά
Lotions for the face	1.8 προσώπου τετανώματα
Unguents to brighten the face	1.9 προσώπου ἐπίχριστα λαμπρυντικά
Cataplasms to brighten the face	1.10 προσώπου καταπλάσματα λαμπρυντικά

Unguents for the brow	1.11 ὀφρύων ἐπίχριστα
Black hair-dyes for the brow	1.12 ὀφρύων μελάσματα
Unguents for the eyes	1.13 ὀφθαλμῶν ἐπίχριστα
Blackening ointments	1.14 στιμμίσματα ἔγχριστα
Against bad smell of the nostrils	1.15 πρὸς δυσωδίαν μυκτήρων
Powders for the teeth	1.16 ὀδόντων ἀποτρίμματα
Chewing gum against bad smell of the arm-pits (?)	1.17 διαμασήματα πρὸς μασχαλῶν δυσωδίαν
Preparations for rubbing	1.18 περιτρίμματα
Chewing gum against bad smell of the mouth	1.19 διαμασήματα πρὸς δυσωδίαν στόματος
Chapters of the second [book]	2. Τοῦ δευτέρου κεφάλαια
Soaps for black spots on the neck	2.1 ῥύμματα πρὸς τὰς ἐπὶ τοῦ τραχήλου μελανίας
Unguents against excessive perspiration of the arm pits	2.2 ἐπίχριστα πρὸς τὰς τῶν μασχαλῶν συνιδρώσεις
Unguents to preserve the breasts	2.3 ἐπίχριστα μαστῶν διαφυλακτικά
Purgatives of the belly	2.4 κοιλίας καθαρτικά
Soaps to brighten the hands	2.5 σμήγματα χειρῶν λαμπρυντικά
[Remedies] against black spots after childbirth	2.6 πρὸς τὰς ἐκ τόκου μελανίας
[Remedies] against wrinkles after childbirth	2.7 πρὸς τὰς ἐκ τόκου ῥυτίδας
[Remedies] against fistulas after childbirth	2.8 πρὸς τὰς ἐκ τόκου ῥαγάδας
Unguents against protruding eyes	2.9 ἐπιχρίσματα πρὸς τὰς τῶν ὀμφαλῶν ἐξοχάς
[Remedies] to retain boys' youth	2.10 ἀνήβων παιδίων διαφυλακτικά
[Remedies] to preserve virginity	2.11 παρθενίας διαφυλακτικά
[Remedies] against women who are excessively wet or excessively cold	2.12 πρὸς τὰς καθύγρους καὶ καταψύχρους γυναῖκας
[Remedies] against black scars	2.13 πρὸς οὐλὰς μελαίνας

Depilatories	2.14 ψίλωθρα τριχῶν
All sorts of thinning soaps	2.15 σμήγματα λεπτυντικὰ παντοῖα
Unguents that destroy the hair	2.16 ἐπίχριστα ἀφανιστικὰ τριχῶν
Soaps for the entire body	2.17 σμήγματα ὅλου τοῦ σώματος
All sorts of brightening purgatives	2.18 καθαρτικὰ λαμπρυντικὰ παντοῖα
Aromatic cataplasms for clothes	2.19 καταπλάσματα ἀρωματικὰ τοῖς ἱματίοις
Sweet-smelling dyes for clothes	2.20 εὐώδεις βαφαὶ ἱματίων
Sprinklings for chambers and covered walks	2.21 ῥάσματα θαλάμων καὶ περιπάτων
Mode of preparation for all sorts of fumigations	2.22 θυμιαμάτων παντοίων σκευασία
Preparations for all sorts of ointments and perfumes [list of perfumes]	2.23 χρισμάτων καὶ μύρων παντοίων σκευασίαι. [list of perfumes]
Chapters of the third [book]	3. Τοῦ τρίτου κεφάλαια
[Remedies] against all sorts of bran-like eruptions	3.1 πρὸς πίτυρα παντοῖα
[Remedies] against cutaneous eruptions on the head	3.2 πρὸς τὰ ἐν τῇ κεφαλῇ ἐκβράσματα
[Remedies] against dandruff	3.3 πρὸς ἀχῶρας
[Remedies] against louses and nits	3.4 πρὸς φθεῖρας καὶ κόνιδας
[Remedies] against alopecia	3.5 πρὸς ἀλωπεκίας
[Remedies] against scabby afflictions on the face	3.6 πρὸς τὰς ἐν τῷ προσώπῳ ψωρώδεις διαθέσεις
[Remedies] against moles	3.7 πρὸς φακούς
[Remedies] against freckles	3.8 πρὸς ἐφήλεις
Cleansing products for the face	3.9 νίμματα προσώπου
[Remedies] against tattoo-marks	3.10 πρὸς στίγματα
[Remedies] against livid-spots	3.11 πρὸς πελιώματα

[Remedies] against black-eyes	3.12 πρὸς ὑπώπια
[Remedies] against eruptions accompanying the first beard	3.13 πρὸς ἰόνθους
[Remedies] against night pustules	3.14 πρὸς ἐπινυκτίδας
[Remedies] against tuberous eruptions on the chin	3.15 τὰ ἐπὶ τῶν γενείων ὀχθώδη
[Remedies] against lichen-like eruptions on the chin	3.16 τοὺς ἐπὶ τῶν γενείων λειχῆνας
Unguents against lichen-like eruption	3.17 λειχήνων ἐπιχρίσματα
Applications which remove skin	3.18 ἐπιθέματα ἐκδόρεια
Fresh applications for after the medications which remove the skin	3.19 χλωραὶ μετὰ τὰς ἐκδορὰς ἐπιτιθέμεναι
White remedies for cicatrisation	3.20 λευκαὶ πρὸς ἀπούλωσιν
Emollient remedies which destroy lichen-like eruptions	3.21 φάρμακα μαλακτικὰ ἐπιτριβέντα λειχήνων
Chapters of the fourth [book]	4. Τοῦ τετάρτου κεφάλαια
[Remedies] against black and white leprosy	4.1 πρὸς ἀλφοὺς μέλανας καὶ λευκούς
[Remedies] for black scars	4.2 πρὸς οὐλὰς μελαίνας
[Remedies] for white [scars]	4.3 πρὸς λεύκας
[Remedies] against leprosy	4.4 πρὸς λέπρας
[Remedies] for itchy nails	4.5 πρὸς ὄνυχας ψωριῶντας
[Remedies] for itching afflictions	4.6 πρὸς κνησμώδεις διαθέσεις
[Remedies] for pimples and excoriations	4.7 πρὸς ψύδρακας καὶ ἐκδάρματα
[Remedies] for spreading afflictions	4.8 πρὸς τὰς ἑρπυστικὰς διαθέσεις
[Remedies] against psoriasis	4.9 πρὸς ψωριάσεις
[Remedies] against warty excrescences	4.10 πρὸς θύμους
[Remedies] against warts	4.11 πρὸς ἀκροχορδόνας
[Remedies] against warts [another type]	4.12 πρὸς μυρμηκίας

[Remedies] against inflammations of the eyes	4.13 πρὸς ὀμφαλοῦ ἐμπνευμάτωσιν
[Remedies] against slipping down of the intestines	4.14 πρὸς ἐντέρων κατολίσθησιν
[Remedies] against hydroceles [water in the scrotum]	4.15 πρὸς ὑδροκήλας
[Remedies] against leipodermos [affliction of the prepuce]	4.16 πρὸς λειποδέρμους
[Remedies] against falling anus	4.17 πρὸς τοὺς προσπίπτοντας ἀρχούς
[Remedies] against chilblains	4.18 πρὸς χείμεθλα
[Remedies] for fractures of the feet	4.19 πρὸς τὰς ἐν ποσὶ ῥωγμάς
In these four books, Crito wrote most carefully almost all the excellent cosmetic remedies, adding to them the adorning remedies, which provide a spurious beauty, not agreeable to truth; for that reason I shall leave them, remembering only those remedies which preserve beauty according to nature	ἐν τούτοις τοῖς τέσσαρσι βιβλίοις ὁ Κρίτων ἐπιμελέστατα σχεδὸν ἅπαντα ἔγραψε τὰ δόκιμα κοσμητικὰ φάρμακα, προσθεὶς αὐτοῖς καὶ τὰ κομμωτικά, νόθον κάλλος, οὐκ ἀληθινὸν ἐκπορίζοντα, διὸ κἀγὼ παραλείψω μὲν αὐτά, μόνον δὲ μνημονεύσω τῶν τὸ κατὰ φύσιν κάλλος φυλαττόντων.

Appendix 2: Recipes Attributed to Cleopatra, Elephantis and Salpe

Cleopatra	Galen, *De comp. med. sec. loc.* I.2 (12.403–5 K.): eight remedies against alopecia from Cleopatra's *Cosmetics*
	Galen, *De comp. med. sec. loc.* I.2 (12.432–4 K.): seven recipes to make the hair grow from Cleopatra's *Cosmetics*
	Galen, *De comp. med. sec. loc.* 1 (12.492 K.): eight recipes against dandruff from Cleopatra's *Cosmetics*
	Metrodora 53: unguent to make the face bright which Berenice, the queen of Egypt, later called Cleopatra, used.
	Aetius, *Tetrabiblon* 6.56 (CMG VIII 2, 205 Olivieri): recipes against falling hair
	Aetius, *Tetrabiblon* 8.6 (CMG VIII 2, 408 Olivieri): unguent of Cleopatra
	Paulus Aegineta 3.2.1 (132 Heiberg): four recipes for curling the hair

Elephantis	Galen, *De comp. med. sec. loc.* 1 (12.416 K.): Elephantis is mentioned as one of the authorities excerpted by Soranus for remedies on alopecia
	Pliny, *H.N.* 28.81: Lais and Elephantis disagree on the which abortives to use
Salpe	Pliny, *H.N.* 28.38: remedy against numbness
	Pliny, *H.N.* 28.66: remedies involving urine
	Pliny, *H.N.* 28.82: remedy against tertian and quartan fever and the bite of a rabid dog used by Lais and Salpe
	Pliny, *H.N.* 28.262: aphrodisiac involving an ass's penis
	Pliny, *H.N.* 32.135: depilatory used by Salpe the *obstetrix* on young slaves
	Pliny, *H.N.* 32.140: remedy so that dogs do not bark

REFERENCES

W. Artelt 1968, *Studien zur Geschichte der Begriffe 'Heilmittel' und 'Gift'* (Darmstadt)

D. Bain 1997, 'Salpe's *Paignia*: Athenaeus 322A and Pliny *H.N.* 23.38', *Classical Quarterly* 48: 262–8.

E. Bartman 2001, 'Hair and the Artifice of Roman Female Adornment', *American Journal of Archaeology*, 105, 1: 1–25

G. W. Bowersock 1969, *Greek Sophists in the Roman Empire* (Oxford)

C. Caballero-Navas 2008, 'The Care of Women's Health and Beauty: An Experience Shared by Medieval Jewish and Christian Women', *Journal of Medieval History*, 34, 2: 146–63

M. Cabré 2000, 'From a Master to a Laywoman: A Feminine Manual of Self-Help' in *Dynamis: Acta Hispanica ad Medicinae Scientiarumque Historiam Illustrandam* 20: 371–93

A. Carson 1990, 'Putting Her in Her Place: Woman, Dirt, and Desire' in D. M. Halperin, J. J. Winkler and F. I. Zeitlin (eds.), *The Construction of Erotic Experience in the Ancient World* (Princeton), 135–67

M. H. Congourdeau 1993, '"Métrodora" et son oeuvre' in E. Patlagean (ed.), *Maladie et société à Byzance: Études présentées au 17e congrès international des sciences historiques* (Madrid), 57–61

J. N. Davidson 1995, 'Don't Try This at Home: Pliny's Salpe, Salpe's *Paignia* and Magic', *Classical Quarterly*, 45: 590–2

M. Dayagi-Mendels 1989, *Perfumes and Cosmetics in the Ancient World*. Catalog Israel Museum (Jerusalem)

L. A. Dean-Jones 1994, *Women's Bodies in Classical Greek Science* (Oxford)

K. Deichgräber 1932, 'Metrodora' in *RE* XV, 2: 1474
G. Del Guerra 1994, *Metrodora: Medicina e cosmesi ad uso delle donne: La antica sapienza femminile e la cura di sé* (Milan)
M. Douglas 1966, *Purity and Danger: An Analysis of Concepts of Pollution and Taboo* (New York)
R. P. Evershed, R. Berstan, M. S. Copley et al. 2004, 'Formulation of a Roman Cosmetic', *Nature*, 432: 35–6
C. Fabricius 1972, *Galens Exzerpte aus älteren Pharmakologen* (Berlin, New York)
M Fissell 2003, 'Hairy Women and Naked Truths: Gender and the Politics of Knowledge in Aristotle's Masterpiece', *The William and Mary Quarterly*, 60, 1: 43–74
R. Flemming 2000, *Medicine and the Making of Roman Women: Gender, Nature, and Authority from Celsus to Galen* (Oxford)
    2007, 'Women, Writing and Medicine in the Classical World', *Classical Quarterly*, 57, 1: 257–79
R. J. Forbes 1965, *Studies in Ancient Technology* (Leiden)
M. Formisano 2001, *Tecnica e scrittura: Le letterature tecnico-scientifiche nello spazio letterario tardolatino* (Rome)
V. Gazzaniga 1997, 'Phanostrate, Metrodora, Lais and the Others: Women in the Medical Profession', *Medicina nei secoli*, 9: 277–90
M. Gleason 1995, *Making Men: Sophists and Self-Presentation in Ancient Rome* (Princeton)
M. H. Green 2000, 'Medieval Gynaecological Texts: A Handlist' in M. H. Green, *Women's Health Care in the Medieval West: Texts and Contexts* (Aldershot), 1–36
    2001, *The Trotula: A Medieval Compendium of Women's Medicine* (Philadelphia)
P. Green 1979, '*Ars gratia cultus*: Ovid as Beautician', *American Journal of Philology* 100, 3: 381–92
B. Grillet 1975, *Les Femmes et les fards dans l'antiquité grecque* (Lyons)
A. E. Hanson 1997, 'Fragmentation and the Greek Medical Writers' in G. W. Most, *Collecting Fragments: Fragmente sammeln* (Göttingen), 289–314
    1998, 'Talking Recipes in the Gynaecological Texts of the Hippocratic Corpus' in M. Wyke, *Parchments of Gender: Deciphering the Bodies of Antiquity* (Oxford), 71–94
    2008, 'The Correspondence between Soranus, Marc Anthony and Cleopatra' in V. Boudon-Millot, V. Dasen and B. Maire, *Femmes en médecine en l'honneur de Danielle Gourevitch* (Paris), 75–104
J. L. Heiberg 1921, *Paulus Aegineta, Libri I–IV, edidit J. L. Heiberg* (Leipzig, Berlin)
E. A. Hemelrijk 1999, *Matrona docta: Educated Women in the Roman Elite from Cornelia to Julia Domna* (London, New York)
M. Hendry 1995, 'Rouge and Crocodile Dung: Notes on Ovid, *Ars* 3, 199–200 and 269–70', *Classical Quarterly*, 45, 2: 583–88
C. P. Jones 1987, 'Stigma: Tattooing and Branding in Graeco-Roman Antiquity', *Journal of Roman Studies*, 77: 139–55
W. H. S. Jones 1957, 'Ancient Roman Folk Medicine', *Journal of the History of Medicine*, 12: 459–72

M. Kilmer 1982, 'Genital Phobia and Depilation', *Journal of Hellenic studies*, 102: 104–12

H. King 1994, 'Sowing the Field: Greek and Roman Sexology' in R. Porter and M. Teich, *Sexual Knowledge, Sexual Science: The History of Attitudes to Sexuality* (Cambridge University Press), 29–46

1998, *Hippocrates' Woman: Reading the Female Body in Ancient Greece* (London, New York)

J. Korpella 1995, '*Aromatarii, pharmacopolae, thurarii et ceteri*: Zur Sozialgeschichte Roms' in P. J. van der Eijk et al., *Ancient Medicine in Its Socio-Cultural Context: Papers Read at the Congress Held at Leiden University, 13–15 April 1992* (Amsterdam), 101–18

A. P. Kousis 1945, 'Metrodora's Work "On the Feminine Diseases of the Womb" according to the Greek Codex 75.3 of the Laurentian Library', *Praktika tes Akademias Athenon*, 20: 46–68

J. Laskaris 2005, 'Error, Loss, and Change in the Generation of Therapies' in P. J. van der Eijk, *Hippocrates in Context: Papers Read at the XIth International Hippocrates Colloquium, University of Newcastle upon Tyne, 27–31 August 2002* (Leiden), 173–89

B. Levick 2007, *Julia Domna: Syrian Empress* (London, New York)

E. Liechtenhan, J. Kollesch, D. Nickel et al. 1968, *Marcelli De medicamentis liber, post M. Niedermann iteratis curis edidit E. Liechtenhan, in linguam Germanicam transtulerunt J. Kollesch et D. Nickel* (Berlin)

G. Marasco 1995, 'Cleopatra e gli esperimenti su cavie umane', *Historia*, 44: 317–25

2003, 'Cléopâtre et les sciences de son temps' in G. Argoud and J.-Y. Guillaumin, *Sciences exactes et sciences appliquées... Alexandrie (IIIe siècle av. J.-C. – Ier siècle ap. J.-C.). Actes du colloque international de Saint-Étienne, 6–8 juin 1996* (Saint-Étienne), 39–53

M. Meilhac-Léonelli 2003, 'La Teinture des cheveux, un geste médical? À propos de Théodore Priscien Eup. 1.5–6 (éd. V. Rose)' in F. Gaide and F. Biville, *Manus medica et gestes de l'officiant dans les textes médicaux latins: Questions de thérapeutique et de lexique* (Aix-en-Provence), 173–81

V. Nutton 1995, 'Roman Medicine, 250 BC to AD 200' in L. I. Conrad, M. Neve, V. Nutton, R. Porter and A. Wear, *The Western Medical Tradition: 800 BC to AD 1800* (Cambridge), 39–70

A. Olivieri 1950, *Aetii Amideni Libri medicinales V–VIII, edidit A. Olivieri* (Berlin)

K. Olson 2009, 'Cosmetics in Roman Antiquity: Substance, Remedy, Poison', *Classical World* 102, 3: 291–310

A. Önnerfors 1964, *Plinii Secundi Iunioris qui feruntur de medicina libri tres* (Berlin)

H. N. Parker 1992, 'Love's Body Anatomized: The Ancient Erotic Handbooks and the Rhetoric of Sexuality' in A. Richlin, *Pornography and Representation in Greece and Rome* (New York, Oxford) 90–111

H. N. Parker 1997, 'Women Doctors in Greece, Rome, and the Byzantine Empire' in L. R. Furst, *Women Healers and Physicians: Climbing a Long Hill* (Lexington) 131–50

2008, 'Metrodora' in P. T. Keyser and G. L. Irby-Massie, *Encyclopedia of Ancient Natural Scientists: The Greek Tradition and Its Many Heirs* (London, New York), 552–3

I. M. Plant 2004, *Women Writers of Ancient Greece and Rome: An Anthology* (Norman, OK)

J. Raeder 1933, *Oribasii Collectionum medicarum reliquiae, libri XLIX–L, libri incerti, eclogae medicamentorum, edidit J. Raeder* (Leipzig, Berlin)

A. Richlin 1983, *The Garden of Priapus: Sexuality and Aggression in Roman Humor* (New Haven, London)

 1995, 'Making up a Woman: The Face of Roman Gender' in H. Eilberg-Schwartz and W. Doniger, *Off with Her Head! The Denial of Women's Identity in Myth, Religion, and Culture* (Berkeley), 185–213

A. M. Riggsby 2007, 'Guides to the Wor(l)d' in J. König and T. Whitmarsh, *Ordering Knowledge in the Roman Empire* (Cambridge University Press), 88–107

E. Rubio Gómez 1996, 'Un capitulo de la ginecologia bizantina: La obra de "Metrodora"' in *Actas del IX congreso Espanol de Estudios Clásicos. VI. Historia y arqueología* (Madrid), 213–18

M. Saiko 2005, *Cura dabit faciem: Kosmetik im Altertum. Literarische, kulturhistorische und medizinische Aspekte* (Trier)

F. Sbordone 1930, 'La morte di Cleopatra nei medici greci', *Rivista indo-greco-italica*, 14: 3–22

J. Scarborough 1985, 'Criton, Physician to Trajan: Historian and Pharmacist' in J. W. Eadie and J. Ober, *The Craft of the Ancient Historian: Essays in Honor of Ch. G. Starr* (Lanham), 387–405

 1991, 'The Pharmacology of Sacred Plants, Herbs, and Roots' in C. A. Faraone and D. Obbink, *Magika hiera: Ancient Greek Magic and Religion* (Oxford), 138–74

 2012, 'Pharmacology and Toxicology at the Court of Cleopatra VII: Traces of Three Physicians' in A. Van Arsdall and T. Graham eds., *Herbs and Healers from the Ancient Mediterranean through the Medieval West: Essays in Honor of John M. Riddle* (Farnham), 7–18

J. Scarborough and A. Touwaide 2008, 'Krito̅n of He̅rakelia Salbake̅, T. Statilius' in P. T. Keyser and G. L. Irby-Massie, *The Encyclopedia of Ancient Natural Scientists: The Greek Tradition and Its Many Heirs* (London, New York), 494–5

G. Sissa 1987, *Le Corps virginal: La virginité féminine en Grèce ancienne* (Paris)

S. Stewart 2007, *Cosmetics and Perfumes in the Roman World* (Stroud)

S. Swain 1996, *Hellenism and Empire: Language, Classicism, and Power in the Greek World* AD *50–250* (Oxford)

L. M. V. Totelin 2007, 'Sex and Vegetables in the Hippocratic Gynaecological Treatises', *Studies in History and Philosophy of Biological and Biomedical Sciences*, 38: 531–40

 2009, *Hippocratic Recipes: Oral and Written Transmission of Pharmacological Knowledge in Fifth- and Fourth-Century Greece* (Leiden)

S. Toulalan 2007, *Imagining Sex: Pornography and Bodies in Seventeenth-Century England* (Oxford)

A. Touwaide 2002, 'Metrodora', *NP* 8: 132

M. Ullmann 1970, *Die Medizin im Islam* (Leiden, Cologne)

H. von Staden 1991, 'Matière et signification: Rituel, sexe et pharmacologie dans le corpus hippocratique', *L'antiquité classique*, 60: 42–61

1992, 'Women and Dirt', *Helios* 19: 7–30

J. Vons 2000, 'Un Vocabulaire médicalisé pour une *ars vivendi*: Dermatologie ou cosmétologie des femmes chez Pline l'ancien' in *Les Textes médicaux latins comme littérature. Actes du VIe colloque international sur les textes médicaux latins du 1er au 3 septembre 1998* (Nantes)

C. A. Williams 2010, *Roman Homosexuality*. 2nd edn. (Oxford)

M. Wyke 1994, 'Woman in the Mirror: The Rhetoric of Adornment in the Roman World' in L. Archer, S. Fischler and M. Wyke, *Women in Ancient Societies: 'An Illusion of the Night'* (Basingstoke, London), 134–51

J. G. Younger 2005, *Sex in the Ancient World from A to Z* (London, New York)

CHAPTER 9

From Discourses to Handbook
The Encheiridion *of Epictetus as a Practical Guide to Life*

Gerard Boter

Introduction

Epictetus (AD c. 50–130) is one of the three major representatives of the so-called New Stoa, the others being Seneca and the Emperor Marcus Aurelius.[1] The bulk of the extant works by Epictetus consists of the four extant books of the *Discourses* (in Latin, *Dissertationes*), which vary in length from eight lines (2.25) to well over thirty pages (4.1).[2] It is certain that there must have been more books, but they have been lost.[3] However, we also have the *Encheiridion* or *Manual*, which is based on the *Discourses*.[4] Although some scholars assume that Epictetus is responsible for at least part of the composition of the written works,[5] it is almost universally accepted that Epictetus did not publish anything himself: everything that remains of his teaching is owing to the work of his pupil Arrian.

The *Discourses* do not give us a full overview of Epictetus' teaching. It is commonly assumed that the *Discourses* represent general and non-technical lectures which could also be attended by people who did not belong to his school: this implies that the technical study of logic is almost absent from the *Discourses*.[6]

According to Epictetus, the world can be divided into two categories: the things that are under our control (τὰ ἐφ' ἡμῖν) and the things that are

---

[1] An excellent recent account of Epictetus' life and philosophy, with an extensive bibliography, is given by Fuentes González 2000.
[2] For the *Discourses (Diss.)* I have used the standard edition by H. Schenkl 1916. Translations of the *Discourses* are borrowed from Oldfather 1925–8 (with adaptations).
[3] See Fuentes González 2000, 119–21. The fragments from the lost books of the *Discourses* are collected in Schenkl 1916², 455–75.
[4] For the *Encheiridion (Ench.)* I have used the edition by Boter (1999 and 2007), in which the line numbers differ somewhat from those in Schenkl's edition. Translations of the *Encheiridion* are borrowed from Boter 1999.
[5] See, for instance, Stellwag 1933, 11–13; Dobbin 1998, xxi–xxiii.
[6] For the role of logic in Epictetus' philosophy, see Barnes 1997, 24–145; Gourinat 2008.

not under our control (τὰ οὐκ ἐφ' ἡμῖν).[7] To the latter category belong such things as health, reputation, possessions. The former category consists of our opinion, choice, desire, aversion etcetera. The category of τὰ ἐφ' ἡμῖν belongs to the domain of our προαίρεσις, which can be rendered as 'moral choice'.[8] The προαίρεσις enables us to distinguish between the things under our control and the things not under our control, thus allowing us to spend all our energy on the first category. We are incessantly exposed to φαντασίαι, 'impressions' or 'appearances', of every kind. It is our task to test our impressions constantly and to decide whether they are true or false. If the result of the test is that an impression is true, we accord our συγκατάθεσις, 'assent', to the impression; if not, we withhold our assent. Because our προαίρεσις is the faculty that agrees with or withholds assent from the impressions, the things under our control are often called προαιρετικά and the things not under our control ἀπροαίρετα.

Epictetus repeatedly speaks about three topics (τόποι) in philosophy (see, for instance, 3.2.1–5). The first topic is concerned with desire (ὄρεξις) and aversion (ἔκκλισις): desire is directed to the things we wish to obtain, aversion to the things we wish to avoid. If we direct our desire and aversion to the things that are not under our control, like wealth, fame and health, or illness, death and poverty, we will sooner or later inevitably be unsuccessful and therefore unhappy. The second topic is concerned with choice (ὁρμή) and refusal (ἀφορμή): it concerns our decision whether or not to do something: it is often designated as our duties (καθήκοντα). The third topic in philosophy deals with the use of impressions and thus with assent; more specifically, it aims at giving a logical foundation and justification for the way we deal with our impressions. Thus it constitutes the theoretical aspect of the basic precept for every man, that is, to make the right distinction between προαιρετικά and ἀπροαίρετα. To be sure, logic is an indispensable part of Stoic philosophy, but one should never forget that it has an ancillary role and that it is not an aim in itself: it is acts, not words, that count.[9]

The pivotal importance of man's actual behaviour is stressed time and again by Epictetus. Perfect theoretical knowledge of philosophy and the ability to solve all philosophical problems are absolutely worthless if the

---

[7] Epictetus also uses the term ἀλλότρια, 'things up to others', for τὰ οὐκ ἐφ' ἡμῖν, while τὰ ἐφ' ἡμῖν are often indicated as τὰ ἐμά, σά, ἡμέτερα: 'the things which are mine, yours, ours'.

[8] For the concept of προαίρεσις in Epictetus see, e.g., Dobbin 1991, Gourinat 2005, Sorabji 2007 and Hofmeister Pich 2010.

[9] For a detailed analysis of Epictetus' attitude towards himself and others with regard to the discrepancy between theory and practice, see Boter 2010, 335–9.

tenets of philosophy are not put into practice in real life. The title of 2.19 runs Πρὸς τοὺς μέχρι λόγου μόνον ἀναλαμβάνοντας τὰ τῶν φιλοσόφων, 'Against those who embrace philosophical opinions only in words'. In 2.19.21–4 he exclaims:

> Στωικὸν δὲ δείξατέ μοι, εἴ τινα ἔχητε. ποῦ ἢ πῶς; ἀλλὰ τὰ λογάρια τὰ Στωικὰ λέγοντας μυρίους... ὡς λέγομεν ἀνδριάντα Φειδιακὸν τὸν τετυπωμένον κατὰ τὴν τέχνην τὴν Φειδίου, οὕτως τινά μοι δείξατε κατὰ τὰ δόγματα ἃ λαλεῖ τετυπωμένον. δείξατέ μοί τινα νοσοῦντα καὶ εὐτυχοῦντα, κινδυνεύοντα καὶ εὐτυχοῦντα, ἀποθνῄσκοντα καὶ εὐτυχοῦντα, πεφυγαδευμένον καὶ εὐτυχοῦντα, ἀδοξοῦντα καὶ εὐτυχοῦντα. δείξατ᾽· ἐπιθυμῶ τινα νὴ τοὺς θεοὺς ἰδεῖν Στωικόν.
>
> But show me a Stoic, if you can. Where or how? But you can show me an endless number who utter small arguments of the Stoics... As we call a statue Phidiac, which is fashioned according to the art of Phidias; so show me a man who is fashioned according to the doctrines which he utters. Show me a man who is sick and happy, in danger and happy, dying and happy, in exile and happy, in disgrace and happy. Show him: I desire, by the gods, to see a Stoic.

In 2.4.10–11 a man caught in adultery, when vituperated by Epictetus for his behaviour, retorts: Ἀλλὰ φιλόλογός εἰμι καὶ Ἀρχέδημον νοῶ, 'But I am a scholar and I understand Archedemus.' (Archedemus was a Stoic philosopher.) Epictetus sarcastically answers: Ἀρχέδημον τοίνυν νοῶν μοιχὸς ἴσθι καὶ ἄπιστος καὶ ἀντὶ ἀνθρώπου λύκος ἢ πίθηκος. τί γὰρ κωλύει; 'Very well then, understand Archedemus and be an adulterer and faithless and a wolf or an ape instead of a man; for what is there to prevent you?'

Epictetus' greatest hero is Socrates exactly because he was ready to sacrifice his life for his philosophical convictions.[10] Socrates' famous dictum from Plato's *Crito* 43d, Ἀλλ᾽, ὦ Κρίτων, εἰ ταύτῃ τοῖς θεοῖς φίλον, ταύτῃ γινέσθω, 'But, Crito, if it pleases the gods this way, it must happen this way', is quoted no fewer than four times in the *Discourses* (1.4.24, 1.29.18, 3.22.95, 4.4.21), and it concludes the *Encheiridion*. Translated into Stoic terms it means that Socrates has managed to live up perfectly to the Stoic ideal of ὁμολογουμένως τῇ φύσει ζῆν, 'living in accordance with nature'.

In comparing the character of the *Discourses* with the character of the *Encheiridion* one might say that the *Discourses* are concerned with philosophical theory and anecdotes that lead up to general remarks on how to

---

[10] For a detailed and excellent study of the figure of Socrates in Epictetus, see Gourinat 2001.

live one's life, whereas the *Encheiridion* gives practical and specific advice based on philosophical theory. As will be amply illustrated in the course of this contribution, the exposition of philosophical theory in the *Encheiridion* is reduced to what is strictly necessary. By the same token, the anecdotes from real life and the frequent references to myth and history with which the *Discourses* are lavishly interspersed are almost completely absent from the *Encheiridion*. The *Encheiridion* exactly gives what the title indicates, namely rules for practical behaviour in all imaginable circumstances in everyday life. The emphasis shifts from the 'words' of the *Discourses* to the 'acts' of the *Encheiridion*. The *Encheiridion* is indeed a practical 'Manual for Life'. In this respect it has a close parallel in the so-called *Golden Verses* attributed to Pythagoras (for which see, e.g., Thom 1995). And, on a more general level, the *Encheiridion* fits in well with the role of philosophy in the first two centuries AD in which philosophy, far from being a purely academic enterprise, is primarily a way of life.[11]

From *Discourses* to *Encheiridion*

In our mediaeval manuscripts the *Discourses* are preceded by a letter from Arrian to Lucius Gellius.[12] Arrian states that he did not spread the *Discourses* among the reading public himself; in fact, he states that the works that have come into circulation are mere lecture notes (ὑπομνήματα) and not fully wrought literary works (συγγράμματα). Whatever Arrian's role in the writing and publishing of the *Discourses* may have been, what interests us here is that, according to Arrian, in delivering his lectures Epictetus 'was clearly aiming at nothing else but to incite the minds of his hearers to the best things' (λέγων αὐτοὺς οὐδενὸς ἄλλου δῆλος ἦν ἐφιέμενος ὅτι μὴ κινῆσαι τὰς γνώμας τῶν ἀκουόντων πρὸς τὰ βέλτιστα). And so we have it in Arrian's own words that, with regard to the *Discourses*, the primary goal of both Epictetus as a teacher and of Arrian as a writer was practical: to make the hearers/readers better men.[13]

The natural starting-point for research on the character of the *Encheiridion* and for an analysis of the relation between the *Encheiridion* and the

---

[11] See, e.g., P. Hadot 1981 and 1995; van Hoof 2010.
[12] For discussions of this letter see, among others, Radt 1990, 364–8, Fuentes González 2000, 121–3, P. Hadot 2000, 30–5, Wehner 2000, 27–36, Long 2002, 39–43.
[13] In the context of Epictetus' philosophy the concept of 'acts', which is one of the central issues in this volume, should be taken in a broad sense. Putting philosophical concepts into practice not only relates to activities involving physical actions (including speaking to others), but also to taking up a stance towards the world that surrounds us.

Discourses is the preface to Simplicius' commentary on the *Encheiridion*.[14] Simplicius states that Arrian 'composed [*or* assembled; συντάξας] the lectures by Epictetus into long books' (P 2–3 ὁ τὰς Ἐπικτήτου διατριβὰς ἐν πολυστίχοις συντάξας βιβλίοις). This phrase obviously refers to the *Discourses*. Simplicius indicates Arrian's activity by means of the verb συντάττω, which in the context of books normally refers to *composing, compiling* or *writing* (LSJ s.v. II 3). I. Hadot 2001, 1 n. 1 argues that in this passage the verb means 'rassembler'.

Simplicius continues his preface as follows (P 4–11):

> Τὸ δὲ βιβλίον τοῦτο, τὸ Ἐπικτήτου Ἐγχειρίδιον ἐπιγεγραμμένον, καὶ τοῦτο αὐτὸς συνέταξεν ὁ Ἀρριανός, τὰ καιριώτατα καὶ ἀναγκαιότατα ἐν φιλοσοφίᾳ καὶ κινητικώτατα τῶν ψυχῶν ἐπιλεξάμενος ἐκ τῶν Ἐπικτήτου λόγων, ὡς αὐτὸς ἐν τῇ πρὸς Μασσαληνὸν ἐπιστολῇ ἔγραψεν ὁ Ἀρριανός, ᾧ καὶ τὸ σύνταγμα προσεφώνησεν, ὡς ἑαυτῷ μὲν φιλτάτῳ, μάλιστα δὲ τὸν Ἐπίκτητον τεθαυμακότι. Τὰ δὲ αὐτὰ σχεδὸν καὶ ἐπ' αὐτῶν τῶν ὀνομάτων σποράδην φέρεται ἐν τοῖς Ἀρριανοῦ τῶν Ἐπικτήτου διατριβῶν γραφομένοις.

> This book, entitled 'The Handbook of Epictetus', was also composed by Arrian himself. In his letter to Massalenus (to whom Arrian dedicated the work because he was a very close friend of his and who also had the greatest admiration for Epictetus), he describes that he picked out from Epictetus' speeches the principal and most necessary elements of this philosophy which are most likely to move the souls of the readers. The same material can be found in practically the same words at various points in Arrian's writings on the discourses of Epictetus.

Several parts of this passage call for comment. In the first place, Simplicius recognizes Arrian as the author of the *Encheiridion*, but the name 'Epictetus' is part of the title. The work is therefore not the *Handbook* of <the author> Epictetus but rather the *Handbook of Epictetus* <by the author Arrian>.[15]

In the second place Simplicius again uses the verb συντάττω. Here, I. Hadot 2001 translates the word as '(qui l'a) composé', stating, however, in n. 5 (on p. 131): 'Arrien a rassemblé les textes contenus dans le *Manuel*, mais il les a aussi choisis et mis en un certain ordre.' In all probability Hadot bases this interpretation on the concluding sentence of the passage quoted above in which Simplicius argues that the elements of the *Encheiridion* are

---

[14] Quotations from Simplicius' commentary are borrowed from I. Hadot's 1996 edition; in the 2001 edition the line numbers differ from those in the 1996 edition. Translations of Simplicius' commentary are based on T. Brennan and Ch. Brittain 2002 (with adaptations).

[15] Cf. Gourinat 1998, 44: 'Faut-il lire alors le *Manuel* comme un *Manuel* d'Épictète ou comme un *Manuel d'Épictète* d'Arrien? À bien des égards, plutôt dans le second sens.' For the use of the word Ἐγχειρίδιον as a book title, see Broccia 1979.

found almost verbatim in the *Discourses* (which, as I shall show, is untrue). To my mind, it would seem a priori implausible that Simplicius uses the same verb (συντάττω) to denote two different activities, and therefore I assume that in both cases the verb is best rendered in the usual general sense of 'compose', 'write'.

In the third place Simplicius quotes from Arrian's dedicatory letter to Massalenus, which has not been preserved for us. According to this letter, Arrian's purpose in composing the *Encheiridion* was to pick out from Epictetus' speeches the elements that are 'the principal and most necessary elements of this philosophy which are most likely to move the souls of the readers'. We have already seen that, in the letter to Lucius Gellius, Arrian expresses the hope that reading the *Discourses* will make the readers better men; his aim in composing the *Encheiridion*, therefore, is to present the reader with a condensed and concentrated version of this corpus of practical pieces of advice.

I shall first investigate the relationship between the *Discourses* and the *Encheiridion*. I shall illustrate that the *Encheiridion* is not just a cento, but an original work based on the *Discourses*. Next I shall discuss the composition of the *Encheiridion*: I shall show that the prevailing opinion, namely that the work by and large follows the tripartition of Epictetus' philosophical system, is wrong. After that I shall tackle the question of the intended readership of the *Encheiridion*: I shall suggest that Arrian composed the *Encheiridion* as a bird's-eye view of the education of the student of Stoic philosophy which is intended not only, or even primarily, as an *aide-mémoire* for the student who has finished his studies, but also, or even principally, for the generally interested reader seeking practical advice based on Stoic philosophy.

The Position of the Addressee in the Encheiridion

A principal distinguishing feature of the *Encheiridion* in comparison with the *Discourses* is constituted by the conversion of the lectures given by Epictetus to his students into a work addressed to the reader himself. In reading the *Discourses* the reader is a spectator, a bystander, who is never addressed personally and who is not involved in the teaching process. In the *Encheiridion*, on the other hand, the reader is constantly apostrophized in the second person. Apostrophized by whom? If the suggestion made above is correct, namely that the title Ἐπικτήτου Ἐγχειρίδιον is to be understood as *The Handbook of Epictetus*, it may have been Arrian's intention that the reader should have the impression of being addressed directly by Epictetus, not by Arrian, who only poses as Epictetus' mouthpiece (or 'prophet' in the literal sense of the word).[16]

Wehner 2000, 251–8 is right in remarking that the lively dialogical character of the *Discourses* is strongly reduced in the *Encheiridion*. But although the reader is usually addressed in the second person,[17] there are cases where the author uses the first person plural, as in 5a.5: ὅταν οὖν ἐμποδιζώμεθα ἢ ταρασσώμεθα ἢ λυπώμεθα, μηδέποτε ἄλλον αἰτιώμεθα, ἀλλ' ἑαυτούς, τουτέστι τὰ ἑαυτῶν δόγματα, 'Therefore, whenever we are hampered or upset or grieved, let us never blame someone else but ourselves, that is, our opinions.' Further, we regularly find the first person singular in soliloquies suggested to the reader, usually introduced by phrases like μελέτα ἐπιλέγειν ('do your best to say', 1.18) or εὐθὺς ἔστω πρόχειρον ὅτι ('you must immediately tell yourself', 16.3–4).

The *Discourses* are reports of actual lectures delivered before actual students: the teacher knows his students and vice versa. Time and again Epictetus gives details of his personal life; and time and again he addresses his pupils in a very straightforward way. In doing so he is often harsh or even rude.[18] The absence of any direct personal bond between the author and the reader of the *Encheiridion* explains the almost complete absence of this reproachful tone in this work. To give one instance: *Ench*. 32, which is devoted to divination, is based on *Diss*. 2.7, which starts with a reproach: Διὰ τὸ ἀκαίρως μαντεύεσθαι πολλοὶ καθήκοντα πολλὰ παραλείπομεν, 'Because we employ divination when there is no occasion for it, many of us neglect many of the duties in life.' And further on Epictetus says (2.7.11): οὕτως ἔδει καὶ ἐπὶ τὸν θεὸν ἔρχεσθαι ὡς ὁδηγόν, 'so also we ought to go to God as a guide',[19] where ἔδει implies that we do not do this at present. In *Ench*. 32, on the other hand, the good and bad ways of dealing with divination are presented in neutral terms: μὴ φέρε οὖν πρὸς τὸν μάντιν ὄρεξιν ἢ ἔκκλισιν ... θαρρῶν οὖν ὡς ἐπὶ συμβούλους ἔρχου τοὺς θεούς, 'Therefore do not bring desire or aversion to the fortune-teller ... full of confidence, then, go to the gods as to counsellors.'[20] In *Diss*. 2.7.12 Epictetus states νῦν δὲ τρέμοντες τὸν ὀρνιθάριον κρατοῦμεν, 'But as it is, we tremble when we grasp at the bird-augur', as an actual fact; in *Ench*. 32.5–6 this is neutrally

---

[16] Cf. P. Hadot 2000, 143.
[17] The reader is very often addressed in the imperative (some 150 occurrences).
[18] See Boter 2010, 339–45.
[19] That is, accepting every piece of advice God gives us through the fortune-teller, either pleasant or unpleasant.
[20] There are a few exceptions. The first one is Chapter 28, which ends with the mildly reproaching οὐκ αἰσχύνῃ τούτου ἕνεκα; In Chapter 52 the author includes himself among those whom he criticizes for paying more attention to the study of logic than to bringing the Stoic precepts into practice.

formulated as a hypothesis: εἰ δὲ μή, τρέμων αὐτῷ πρόσει, 'otherwise, you will come to him full of fear'.[21] Instead of the negative approach in 2.7.9, Τί οὖν ἡμᾶς ἐπὶ τὸ οὕτω συνεχῶς μαντεύεσθαι ἄγει; ἡ δειλία, τὸ φοβεῖσθαι τὰς ἐκβάσεις, 'What, then, induces us to employ divination so constantly? Cowardice, fear of the outcome', Arrian strikes a positive note in *Ench.* 32.6, by means of the word θαρρῶν, 'full of confidence'.

The Discourses *as the Source of the* Encheiridion

As we have seen, Simplicius states that in composing the *Encheiridion* Arrian picked out the essential elements from the *Discourses*; he further states that many phrases in the *Encheiridion* are also found in the *Discourses*. Simplicius' statement has undoubtedly contributed to the generally accepted opinion that the *Encheiridion* is no more than an excerpt from the *Discourses*.[22] Of course, I do not wish to deny that the *Discourses* are the source of the *Encheiridion*, but Arrian did not content himself with picking out passages from the *Discourses* and putting them together.[23]

To start with, there are only a very few passages in the *Encheiridion* where passages from the *Discourses* recur (almost) verbatim; in addition, such passages are hardly ever longer than just a few words.[24] Thus the opening phrase of the *Encheiridion* (1.1), Τῶν ὄντων τὰ μέν ἐστιν ἐφ' ἡμῖν, τὰ δὲ οὐκ ἐφ' ἡμῖν, 'There are two classes of things: those that are under our control and those that are not', is found in exactly the same form in 1.22.10; the phrase τὴν ἐμαυτοῦ προαίρεσιν κατὰ φύσιν ἔχουσαν τηρῆσαι

---

[21] My text of *Ench.* 32 differs considerably from the text in previous editions, especially in section 2. Here is my text of section 2:

> μὴ φέρε οὖν πρὸς τὸν μάντιν ὄρεξιν ἢ ἔκκλισιν (εἰ δὲ μή, τρέμων αὐτῷ πρόσει), ἀλλὰ διεγνωκὼς ὅτι πᾶν τὸ ἀποβησόμενον ἀδιάφορον καὶ οὐδὲν πρός σέ, ὁποῖον δἂν ᾖ (ἔσται γὰρ αὐτῷ χρήσασθαι καλῶς καὶ τοῦτο οὐδεὶς κωλύσει) – θαρρῶν οὖν ὡς ἐπὶ συμβούλους ἔρχου τοὺς θεούς, καὶ λοιπὸν ὅταν τί σοι συμβουλευθῇ, μέμνησο τίνας συμβούλους παρέλαβες καὶ τίνων παρακούσεις ἀπειθήσας.

[22] See for instance Colardeau 2004, 27: 'Il choisit dans ceux-ci [les *Entretiens*] les passages qui lui parurent les plus utiles'; Reale 1982, 522: 'un semplice procedimento rapsodico'; P. Hadot 1997, 76: 'Le *Manuel* est un choix de textes tirés des *Propos* d'Épictète'; Fuentes González 2000, 118: 'Un abrégé'; Wehner 2000, 251: 'ein Kompendium'; Seddon 2005, 7: 'Some of the text is taken from the *Discourses*'; Sellars 2007, 136: 'It takes the form of a collection of material from the *Discourses*.'
[23] It might perhaps be argued that much of the material in the *Encheiridion*, for which there is no direct source in the *Discourses*, was taken from the lost books of the *Discourses*. This can neither be proved nor disproved, but the general picture that arises from the following analysis is so clear that I regard this hypothesis as antecedently improbable.
[24] The source places from the *Discourses* are indicated in a separate apparatus in my editions. The symbol = indicates (almost) verbatim borrowings; 'sim.' is used for passages where the text of the *Encheiridion* is a free adaptation of the source text in the *Discourses*.

('to keep my moral choice in accordance with nature', 4.6, 4.9–10) is directly borrowed from 2.2.2 and 3.4.9; 'οἴμοι' καὶ 'τάλας ἐγώ' (" "Alas!" and "Poor me!"', 26.7–8) comes from 1.4.23; ἔδοξεν αὐτῷ ('So it seemed best to him', 42.8) is taken from 1.11.31.

Passages of some length, when taken from the *Discourses*, are usually moulded into a different form; see, for instance, *Ench.* 47.4–5 ἀλλὰ διψῶν ποτε σφοδρῶς ἐπίσπασαι ψυχροῦ ὕδατος καὶ ἔκπτυσον καὶ μηδενὶ εἴπῃς, 'but on occasion when you are very thirsty, take a mouthful of cold water and spew it out, and do not tell anyone', which echoes 3.12.17 ὅταν θέλῃς σαυτῷ ἀσκῆσαι, διψῶν ποτε καύματος ἐφέλκυσαι βρόγχον ψυχροῦ καὶ ἔκπτυσον καὶ μηδενὶ εἴπῃς, 'When you wish to train for your own sake, then when you are thirsty some hot day take a mouthful of cold water, and spit it out – and don't tell anybody about it!'

Chapter 47, which has only five lines, may well serve to illustrate how Arrian used elements taken from several places of the *Discourses* in composing the *Encheiridion*. Chapter 47 forms a diptych with Chapter 46: both chapters state that one should not show off to others. Chapter 46 deals with this topic with regard to philosophical tenets: 'do not talk much about philosophic principles but practise what follows from these principles.' Chapter 47 continues the topic of not showing off with regard to bodily exercise: Ὅταν εὐτελῶς ἡρμοσμένος ᾖς κατὰ τὸ σῶμα, μὴ καλλωπίζου ἐπὶ τούτῳ, 'When you have become adapted to simple living with regard to your body do not make a show of it'. Arrian continues with the issue of drinking water (instead of wine): lines 1–2 μηδ' ἂν ὕδωρ πίνῃς, ἐκ πάσης ἀφορμῆς λέγε ὅτι ὕδωρ πίνεις, 'and when you drink water do not say on every occasion that you are drinking water'; this phrase is a very condensed version of 3.14.4–6, in which drinking water (without telling others) is mentioned as an instance of asceticism. This mentioning of asceticism in 3.14.4 may have led to the topic of not showing off asceticism in general, which is mentioned in lines 3–4: κἂν ἀσκῆσαί ποτε πρὸς πόνον θέλῃς, σαυτῷ καὶ μὴ τοῖς ἔξω, 'and if ever you want to train yourself in enduring physical discomfort, do it for yourself and not to impress outsiders'; the latter part of this phrase, σαυτῷ καὶ μὴ τοῖς ἔξω, appears to be based on 3.12.17, which will return at the end of the chapter. Next, Arrian moves on to another item of bodily exercise, embracing statues (for which see D.L. 6.23): μὴ τοὺς ἀνδριάντας περιλάμβανε, a practice which is mentioned in 3.12.2,10 and 4.5.14. A third method of bodily exercise is abstaining from drinking when being thirsty, for which see the end of the preceding paragraph.

Another good instance of Arrian's method is furnished by Chapter 3, where the reader is advised to train himself in acquiescing in the loss of the things or people he loves by starting with small and insignificant losses. The whole chapter is ultimately based on 3.24.84–8, but verbatim echoes are hardly found. The opening phrase, 'Εφ' ἑκάστου τῶν ψυχαγωγούντων ἢ χρείαν παρεχόντων ἢ στεργομένων, 'With all the things that attract you or that are useful or that are appreciated', is not found as such in the *Discourses*, but the three elements are taken from different passages.[25] The advice to start with small things, ἀπὸ τῶν σμικροτάτων ἀρξάμενος, comes from 1.18.18 and 4.1.111. The earthenware pot and the child as objects of affection are taken from 3.24.84, where, however, the wife is not mentioned; elsewhere the wife is mentioned in one breath with the child, e.g. 4.1.67,100,111. The phrase καταγείσης γὰρ αὐτῆς οὐ ταραχθήσῃ, 'for if it gets broken you will not be upset', is an adaptation of 3.24.84 ἵν' ὅταν καταγῇ, μεμνημένος [μὴν] μὴ ταραχθῇς; in the next sentence, the phrase ἂν παιδίον σαυτοῦ καταφιλῆς, 'if you kiss your child', is taken over verbatim from 3.24.85, while ἄνθρωπον καταφιλεῖς, 'you kiss a human being', is an adaptation of 3.24.86 θνητὸν φιλεῖς, 'you love a mortal'. An element for which there is no precedent in the *Discourses* is the similar structure of καταγείσης γὰρ αὐτῆς οὐ ταραχθήσῃ and ἀποθανόντος γὰρ οὐ ταραχθήσῃ: this betrays Arrian the accomplished stylist, striving at well-balanced phrasing.[26]

As an instance of Arrian's way of composing a longer chapter, I shall briefly discuss *Ench.* 32 (already mentioned above), which is based on *Diss.* 2.7.[27] To start with, verbal echoes are few and far between: 32.6 τρέμων, 'full of fear', comes from 2.7.12 τρέμοντες; 32.10 συμβούλους, 'counsellors', is taken from 2.7.14 τὸν σύμβουλον; 32.15 ὅταν δεήσῃ συγκινδυνεῦσαι φίλῳ ἢ πατρίδι, 'whenever it is necessary to share a danger with a friend or with your country', is taken from 2.7.3 ἂν οὖν δέῃ κινδυνεῦσαι ὑπὲρ τοῦ φίλου; that is all. Wehner 2000, 257 rightly points out the presence of the numerous imperatives in *Ench.* 32 and, in general, the variegated presentation of 2.7, with discussion with fictive interlocutors, anecdotes etc. being replaced by a continuous exposition; this presentation suits the practical purpose of the *Encheiridion*. In 2.7 Epictetus starts negatively

---

[25] ψυχαγωγεῖν occurs in 2.16.38, 3.21.23 and 4.4.4; χρείαν παρέχειν in 3.26.26; τὰ στεργόμενα in 3.24.28.
[26] There are many more instances of Arrian's attempts at literary embellishment. Thus we find a number of aphorisms in the *Encheiridion* for which there is no parallel in the *Discourses* (5b, 8, 27, 48a).
[27] See also Wehner 2000, 258–9.

by stating that we make use of divination in a perverted manner, which initially leads him to reject it altogether. It is only at the end of the lecture (2.7.10–14) that he explains the right way to make use of divination. In *Ench.* 32 Arrian introduces the subject neutrally: Ὅταν μαντικῇ προσίῃς, 'Whenever you make use of divination'. He then systematically opposes the two ways of consulting fortune-tellers by contrasting the question of what will happen (τί μὲν ἀποβήσεται) and the question of the character of what will happen (ὁποῖον δέ τί ἐστιν). The latter, he argues, is clear in advance: everything the fortune-teller can tell necessarily belongs to the domain of τὰ οὐκ ἐφ' ἡμῖν: 32.6 πᾶν τὸ ἀποβησόμενον ἀδιάφορον καὶ οὐδὲν πρὸς σέ. Arrian adds that we are completely free to deal with any events the right way (ἔσται γὰρ αὐτῷ χρήσασθαι καλῶς καὶ τοῦτο οὐδεὶς κωλύσει): this is common Epictetean doctrine, which is, however, not explicitly mentioned in *Diss.* 2.7. In order to illustrate how we should consult fortune-tellers, he refers to Socrates' practice with regard to divination, namely to consult the fortune-teller only for matters about which there is no other method of getting certainty; this anecdote, which we also find in Xen. *Mem.* 1.1.7–9, is not found in *Diss.* 2.7 or elsewhere in the extant books of the *Discourses*: it may have been in one of the lost books, but it may also have been added by Arrian himself (who regarded himself as a second Xenophon). Arrian ends the chapter by explicitly pointing out what is stated at the beginning of *Diss.* 2.7, namely that we should always fulfil our duty towards our friends, even if this means our death; once more, he adds an anecdote that is not found in the extant *Discourses*: Apollo chased from his temple a man who had failed to help his friend when he was being attacked.

Another text-type which we encounter in the *Encheiridion* is the aphorism, which is practically absent from the *Discourses*. As an instance of Arrian's way of turning Epictetus' thought into the form of an aphorism, let us have a look at Chapter 19a Ἀνίκητος εἶναι δύνασαι, ἐὰν εἰς μηδένα ἀγῶνα καταβαίνῃς ὃν οὐκ ἔστιν ἐπὶ σοὶ νικῆσαι, 'You can be invincible if you never enter any contest in which victory is not under your control.' The aphorism is inspired by *Diss.* 3.6.5 Ὁ σπουδαῖος ἀήττητος, 'The good man is invincible', which provides the basis for the opening words of the aphorism, and 3.22.102 οὐδέποτ' οὖν εἰς τοῦτον καταβαίνει τὸν ἀγῶνα, ὅπου δύναται νικηθῆναι, 'Therefore, he never enters this contest where he can be beaten', which is the model for the rest of the aphorism. The original ἀήττητος is turned into ἀνίκητος in order to balance the final word νικῆσαι. The third person in the source passage (that is, the ideal Cynic) is replaced by the second person and thus addressed to the reader.

Finally, the passive and therefore negative νικηθῆναι is substituted by the active and positive νικῆσαι.

The analyses of Chapters 3, 19a, 32 and 47 show that in composing the *Encheiridion* Arrian did not just make an excerpt or an abridged version of the *Discourses*. When dealing with a theme, he selected passages scattered throughout the *Discourses* and incorporated them in various ways (Chapters 3, 19a, 47) or he used the text of the *Discourses* as the starting-point for a chapter of his own invention (Chapter 32). This is not just copy-and-paste: this is creative writing. The gist of Epictetus' thought on a particular subject is presented in a clear-cut and systematic way so that the reader immediately knows what he must do and why he must do it. Every chapter teaches a clear lesson, usually accompanied by philosophical justification, without going into too much theoretical detail. The thought is Epictetus'; the composition and the phrasing is Arrian's, with scattered verbal echoes from the *Discourses*.

The Structure of the *Encheiridion*

Traditionally, the *Encheiridion* is divided into fifty-three chapters, the length of which varies from slightly more than one line (Chapter 27) to almost fifty lines (Chapter 33).[28] As we have seen in the previous section, each chapter is a self-contained unit. I shall now turn to the composition of the *Encheiridion* as a whole. Pohlenz (1990, 162) was the first to suggest a structuring principle for the *Encheiridion*. Pohlenz states that the *Encheiridion* is composed according to the three topics of Epictetus' philosophy (for which see above): the first topic, that of ὄρεξις and ἔκκλισις, is dealt with in Chapters 1–29;[29] the second topic, the one of the καθήκοντα, is the subject of Chapters 30–51;[30] the third topic, concerned with logic, is treated in Chapter 52; the last chapter is an epilogue, consisting of four maxims. This analysis is accepted by, among others, Stadter 1980, 29; P. Hadot 1997, 115 with n. 30; I. Hadot 1996, 149–51; Wehner 2000, 251–2.[31] I believe that this

---

[28] This chapter division was introduced by J. Schweighäuser, in his late eighteenth-century *editio maior* of the *Encheiridion* (Schweighäuser 1798). In my edition, four chapters are split into two: 5a/b, 14a/b, 19a/b, 48a/b, which corresponds to the division in the manuscripts of the *Encheiridion*.
[29] I regard Chapter 29 as an interpolation; see Boter 1999, 127.
[30] According to Pohlenz, the opening line of Chapter 30, Τὰ καθήκοντα ὡς ἐπίπαν ταῖς σχέσεσι παραμετρεῖται, 'Our duties are in general measured by relationships', is programmatic for the following chapters as well ('das Stichwort καθήκοντα').
[31] Wehner 2000, 251 discerns a form of ring composition: the division between the things under our control and not under our control, dealt with in the first chapter and returned to in the final chapter. This is true, but to my mind all the intervening chapters are concerned with this distinction as well.

structure is an oversimplification. I agree with Pohlenz on the third topic, the theory of logic, which is discussed, or rather dismissed, in Chapter 52 (although Chapter 49 too is concerned with the interpretation of the texts of Chrysippus and therefore with theory). But it is principally wrong to draw such a sharp distinction between the first and the second topics. On the one hand, there are many passages dealing with behaviour in Chapters 1–28,[32] and, on the other, the topic of ὄρεξις and ἔκκλισις is also found after Chapter 30.[33]

In later publications[34] P. Hadot offers a refinement of Pohlenz' structure. He argues that Chapters 3–6 are devoted to the discipline of judgement; further, he introduces the category of 'conseils au progressant', which he identifies in Chapters 12–13, 22–5 and 46–53. This is not an improvement. For one thing, it is patently wrong to attribute the third topic, that of judgement, to Chapters 3–6 because in these chapters the *practice* of judgement is central, while the third topic deals with the *theory* of judgement, that is, with logic; moreover, there is hardly any passage in the *Encheiridion* where the practice of judgement plays no role at all.[35] For another thing, it is not helpful to set apart a number of chapters as being especially addressed to the προκόπτων: the themes treated in these chapters do not essentially differ from those in other chapters, and in some cases the same themes recur elsewhere.[36]

Gourinat 1998, 45–53, while basically accepting Pohlenz' division, divides the category of the καθήκοντα into several subcategories; thus he claims that Chapters 42–5 are concerned with logic (49–50). This interpretation founders on the same objection that was raised against Hadot's claim that logic is dealt with in Chapters 3–6: in both passages the *practice* of judgement is at stake, not the theory.

I do not wish to deny that the first two topics play important roles in the *Encheiridion*; on the contrary, these topics are essential. But the claim

---

[32] See Chapters 4, 7, 10, 12, 13, 15, 16, 17, 22, 24, 25.
[33] See Chapters 30, 31, 32, 34, 42, 45, 48b, 49.
[34] P. Hadot 2000, 140–2; I. Hadot and P. Hadot 2004, 34–40.
[35] P. Hadot 2000, 142 admits that the discipline of judgement 'reparaît plusieurs fois'. In fact, the practice of judgement is essential for the practice of the first two topics: cf. Bonhöffer 1890, 24 n. 1, and see below. With regard to Chapter 17, in which life is compared to a theatre piece in which everyone has to play the role assigned to him by the producer, I. Hadot and P. Hadot 2004, 36 state that it is 'un peu erratique': this is an understatement, because the chapter deals with the task imposed on us by God and therefore to the καθήκοντα. In the *Discourses* Epictetus repeatedly says that one should confront the third topic only after having fully mastered the first two: see, for instance, 3.2.5–8, 4.10.13.
[36] For instance, the theme of not paying heed to the opinion of others is found in Chapters 13, 22, 24, 28, 35, 42, 46, 50.

that the work is split into two main parts, each corresponding to one of the topics, is untenable because, as I have indicated, there are chapters dealing with καθήκοντα in the first part of the *Encheiridion* and chapters dealing with desire and aversion in the second part; and the practical application of the faculty of judgement is not confined to just a few chapters but it is at work practically everywhere. What is more, in a number of cases the three topics are inextricably related, which makes it undesirable to try to separate them. I shall give a few instances.

Chapter 31, which Pohlenz and others assign to the part about the καθήκοντα, deals with our attitude towards the gods.[37] The chapter starts with the remark that we should have the right opinions about them, namely that they exist and rule the universe in the best possible way. Thus from the start the faculty of judgement plays a crucial role. In section 2 Arrian continues, stating that it is imperative to place good and evil (the objects of ὄρεξις and ἔκκλισις respectively) in the domain of τὰ ἐφ' ἡμῖν exclusively: this is the central issue of the first topic. At the end of the chapter the two topics of desire/aversion and piety (that is, our duty towards the gods) are even identified: ὥστε ὅστις ἐπιμελεῖται τοῦ ὀρέγεσθαι ὡς δεῖ καὶ ἐκκλίνειν, ἐν τῷ αὐτῷ καὶ εὐσεβείας ἐπιμελεῖται, 'accordingly, whoever takes care to desire and avoid as he should do, takes care of piety at the same time'. The situation in Chapter 32 is comparable. For the right attitude towards divination it is necessary to leave aside ὄρεξις and ἔκκλισις and to realize that the fortune-teller can only tell us things that belong to the domain of τὰ οὐκ ἐφ' ἡμῖν. Only thus shall we be ready to fulfil our duty irrespective of the outcome, for instance by helping our friends or our country even when we know that this will end in our death.

I conclude that any attempt to find a rough and ready structure for the *Encheiridion* according to the three topics of philosophy is not sustainable. But is there an alternative? The opening sentence of the *Encheiridion*, Τῶν ὄντων τὰ μέν ἐστιν ἐφ' ἡμῖν, τὰ δὲ οὐκ ἐφ' ἡμῖν, 'There are two classes of things: those that are under our control and those that are not', sets the tone for the whole work. The rigid dichotomy between the things under our control and the things not under our control extends to all aspects of life: real good and evil versus apparent good and evil, real value versus apparent value, mind versus body, primary activities versus secondary activities,

---

[37] P. Hadot 2000, 151 argues that the gods play only a very minor role in the *Encheiridion*; according to Hadot, they appear practically exclusively in Chapters 31–2. This view is wrong: the gods are also present, though sometimes mentioned only indirectly, in Chapters 7 (ὁ κυβερνήτης), 11 (ὁ δούς), 17 (ὁ διδάσκαλος), 22 (ὑπὸ θεοῦ τεταγμένος εἰς ταύτην τὴν χώραν); moreover, the first three of the four maxims in the final chapter are concerned with the gods.

being a philosopher versus appearing to be one. Therefore, the starting-point for the good life is the choice of the correct alternative. The choice of the wrong alternative or the confusion of the two alternatives results in a lack of freedom and in pain, sorrow, frustration and so on. Accordingly, the first chapter aptly ends with the κανών, 'standard', we must apply in our use of impressions. Every impression must be tested; in every case the test should start with the application of the κανών, which consists of the question 'Is it under my control or not under my control?'[38] The answer to this question determines our attitude towards the first two topics: do ὄρεξις and ἔκκλισις apply or not? Is this a situation involving a duty or not? The second chapter opens with elaborating the concepts of ὄρεξις and ἔκκλισις, which leads to the conclusion that ὄρεξις should be suppressed altogether for the time being, while ἔκκλισις should only be applied to the unnatural things under our control, that is, to negative emotions (like rage; see below on Chapter 5a); the chapter ends with the remark that we should use our impulse to act or not to act 'lightly and with reservation and without straining'.

The *Encheiridion* illustrates the practical application of this theoretical model, with the κανών as the key instrument, in a variety of situations and aspects.[39] These situations and aspects are not presented in separate blocks never to be repeated; rather, they are recurrent and intertwined motifs, and as such the *Encheiridion* might be compared to a theme with variations, or to a fugue with several themes. Further, the *Encheiridion* leads the reader to a well-marked goal, as I shall illustrate: the way to this goal is marked by ἄσκησις, philosophical training.[40] I shall not give a detailed analysis of all the chapters of the *Encheiridion*; instead, I shall point out the various motifs, indicating where they are found for the first time and where they recur. This discussion is not meant to be exhaustive, because in many cases several motifs play a role in one chapter.

---

[38] The pivotal importance of the κανών as the starting point for daily meditation in Epictetus' *Discourses* is duly emphasized by Newman 1989, 1498–500. Newman states (1498): 'Epictetus reduced the κανών to its basic elements in 4.4.29 to fit it into a meditation: τί ἐμόν, τί οὐκ ἐμόν; τί μοι δίδοται; τί θέλει με ποιεῖν ὁ θεὸς νῦν, τί οὐ θέλει; ["What is mine? What is not mine? What has been given me? What does God will that I do now, what does he not will?"]' Newman, in his discussion of Epictetus, confines himself to discussing passages from the *Discourses*, mentioning the *Encheiridion* only once, in passing (1503): 'Besides minor characters, such as Zeno (*Ench.* 33.12) and Helvidius Priscus.'
[39] Cf. Newman 1989, 1499: 'Epictetus' innovation lay chiefly in his use of the κανών as an integral and omnipresent part of the *meditatio*.'
[40] The pilot study on ἄσκησις in Epictetus is Hijmans 1959. It is remarkable that many scholars neglect the concept in their analyses of the *Encheiridion*. Sellars 2007 is a positive exception to the rule.

Chapters 3 and 4 deal with the loss of pleasant ἀλλότρια ('things which are up to others') and the occurrence of unpleasant ἀλλότρια respectively. The reader is advised to start his ἄσκησις, 'philosophical training',[41] with small things, like the loss of an earthenware jug or the theft of clothes in the baths. The motif of starting with small things returns in Chapters 12 and 26.[42]

Chapter 5a demonstrates the consequences of the wrong application of the κανών. Fear of death serves as an example of negative emotions, which belong to τὰ παρὰ φύσιν τῶν ἐπὶ σοί (2.7): they are παρὰ φύσιν, 'unnatural', because they confuse τὰ ἐφ' ἡμῖν and τὰ οὐκ ἐφ' ἡμῖν – they are ἐπὶ σοί, 'up to you', because we can apply the κανών correctly and thus dispel them.[43]

Chapter 5b, the first aphorism in the *Encheiridion*, is closely linked to Chapter 5a: the uneducated blames others for the trouble he finds himself in, the man who is making progress blames himself, the wise man blames neither others nor himself. The content of this chapter is taken up again in Chapters 48a and 48b (see below).[44]

Chapter 6 introduces another type of ἀλλότρια, namely our material possessions, a motif that recurs in Chapter 44.

The enigmatic comparison of Chapter 7 states that we should readily give up ἀλλότρια and concentrate on our principal ἔργον, that is, the task that God, 'the captain of the ship', imposes upon us; and thus this chapter refers to the topic of the καθήκοντα.[45] The theme of one's proper task recurs in Chapters 17 and 37.[46]

Chapter 9 introduces the body as another ἀλλότριον in opposition to our προαίρεσις, 'moral choice'. In Chapters 39–41 the care of the body is taken up again. There it is argued that we should take care of our bodies as a side issue; exaggerated care of the body conflicts with the primary goal – that is, taking care of our προαίρεσις.[47]

---

[41] To all practical intents and purposes this term is equivalent to τὰ οὐκ ἐφ' ἡμῖν, 'the things which are not under our control', and to ἀδιάφορα, 'the things which are unimportant'. See the introduction.

[42] For further instances of ἄσκησις, see Chapters 10, 11, 21, 30, 34, 47. For further instances of dealing with the loss of positive ἀλλότρια, see Chapters 6, 7, 11, 14a, 15, 16, 19b, 24, 25, 31, 34; for the occurrence of negative ἀλλότρια, see Chapters 9, 14a, 16, 18, 20, 21, 24, 30, 31, 32, 42, 43.

[43] Further discussions of wrong opinions as the source of negative emotions are found in Chapters 16, 18, 20, 30, 42.

[44] The theme of blaming others also recurs in Chapters 20 and 31. Other aphorisms are found in Chapters 8, 19a, 27, 48a.

[45] The chapter is interpreted differently by I. Hadot and P. Hadot 2004, 127–38 and I. Hadot and P. Hadot 2005, 427–36: following Simplicius' interpretation (Simp. *In Epict. Ench.* ch. 13), they argue that the comparison teaches us that we should readily accept death when God calls us.

[46] For καθήκοντα in general, see also Chapters 10, 15, 16, 22, 24, 30, 31, 32, 36, 42, 50.

[47] See also Chapters 28 and 33[7].

Chapter 13 is the first time an issue is raised that often returns in the *Encheiridion*, namely the opinion of others. On the one hand, we should not be afraid to incur other people's contempt, and on the other we should not strive for their admiration. The first issue is also dealt with in Chapters 22, 23, 24, 46 and 47; the second returns in Chapters 28, 35, 42 and 50. Above all, we should stick to our principles, whatever opinion others may have about them.

Chapter 14b picks up the theme of freedom and slavery, first mentioned in 1² as directly related to the right or wrong application of the κανών. The aphorism of Chapter 19a expresses the same thought by means of the concept of invincibility.

Chapters 16 and 26 are about other people's sorrow, but both chapters approach the concept from a different angle: in Chapter 16 the addressee is advised to show compassion to someone who is suffering on the condition that he does not really suffer himself; in Chapter 26 the addressee is told to take up the same attitude towards his own sorrow as he does in the case of other people's sorrow.

Chapters 18 and 32 are about our attitude towards divination. Here too the correct application of the κανών teaches us that everything we learn about the future is outside our control and therefore does not affect our προαίρεσις.

Chapter 19b teaches us that we should never envy others for external goods. Chapter 25 expands this thought by pointing out that nobody gets anything for free: being accepted as a client by an influential man is paid for by having to flatter this man and by having to endure the humiliations of his servants.

Chapters 24 and 25 form a diptych. They both deal with real and apparent τιμή, which is taken in the sense of both 'honour' and 'value'. Chapter 24 demonstrates that being valued by others is unimportant and that having real value is important. One achieves real value by having self-respect, being faithful and high-minded: an attitude that can only be achieved by disregard for externals.[48] Chapter 25, as already stated above, deals with other people's τιμή, which is apparent because it is bought for the high price of selling one's self-respect.

Chapter 31, dealing with our attitude towards the gods, has already been discussed above, and the same goes for Chapter 32 on divination.

Chapter 33 is the only passage in the *Encheiridion* where the reader is given concrete advice for behaviour in concrete situations, many of them in

---

[48] For a discussion of the ambiguity of the concept of τιμή in this chapter, see Boter 1999, 124–5.

a very lapidary form: 'do not laugh', 'don't swear', 'don't have sex before marriage', and so forth. This is the only part of the *Encheiridion* where the application of the κανών is not directly relevant: the advice concerning good manners might apply to anyone of any philosophical conviction.

Chapter 45 warns against jumping to conclusions about other people's behaviour and thus exposes the danger of a rash application of the συγκατάθεσις: only by learning other people's motives can we judge whether their actual behaviour is right or wrong.

In Chapters 48a–48b what we might call the epilogue of the *Encheiridion* begins. Chapter 48a resembles Chapter 5b in its aphoristic form. The ἠργμένος παιδεύεσθαι of Chapter 5b is discussed at length in Chapter 48b, where he is designated as the προκόπτων, 'the person who is making progress'. This sketch of the προκόπτων contains elements that have been dealt with separately in the preceding part of the *Encheiridion*.[49] It is apparently a checklist of the things a student should have mastered by the end of his education. Taken this way, Chapter 48b forms a ring composition with Chapter 5b: the goal that has been suggested for the beginner student in 5b has been reached by the προκόπτων of 48b.

Chapters 49–53, too, can be interpreted as final advice to the student at the end of his education. Chapter 49 warns that it is deeds that count, not words; Chapter 50 emphasizes that one should stand by one's principles. Chapter 51 might even be characterized as a schoolmaster's farewell speech, urging the departing students to put into practice everything they have learned during their education. After these chapters, Chapter 52 comes as rather a shock, not so much because of its message (which is roughly the same as that of Chapter 49) but because of its form. Chapter 49 is moulded into the form of a soliloquy in the first person singular, in which the I advises himself not to give preference to theory over practice. Chapter 52, on the contrary, is formulated in the first person plural: the speaker states as a fact that we (including himself) put all our energy into elaborating

---

[49] Here is a list of parallel passages: 48b.1 οὐδένα ψέγει, οὐδένα ἐπαινεῖ = 33²; 48b.2 οὐδένα μέμφεται, οὐδενὶ ἐγκαλεῖ = 1.10; 48b.2–3 οὐδὲν περὶ ἑαυτοῦ λέγει ὡς ὄντος τι ἢ εἰδότος τι: cf. 13.2–3, 22, 46; 48b.3 ὅταν ἐμποδισθῇ τι ἢ κωλυθῇ, ἑαυτῷ ἐγκαλεῖ = 5b2; 48b.3–4 κἄν τις αὐτὸν ἐπαινῇ, καταγελᾷ τοῦ ἐπαινοῦντος αὐτὸς παρ' ἑαυτῷ: cf. 13.2–3; 48b.4 κἄν ψέγῃ, οὐκ ἀπολογεῖται: cf. 33⁹, 42; 48b.5–6 περίεισι δὲ καθάπερ οἱ ἄρρωστοι, εὐλαβούμενός τι κινῆσαι τῶν καθισταμένων, πρὶν πῆξιν λαβεῖν: this image does not occur elsewhere in the *Encheiridion*, but the thought resembles 46.7–13; 48b.6–7 ὄρεξιν ἅπασαν ἦρκεν ἐξ ἑαυτοῦ = 2.7–8; 48b.7–8 τὴν δὲ ἔκκλισιν εἰς μόνα τὰ παρὰ φύσιν τῶν ἐφ' ἡμῖν μετατέθεικεν = 2.6–7; 48b.8 ὁρμῇ πρὸς ἅπαντα ἀνειμένῃ χρῆται = 2.10–11; 48b.8–9 ἂν ἠλίθιος ἢ ἀμαθὴς δοκῇ, οὐ πεφρόντικεν = 13.1–2. The concluding sentence, ἑνί τε λόγῳ, ὡς ἐχθρὸν ἑαυτὸν παραφυλάσσει καὶ ἐπίβουλον, has no predecessor: the introduction by means of ἑνί τε λόγῳ shows the resumptive character of the phrase.

the theory of ethics (or rather the meta-theory),[50] while utterly failing to put this theory into practice. How are we supposed to take this chapter? It would be self-destructive and self-contradictory for Arrian if he intended his readers to take this chapter at face value and to acquiesce in the situation that is sketched out here. I would therefore rather read this chapter as a kind of self-denying prophecy, presenting to the προκόπτων the desired way of life *e contrario*. By the same token, the layman who reads this chapter is warned of the risks he runs when confronting the professional study of philosophy.

Finally, Chapter 53 provides the reader with four maxims which we should have ready to hand in all situations; it might not be exaggerated to style these four maxims as four ἐγχειρίδια in the sense of 'daggers', weapons with which to protect oneself against the attacks of life.[51]

The picture that emerges from the above analysis is clear. The goals, and the instrument to reach these goals, are set out in the first two chapters; the various themes related to these goals are treated in a kaleidoscopic way, culminating in the epilogue of Chapters 48a–53. The aspect of ἄσκησις, including preparation for and anticipation of unpleasant events that may befall us, is relevant in almost every chapter, the answer to the question of how to deal with such events always being the same: apply the κανών under all circumstances and apply it well. The *Encheiridion* is not a guide that one only consults in case of actual problematic situations, like the help site in a computer program; rather, it aims at sketching the way of life in accordance with Stoic doctrine with which one should get acquainted through incessant meditation and spiritual exercise and by reading and re-reading the work as one coherent whole. This brings us to the next question: what type of readers did Arrian have in mind when composing the *Encheiridion*?

The Intended Readers of the *Encheiridion*

P. Hadot has repeatedly and convincingly argued that, in antiquity, philosophy is primarily a way of life, not a purely intellectual endeavour.[52] With regard to the *Encheiridion* he aptly states (P. Hadot 2000, 142–3) that it is not a 'petit traité de «morale»': it is concerned with the problem of *comment devenir philosophe*. Hadot argues that the *Encheiridion* is not addressed to

---

[50] Cf. P. Hadot 2000, 136: 'ce que l'on pourrait appeler la théorie de la théorie'.
[51] For a detailed analysis of this chapter, see Carlini 1995.   [52] See especially P. Hadot 1981, 1995.

outsiders.⁵³ The work, according to Hadot, contains too much technical philosophical language to be accessible to someone who is totally ignorant of Epictetus' philosophy. Hadot further draws attention to formulas like Εἰ προκόψαι θέλεις ('If you want to make progress', Chapters 12–13) and Εἰ φιλοσοφίας ἐπιθυμεῖς ('If you long for philosophy', Chapter 22): with such phrases, Hadot argues, Arrian addresses himself to former students of Epictetus or to readers of the *Discourses*. He therefore concludes that it is addressed to the προκόπτοντες, 'those who are making progress', and this opinion is shared by the majority of scholars.⁵⁴ Maltese (1990, xviii) goes still further, claiming that the *Encheiridion* is intended exclusively for the accomplished philosopher. Others, on the contrary, follow Simplicius in believing that the work is written for beginners.⁵⁵

In contradiction to Hadot's argument concerning the technical language in the *Encheiridion* it could be argued that Arrian gives an explanation of the crucial terms in the first two chapters, which would be superfluous for those already thoroughly acquainted with Epictetus' philosophical system (this also answers the position of Sellars, for which see n. 54). Moreover, in the *Discourses* too there is technical language that is frequently not explained, so that the reader of the *Discourses* has to cope with the same problems as the reader of the *Encheiridion*.⁵⁶ As to Hadot's second argument: the two phrases cited above (Εἰ προκόψαι θέλεις and Εἰ φιλοσοφίας ἐπιθυμεῖς) are not directed to those who *are* making progress but to those who *want to* make progress. On the other hand, against the thesis that the work is directed at beginners it could be argued that Chapters 49 and 52 (about

---

⁵³ P. Hadot 2000, 145: 'Il est sûr que l'ouvrage n'est pas destiné à être lu par des profanes qu'Arrien aurait voulu convertir.' Cf. I. Hadot and P. Hadot 2004, 40–6.

⁵⁴ See, for instance, I. Hadot 1996, 147, Wehner 2000, 259, Colardeau 2004, 28. Sellars 2007, 135–8 argues that the *Encheiridion* must be intended for people already well acquainted with philosophy because 'Epictetus proposes first the study of *theories* and then, only once these have been mastered, a series of *exercises* designed to digest those theories' (138). Now, it is true that one should have knowledge of the Stoic doctrines before trying to put these into practice, but that does not mean that one has to be an advanced student of philosophy before one tries to live according to the Stoic tenets. In fact, the gist of Stoic ethics is very easy to grasp, especially in Epictetus: 'make the right distinction between the things under our control and the things not under our control'. It is the study of the logical foundations of these tenets which takes up much time and energy; and this study of logic should only be addressed when one is able to bring the Stoic tenets into practice in one's own life, as Epictetus argues in 3.2.5–8. Moreover, the *Discourses* (which form the basis of the *Encheiridion*) do not contain much theoretical instruction either.

⁵⁵ Simplicius, *In Epict. Ench.* P 58–60: καὶ τῶν φιλομαθῶν οἱ πρὸς λόγους ἀσυνηθέστεροι ἴσως ἕξουσί τινα χειραγωγίαν ἐκ τῆς ἑρμηνείας αὐτῶν, 'and students who are less accustomed to such speeches will perhaps receive some assistance from their interpretation'. See, for instance, Schenkl 1916², lii, Erler 1997, 574, Gourinat 1998, 77.

⁵⁶ In the letter to Lucius Gellius which precedes the *Discourses* in our manuscripts (for which see above) Arrian does not state that the *Discourses* are directed to readers with prior knowledge of Epictetus' philosophy.

interpreting Chrysippus and about logic) cannot possibly be exclusively directed to the layman.

I think that Arrian had both types of reader in mind.[57] In my analysis I have suggested that Chapter 48b recapitulates all the major elements already discussed in detail in various passages; I have interpreted the rest of the work as an epilogue to the student leaving school, including a 'farewell speech' in Chapter 51. Accordingly, the *Encheiridion* provides a kind of blueprint of what a full course in the practice of Stoic philosophy includes, not by presenting a detailed schedule of the various subjects, but by repeatedly bringing to one's attention the practical consequences of Stoic doctrine. For the student of Epictetus (or, for that matter, for the reader of the *Discourses*) the *Encheiridion* may therefore serve as an *aide-mémoire*, recapitulating the essential elements of Epictetus' teaching; it can also be used as a kind of checklist to see whether the student is living up to the philosophical standards he has learned. For the layman, on the other hand, reading the *Encheiridion* may give an impression of the aims and goals Epictetus set himself and his students.[58]

The absence of any treatment of logic (apart from the brief mention of the subject in Chapters 49 and 52) and the rare occurrence of references to philosophical theory in general are in full accordance with the aim Arrian set himself in composing the *Encheiridion*, as stated in his lost dedicatory letter to Massalenus quoted by Simplicius (see above), namely to pick out from Epictetus' speeches 'the principal and most necessary elements of this philosophy which are most likely to move the souls of the readers'. Here I take 'the principal and most necessary elements of this philosophy' to refer to the bare minimum of theory, such as it is presented in the two opening chapters, and 'which are most likely to move the souls of the readers' to refer to the bulk of the work, which is concerned with the application of the κανών. The logical, cosmological and theological foundations of Epictetus' system are not expounded. By repeatedly reading the *Encheiridion* the layman will be able to train himself in appropriating the Stoic attitude to life; at the same time he will be aware that it takes a lot of serious study to provide a logical basis for the Stoic tenets.[59]

---

[57] This possibility is mentioned by Sellars 2007, 138 n. 110, who rejects it because 'if it [the *Encheiridion*] were used by a beginner who had not yet studied philosophical theory, it would in effect be a series of *exercises* without *theory*'. Cf. my remark on Sellars' views in n. 54.
[58] Gourinat 1998, 77 aptly writes: 'Ainsi, le *Manuel* n'est pas loin de s'achever par le portrait d'un homme qui a accompli le programme que les premiers chapitres proposent au néophyte.'
[59] I wish to thank the members of the Amsterdam Hellenist Club for discussing an earlier draft of this chapter with me and Nina King for correcting my English.

REFERENCES

J. Barnes 1997, *Logic and the Imperial Stoa* (Leiden, New York, Cologne)
M. Billerbeck 1978, *Epiktet, Vom Kynismus* (Leiden)
A. Bonhöffer 1890, *Epictet und die Stoa* (Stuttgart)
G. J. Boter 1993, 'The Greek Sources of the *Translations* by Perotti and Politian of Epictetus' *Encheiridion*', *Revue d'Histoire des Textes*, 23: 159–88
  1999, *The Encheiridion of Epictetus and Its Three Christian Adaptations* (Leiden, New York)
  2007, *Epictetus: Encheiridion* (Berlin, New York)
  2010, 'Evaluating Others and Evaluating Oneself in Epictetus' *Discourses*' in R. M. Rosen and I. Sluiter (eds.), *Valuing Others in Antiquity* (Leiden, Boston), 323–52
T. Brennan and Ch. Brittain 2002, *Simplicius: On Epictetus' Handbook 1–26* (Ithaca, NY)
G. Broccia 1979, *Enchiridion: Per la storia di una denominazione libraria* (Rome)
A. Carlini 1995, 'Osservazioni sull'epilogo del *Manuale* di Epitteto', *Studi Italiani di Filologia Classica*, 13: 214–25
Th. Colardeau 2004, *Étude sur Épictète: Avant-propos, traduction nouvelle des textes grecs et latins par Jean-Baptiste Gourinat. Préface de Pierre Hadot* (Paris [1903])
  [NB the page numbers in the 2004 edition differ from those in the original 1903 edition]
R. Dobbin 1991, 'Prohairesis in Epictetus', *Ancient Philosophy*, 11: 111–35
  1998, *Epictetus, Discourses Book I* (Oxford)
M. Erler 1997, 'Römische Philosophie' in F. Graf (ed.), *Einleitung in die lateinische Philologie* (Stuttgart, Leipzig), 537–98
P. P. Fuentes González 2000, 'Épictète' in R. Goulet (ed.), *Dictionnaire des philosophes antiques*, vol. III (Paris), 106–51
J.-B. Gourinat 1998, *Premières leçons sur le Manuel d'Épictète* (Paris)
  2001, 'Le Socrate d'Épictète', *Philosophie antique: Problèmes, renaissances, usages* 1: 137–65
  2005, 'La «prohairesis» chez Épictète: Décision, volonté ou «personne morale»?', *Philosophie antique: Problèmes, renaissances, usages*, 5: 93–134
  2008, 'Épictète et la logique: Une variation sur le thème des vertus dialectiques du sage' in M. Broze, B. Decharneux and S. Delcomminette (eds.), *Mélanges de philosophie et de philologie offerts à Lambros Couloubaritsis* (Paris), 435–46
I. Hadot 1996, *Simplicius: Commentaire sur le Manuel d'Épictète. Introduction et édition critique du texte grec* (Leiden)
  2001, *Simplicius: Commentaire sur le Manuel d'Épictète*, vol. I, chapters 1–29 (Paris)
I. Hadot and P. Hadot 2004, *Apprendre à philosopher dans l'Antiquité: L'Enseignement du "Manuel d'Épictète" et son commentaire néoplatonicien* (Paris)
  2005, 'La Parabole de l'escale dans le Manuel d'Épictète et son commentaire par Simplicius' in G. Romeyer Dherbey and J.-B. Gourinat (eds.), *Les Stoïciens* (Paris), 427–49

P. Hadot 1981, *Exercices spirituels et philosophie antique* (Paris)
1995, *Qu'est-ce que la philosophie antique?* (Paris)
1997, *La Citadelle intérieure: Introduction aux Pensées de Marc Aurèle* (Paris [1992])
2000, *Manuel d'Épictète* (Paris)
B. L. Hijmans 1959, *ΑΣΚΗΣΙΣ: Notes on Epictetus' Educational System* (Assen)
R. Hofmeister Pich 2010, 'Prohairesis und Freiheit bei Epiktet: Ein Beitrag zur philosophischen Geschichte des Willensbegriffs' in R. Hofmeister Pich and J. Müller (eds.), *Wille und Handlung in der Philosophie der Kaiserzeit und Spätantike* (Berlin, New York), 95–129
A. A. Long 2002, *Epictetus: A Stoic and Socratic Guide to Life* (Oxford)
E. V. Maltese 1990, *Epitteto, Manuale* (Milan)
R. J. Newman 1989, 'Cotidie meditare: Theory and Practice of the meditatio in Imperial Stoicism', *ANRW* II 36.3 (Berlin, New York), 1473–517
W. A. Oldfather 1925-8, *Epictetus: The Discourses as Reported by Arrian, the Manual, and Fragments* (Cambridge, MA, London)
1927, *Contributions toward a Bibliography of Epictetus* (Urbana)
M. Pohlenz 1990, *Die Stoa: Geschichte einer geistigen Bewegung*, vol. II. 6th edn (Göttingen [1949])
S. L. Radt 1990, 'Zu Epiktets Diatriben', *Mnemosyne*, 43: 364–73
G. Reale 1982, *Epitteto: Diatribe, manuali e frammenti* (Milan)
H. Schenkl 1916, *Epictetus: Dissertationes ab Arriano digestae* 2nd edn (Leipzig)
J. Schweighäuser 1798, *Epicteti Manuale et Cebetis Tabula* (Leipzig)
J. Sellars 2007, 'Stoic Practical Philosophy in the Imperial Period' in R. Sorabji and R. W. Sharples (eds.), *Greek and Roman Philosophy 100 BC–AD 200*. Bulletin of the Institute of Classical Studies, Supplement 94 (London), 115–40
R. Sorabji 2007, 'Epictetus on proairesis and Self' in Th. Scaltsas and A. S. Mason (eds.), *The Philosophy of Epictetus* (Oxford), 87–98
P. A. Stadter 1980, *Arrian of Nicomedia* (Chapel Hill)
H. W. F. Stellwag 1933, *Epictetus, Het eerste boek der Diatriben* (Amsterdam, Paris)
J. C. Thom 1995, *The Pythagorean 'Golden Verses'* (Leiden)
L. van Hoof 2010, *Plutarch's Practical Ethics: The Social Dynamics of Philosophy* (Oxford, New York)
B. Wehner 2000, *Die Funktion der Dialogstruktur in Epiktets Diatriben* (Stuttgart)

CHAPTER 10

The Problem of Practical Applicability in Ptolemy's Geography

Klaus Geus

εὔχρηστον and εὔλογον

When one wants to re-translate the term 'practical applicability' into Greek, one will probably come up with the word εὐχρηστία or simply τὸ εὔχρηστον. Literally, it means 'good use' or 'serviceableness.'*[1]

This word is quite common in philosophical and historical contexts but rarely used in scientific or practical treatises. For example, Aelian, the tactician, uses it to declare that the first rule in tactics is to 'determine the number of troops "in a way that is appropriate and serviceable" for (the requirements of) war' (ἀριθμόν τε τοῦ παντὸς πλήθους ὁρίσαι συμμέτρως καὶ εὐχρήστως πρὸς τὰ κατὰ τὸν πόλεμον).[2] Aelian means here that the military commander shall not to stick to an abstract or theoretical principle, but act according to the needs of the specific situation, e.g. the time and place of battle.

Of course, the meaning of εὔχρηστος and εὐχρηστία is not limited to 'practical applicability' – the concept that concerns us here – but it is interesting to see that the famous geographer and astronomer Ptolemy (c. AD 150) makes use of this word εὔχρηστος in the theoretical chapters of his Γεωγραφικὴ ὑφήγησις, his *Introduction to Geography*. The following discussion will try to show what exactly Ptolemy means by this, and why he attaches great importance to the principle of εὔχρηστος.

First of all, it is worth noting here that Ptolemy uses this word thrice in his *Geography*. This makes the question of its exact meaning and usage all the more important.[3]

---

* I would like to thank Markus Asper, Elton Barker, Anca Dan, Philip van der Eijk, Marco Formisano, Andrea Ladány, Søren Lund Sørensen and Edgar Reich for their valuable suggestions, and Irina Tupikova for drafting the images.
[1] Since the literature on Ptolemy is vast, I have limited myself in the following to standard and recent works. For a good, modern overview, see Jones 2010 (with ample bibliography).
[2] Aelian. *Tact.* (*Tactica*) 3.2.266 Köchly-Rüstow.
[3] In contrast, εὔχρηστος and εὐχρηστία are attested twenty times in his *Syntaxis mathematikē* (*Almagest*).

In 1.6.2 Ptolemy informs his reader why he felt compelled to write a new handbook on geography despite the fact that shortly before him the geographer Marinos of Tyre had already done the same.

Ἀλλ' εἰ μὲν ἑωρῶμεν μηδὲν ἐνδέον αὐτοῦ τῇ τελευταίᾳ συντάξει, κἂν ἀπήρκεσεν ἡμῖν ἀπὸ τούτων μόνων τῶν ὑπομνημάτων ποιεῖσθαι τὴν τῆς οἰκουμένης καταγραφήν, μηδέν τι περιεργαζομένοις. Ἐπεὶ δὲ φαίνεται καὶ αὐτὸς ἐνίοις τε μὴ μετὰ καταλήψεως ἀξιοπίστου συγκατατεθειμένος καὶ ἔτι περὶ τὴν ἔφοδον τῆς καταγραφῆς πολλαχῇ μήτε τοῦ προχείρου μήτε τοῦ συμμέτρου τὴν δέουσαν πρόνοιαν πεποιημένος, εἰκότως προήχθημεν, ὅσον ᾠόμεθα δεῖν, τῇ τἀνδρὸς πραγματείᾳ συνεισενεγκεῖν ἐπὶ τὸ εὐλογώτερον καὶ εὐχρηστότερον.

Now if we saw no defect in his [Marinos'] final composition, we could content ourselves with making the map of the *oikoumenē* on the basis of these writings alone, without taking any more trouble about it. Since, however, even he turns out to have given assent to certain things that have not been creditably established, and in many respects not to have given due thought to the method of map-making, with a view either to convenience or to the preservation of proportionality, we have justifiably been induced to contribute as much as we think necessary to the man's work *to make it more logical and easier to use* (ἐπὶ τὸ εὐλογώτερον καὶ εὐχρηστότερον). (trans. Berggren and Jones 2000, with slight modifications)

In short, Ptolemy took up the burden not only to draw a map of the *oikoumenē*, the inhabited world, but also to make Marinos' basic work more logical[4] and *easier*, *i.e. more practical to use*. Despite some claim to the contrary,[5] this passage is positive proof that Ptolemy did not plan to write a wholly new book on geography. He settled for making a revision of Marinos' work by correcting his astronomical and mathematical errors and making his vast material εὐχρηστότερον.[6]

In the following, Ptolemy gives many examples where he corrects Marinos' mistakes and where he gives ample advice for making the data 'user-friendly'. While modern scholars in their discussions of these passages regularly treat the εὐλογώτερον, especially Ptolemy's amendments and

---

[4] It seems that εὔλογος here has both the literary and mathematical meaning of a 'good' λόγος, i.e. 'word' and 'reckoning'.
[5] But see Herrmann 1930, especially 48.
[6] Maybe it is useful to recall the fact that nobody in antiquity wrote a 'new' geography. All the descriptions of the world we have are 'corrections of the map'. It is clear in Herodotus, who constructs his *Mappa mundi* by correcting the Ionians; it is also obvious in Hipparchus and Strabo (less clear with Eratosthenes, but see Geus 2002, 261–88). The 'correction of the map' (διόρθωσις πίνακος) was an important principle of ancient geography: all the authors correct explicitly or implicitly the previous scientific works and sometimes present this as the only purpose of their work. (I am indebted to Anca Dan for this note.)

improvements on Marinos, the εὐχρηστότερον tends to get ignored.[7] But, we may ask, what exactly is the difference between εὔλογον and εὔχρηστον here? Why is εὔχρηστον important at all – and Ptolemy's wording makes it sound as if it is of the same importance as εὔλογον – when cartography is all about correctness and exactness? What level of knowledge and expertise is he envisioning in his audience – explicitly and implicitly? Or, to paraphrase it: what kind of procedure is Ptolemy expecting from his reader?

A Paradigm for the Reader

In the following, I shall discuss one example in order to show that both aspects, εὔλογον and εὔχρηστον, are closely entwined in Ptolemy's thinking.[8] This example concerns his main goal: drafting a world map according to a certain kind of projection on the basis of simple measurements transmitted by sailors, travellers and earlier geographers.

In 1.13.1–4 he writes:

> Ἀπὸ γὰρ τοῦ μετὰ τὸν Κολχικὸν κόλπον ἀκρωτηρίου, ὃ καλεῖται Κῶρυ, τὸν Ἀργαρικὸν κόλπον φησὶ [i.e. Μαρῖνος] διαδέχεσθαι σταδίων ὄντα μέχρι Κούρουλα πόλεως τρισχιλίων τεσσαράκοντα, καὶ κεῖσθαι τοῦ Κώρυ τὴν Κούρουλα πόλιν ὡς ἀπὸ βορέου. (2) Συνάγοιτ' ἂν οὖν ἡ διαπεραίωσις ὑφαιρουμένου τοῦ τρίτου κατὰ τὸ ἀκόλουθον τῷ Ἀργαρικῷ κόλπῳ δισχιλίων τριάκοντα ἔγγιστα σταδίων μετὰ τῆς ἀνωμαλίας τῶν δρόμων. (3) Ἐξ ὧν εἰς τὴν συνέχειαν ὑπολογισθέντος ἔτι τοῦ τρίτου, καταλειφθήσονται στάδιοι χίλιοι τριακόσιοι πεντήκοντα ἔγγιστα κατὰ τὴν βορρᾶν θέσιν. (4) Ἧς μεταφερομένης ἐπὶ τὴν τῷ ἰσημερινῷ παράλληλον καὶ ὡς πρὸς ἀπηλιώτην, μειώσει τοῦ ἡμίσεως ἀκολούθως τῇ μεταλαμβανομένῃ γωνίᾳ ἕξομεν τὴν μεταξὺ τῶν δύο μεσημβρινῶν διάστασιν, τοῦ τε διὰ τοῦ Κῶρυ ἀκρωτηρίου καὶ τοῦ διὰ τῆς Κούρουλα πόλεως, σταδίων μὲν ἑξακοσίων οε', μοίρας δὲ μιᾶς ἔγγιστα καὶ τρίτου, διὰ τὸ τοὺς κατὰ τοῦτον τὸν τόπον παραλλήλους μηδενὶ ἀξιολόγῳ διαφέρειν τοῦ μεγίστου κύκλου.

Thus [Marinos] says after the cape marking the end of the Bay of Kolchoi, which is called Kory, follows the Bay of Argaru, and this is 3,040 stades as far as the city of Kurula; and the city of Kurula is in the direction of the *Boreas* [northnortheast] wind from Kory. Hence if a third is subtracted to account for following [the arc of] the Bay of Argaru, the crossing will amount to approximately 2,030 stades, with the irregularities of the daily sails [still incorporated in the total]. If a third is again subtracted from these [2,030 stades] to get the total distance, approximately 1,350 stades will remain in

[7] But see Stückelberger and Grasshoff 2006, 1.18–24.
[8] I have treated this example from another perspective in Geus 2007 and Geus 2013.

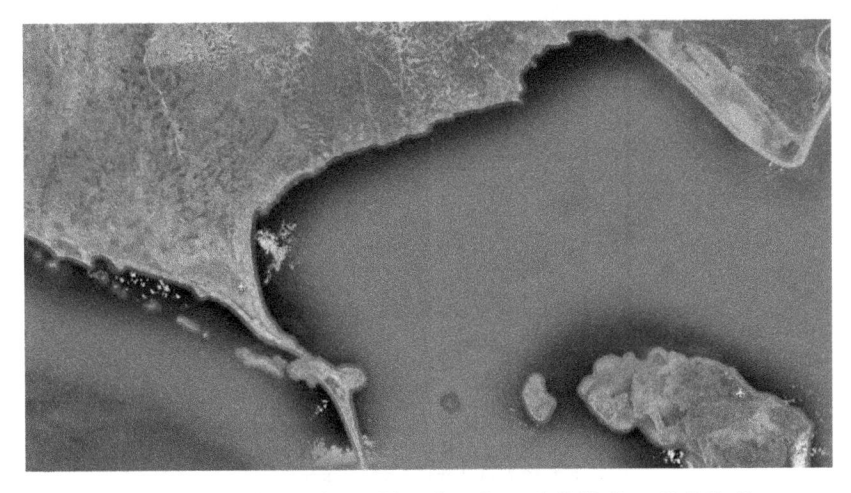

Fig. 10.1 *The Palk Strait* https://de.wikipedia.org/wiki/Palkstra%C3%9Fe

the direction of the *Boreas* wind. When this has been transferred to the [circle] parallel to the equator, and to the direction of the *Apeliotes* [east] wind, by subtracting half in accordance with the subtended angle, we shall get the distance between the two meridians through Cape Kory and the city of Kurula as 675 stades. This is approximately 1 ⅓° since the parallels through these places do not differ significantly from the great circle [i.e. the equator]. (trans. Berggren and Jones 2000, with slight modifications)

The said Cape Kory is the cape that lies at the so-called Palk Strait (see Fig. 10.1). It is the strait between South India and the northernmost part of Sri Lanka, and connects the Bay of Bengal in the northeast with the Palk Bay and thence with the Gulf of Mannar in the southwest.[9]

Marinos knew that the voyage from Kory to Kurula along the coast of Argaru amounts to 3,040 stades. Also known was the direction of the voyage to the Boreas wind, i.e. northnortheast.[10]

Ptolemy makes use of these two statements for calculating the distance between the cities and also for reckoning the distance between the meridians of the two cities. The latter is important for drawing a map according to his own cone projection.

Ptolemy solves the problem in three steps (see Figs. 10.2, 10.3 and 10.4).

---

[9] The strait is 33 to 50 miles (53 to 80km) wide. Several rivers flow into it, including the Vaigai River of Tamil Nadu. The strait is named after Robert Palk, who was a governor of the Madras Presidency (1755–63) during the John Company Raj period.
[10] Both pieces of information stem surely from an older Greek account, probably from the *Periplous* of a Greek merchant used by Marinos for his geographical treatise.

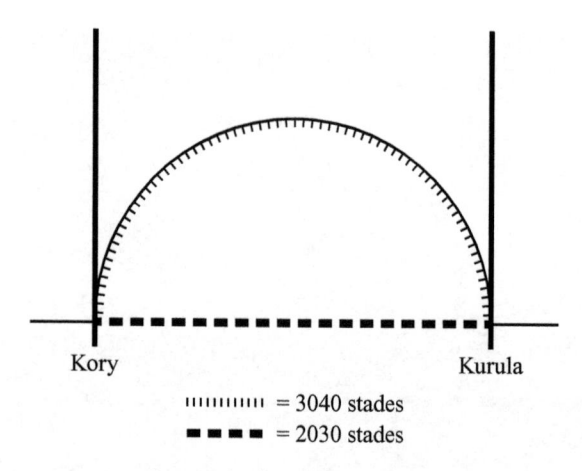

Fig. 10.2 Ptolemy's first reduction

Kory — Kurula

‖‖‖‖‖‖‖‖ = 3040 stades
■ ■ ■ ■ = 2030 stades

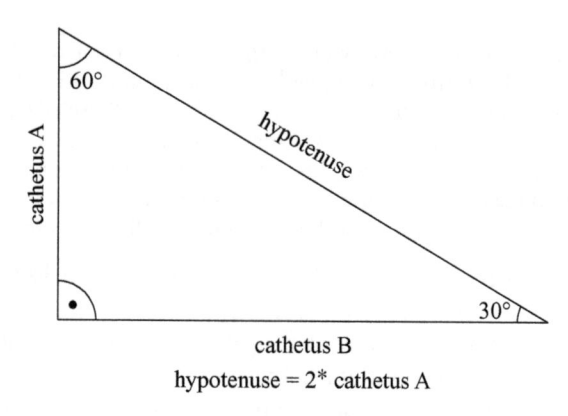

Fig. 10.3 Ptolemy's triangle

cathetus A, cathetus B, hypotenuse
60°, 30°
hypotenuse = 2* cathetus A

Step 1: Ptolemy subtracts one-third from the transmitted 3,040 number 'for following [the arc of] the Bay of Argaru'. Obviously, Ptolemy compares the voyage along the coastline of a bay to an arc of a full circle. His scant explanation harkens back to a mathematical rule of thumb: the ratio between the circumference of a circle and its diameter equals approximately 3:1 (π). Therefore, a straight line between two points that are the extremities of a half circle is two-thirds shorter (2:3) than the whole circumference and one-third shorter than the circumference of one half of the circle (1:3). Since Ptolemy compared the voyage to the length of the arc of a half-circle, he subtracted one-third from the original number.

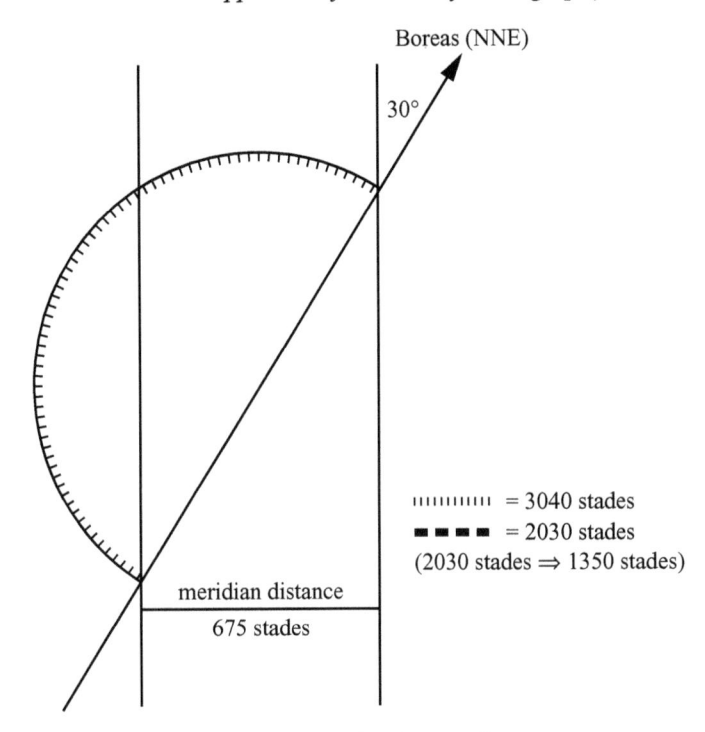

Practical Applicability in Ptolemy's Geography

Fig. 10.4 Ptolemy's reduction, Kory

The first step reduces the 3,040 stades to 2,030 stades.[11]

Step 2: Again, Ptolemy subtracts one-third, because of 'the irregularities of the daily sails', which are incorporated in the total so far. Obviously, Ptolemy thinks of the transmitted number of days as too big. He does not inform us what kind of irregularities he had in mind. Stops, bad weather, counter-currents are the most likely. The reason for subtracting one-third is nevertheless clear enough: again, Ptolemy tries to cancel out irregularities with a mathematical rule of thumb – 'Reduce a transmitted number by one-third if there are indications that the sailing was not according to a certain norm.' Ptolemy also uses this principle in other instances in his work. Thus we may describe it as his guiding rule.[12]

The second step has brought down the 2,030 stades to 1,350 stades.

Step 3: Finally, Ptolemy reduces the number for a third time by cutting the 1,350 in half. He tries to find the distance between the meridians of

---

[11] Strictly speaking 2,026,666.
[12] Cf., e.g. *Ptol. geogr.* 1.10.2 the return journey from Garame to Leptis Magna took only twenty days, while the first journey was thirty days long 'because of the diversions'.

both cities, Kory and Kurula. Since Kurula is not in the east of Kory but in direction of the Boreas wind, he has to tamper with the data again. According to the compass rose of Timosthenes of Rhodes, which Ptolemy uses throughout his work, each of the twelve winds comprises 30°.[13] The Boreas wind lies 30° degrees east of the north direction, indicated by the Aparktias. Again, Ptolemy applies a mathematical procedure here. Probably, what he has in mind is that the hypotenuse in a right-angled triangle is twice the one cathetus, which has a 60° angle with this hypotenuse.[14]

In other words: if you turn your triangle in such a way that the hypotenuse matches the direction of the Boreas wind, you will find the (modified) sailing distance as twice the distance between the meridians of Kory and Kurula.

Step 3 has finally reduced the 1,350 stades to 675 stades. This third reduction allows Ptolemy at last to draft the meridians of Kory and Kurula on this map.

Ptolemy is content with his result here. In principle, he should have proceeded with a fourth reduction. His 675 stades are only true for the parallel of Kurula and Kory. Hence, if one wants to plot the meridian distance onto Ptolemy's map drawn in a cone projection, one has to adapt the number. But as Ptolemy himself writes: 'the parallels through these places do not differ significantly from the great equator'. Thus he feels entitled to leave out this fourth step.

Ptolemy calls the whole procedure 'the reckoning of positions [of localities] by reducing [literally, approaching] the excesses'.[15] In our judgement, these steps may seem bold or even arbitrary to some extent, but they are consistent, mathematically verifiable and – easy to apply. Every reader who understands this procedure will be able to use it as a model and adapt it to other, similar, tasks.

Obviously, Ptolemy takes a lot as given on the part of his reader. Without some mathematical expertise one will probably have trouble understanding *Step 1* and *Step 3*, where the mathematical principle is not cited or even explained. For sure, the mathematics in this example is basic, and probably

---

[13] Ptolemy makes use of the compass rose of Timosthenes of Rhodes, which comprises twelve winds: Aparktias (N), Boreas (30°, NNE), Kaikias (60°, NEE), Apeliotes (90°, E), Euros (120°, SEE), Euronotos (150°, SES), Notos (180°, S), Libonotos (210°, SWS), Lips (240°, SWW), Zephyros (270°, W), Argestes (300°, NWW), Thraskias (330°, NWN). See, e.g., Nielsen 1945, especially 41–4.

[14] See Wilberg's edition in Ptolemaeus 1838, 45–6 ('*observationes*'), and Mžik and Hopfner 1938, 32 nn. 2, 44, 5.

[15] Ptol. *geogr.* 1.13.1 ἐὰν τὸ παρὰ . . . τὰς θέσεις ἐπιλογίζηται κατὰ συνεγγισμὸν τῶν ἐπιβολῶν. For a different view, see Berggren and Jones 2000, 49: 'We take *epibolē* here to mean "landfall", as it does in *Periplous* 55.'

all Greeks must have learned these rules in school. Nevertheless, the fact that they are used here in a context not of pure, but applied, mathematics, like geography, makes it hard on the reader.

This example illustrates neatly the εὔλογον of Ptolemy's strategy in making Marinos' work accessible to his readers. But what about the εὔχρηστον, the practical applicability?

Practical Procedure and Practical Advice

Let us look again at *Step 1*: strictly speaking, Ptolemy's mathematics is loose or sketchy. The ratio between the half-circle and the diameter is not 3:2. As we all know from school, the circumference of a circle equals 2π*r, so a half-circle is 2π*r/2 = p*r. The diameter equals 2*r. Thus the ratio between a half-circle and a diameter equals π*r:2*r or π:2.

Hence Ptolemy's calculation is only true for the approximation of π = 3. This, of course, had been the standard approximation in antiquity since Babylonian times, often used for practical purposes.[16] Ptolemy surely knew of and could have used a much better approximation for π, like that of Archimedes, but clearly chose not to. Had he done this, the result (1,935 stades) would not have been a real improvement, given the bold steps he has made during his procedure. So he was content with the easier or 'more practical' reckoning of π = 3.

To provide some intermediary findings at this point: while writing his *Introduction to Geography*, Ptolemy surely has a reader in mind who is competent in mathematics way above the average. This reader has not only to add mentally mathematical propositions which Ptolemy nowhere explicitly states, but must also have some implicit or tacit knowledge about applied mathematics, like 'π equals 3'. On the other hand, this reader is not required to carry out the mathematical operations to the ultimate degree of accuracy. Obviously, a handy guideline is sufficient here.

In his introduction, Ptolemy gives a lot of practical advice for drawing a map. Let us now go over to the passage where Ptolemy uses the word εὔχρηστον for the second time:

2, 1, 4–5: Προειλόμεθα δὲ τάξιν τοῦ περὶ τὴν καταγραφὴν εὐχρήστου πανταχῇ ποιούμενοι πρόνοιαν, τουτέστι καθ'ἣν ἐπὶ δεξιὰ ποιησόμεθα τὰς μεταβάσεις, ἀπὸ τῶν ἤδη κατατεταγμένων ἐπὶ τὰ μηδέπω τῆς χειρὸς

---

[16] If Ptolemy had used a better approximation of π, such as 3.1415, the result would have been 1,935 stades, not 2,030 stades. This is evident proof that Ptolemy used the standard approximation of π and rounded up 2,026,666 to 2,030. For π = 3 in ancient cultures, cf. now Brunke 2011.

ἐκλαμβανομένης. (5) Τοῦτο δὲ γένοιτ' ἄν, εἰ γράφοιτο τά τε βορειότερα πρότερα τῶν νοτιωτέρων καὶ τὰ δυσμικώτερα τῶν ἀπηλιωτικοτέρων, ὅτι πρὸς τὰς τῶν ἐγγραφόντων ἢ ἐντυγχανόντων ὄψεις ἄνω μὲν ἡμῖν ὑπόκειται τὰ βορειότερα, δεξιὰ δὲ τὰ ἀπηλιωτικώτερα τῆς οἰκουμένης ἐπί τε τῆς σφαίρας καὶ τοῦ πίνακος.

We have chosen an order [of presentation] with forethought to convenience in the drawing of the map in every respect, namely progressing toward the right, with the hand proceeding from the things that have already been inserted to those that have not yet [been inscribed]: this would be achieved by having the more northern [places] drawn before the more southerly ones, and the more western before the more eastern, because our convention is that 'up' with respect to the map-makers' or spectators' view means 'north', and 'right' means 'east' in the *oikoumenē*, both on a globe and on a planar map. (trans. Berggren and Jones 2000, with slight modifications)

First, it is interesting to see that Ptolemy again uses εὔχρηστος not as a simple but as a nominalized adjective: it is τὸ εὔχρηστον, not simply εὔχρηστον. This underlines his aim to provide an abstract principle for his operation.

The εὔχρηστον of this advice, which Ptolemy underlines, is that the map-maker's hand shall progress from the upper left side and move on to the right side and downwards. Accordingly, the data in his catalogue starts with the northwest of Europe and ends with the Island of Taprobane, Sri Lanka, in the southeast of Asia.

At first sight, this advice seems quite petty: we all start writing (or drawing) – at least if we are right-handers – on the left side at the top of a page. But what Ptolemy assigns a reason for is far from trivial. That the northern parts are 'up' is not simply a 'convention', as the translation of Berggren and Jones suggests. In fact, ancient maps were often oriented to the east.[17] Thus Ptolemy's statement is also an anthropological constant as well as a practical hint.

Ptolemy's reasoning or line of thought in this section is clearly an astronomical one, as the Greek text clearly suggests. The word ὑπόκειται has to be translated simply as 'lie under', which means that the north pole of our earth lies exactly under the celestial north pole. Every spectator, at least in the northern hemisphere, will see the heavens move around this axis point above his head. Therefore, 'north' is 'up' and 'east' is 'right'. What Ptolemy does here is connect micro- and macrocosm, intuitive or naive apperception with abstract or universal orientation.

---

[17] The term 'orientation' itself stems from the Latin word *oriens*, east. Cf. Podossinov 1991, 233.

Following His Own Advice?

In order to ascertain whether Ptolemy has actually followed his own advice or paid only lip-service to some rhetorical claims of practical applicability, we turn now to another example.

It concerns the notorious problem of finding the longitude of localities in antiquity. Ptolemy pays great attention to it and complains of the rareness of reliable data like that of lunar eclipses in older sources (1.4.2):

> καὶ διὰ τὸ μὴ πλείους τῶν ὑπὸ τὸν αὐτὸν χρόνον ἐν διαφόροις τόποις τετηρημένων σεληνιακῶν ἐκλείψεων – ὡς τὴν μὲν Ἀρβήλοις πέμπτης ὥρας φανεῖσαν, ἐν δὲ Καρχηδόνι δευτέρας – ἀναγραφῆς ἠξιῶσθαι κτλ.

> because no one bothered to record more lunar eclipses that were observed simultaneously at different localities (such as the one that was seen at Arbēla at the fifth hour and at Carthage at the second hour) (trans. Berggren and Jones 2000, 63)

The lunar eclipse Ptolemy mentions here is the one experienced by Alexander the Great on 20 September 331 BC at Erbil near the Tigris. The time difference between Carthage and Arbela is not very accurately given. Instead of the three hours mentioned it is in fact only 2h 14min (or, converted to degrees, 33° 50'). But this very imprecision enables us to check on Ptolemy's procedure. A review of the co-ordinates in his catalogue shows that Ptolemy actually reckoned with three hours (or 45°) as the distance between Carthage and Arbela:

Καρχηδὼν μέγα ἄστυ	λδ° L° ′γ	λβγο
Karchedon/Carthage, a large city (geogr. 4, 3, 7)	34° 50 ′	32° 40'
Ἄρβηλα	π°	λϚ°δ'
Arbela/Erbil (geogr. 6, 1, 5)	80 °	37° 15'

The longitudinal difference between Carthage (34° 50') and Arbela (80°) is 45° 10', or nearly the three hours Ptolemy has spoken of. This is ample proof for believing that Ptolemy followed his own counsel in assembling reliable data.

Has Ptolemy also succeeded in presenting the data in a user-friendly way? While modern scholars would doubt that and compare Ptolemy's huge work with a telephone book, ancient authors may have had a different view. At least, Cassiodorus in his *Institutiones* (I 25) writes enthusiastically that Ptolemy's *Geography* shows all localities 'plainly' (*evidenter*).[18]

---

[18] *Tum si vos notitiae nobilis cura flammaverit habetis Ptolemaei codicem, qui sic omnia loca evidenter expressit, ut eum cunctarum regionum paene incolam fuisse iudicetis. Eoque fiat ut uno loco positi, sicut monachos decet, animo percurratis, quod aliquorum (v.l. aliorum) peregrinatio plurimo labore collegit.*

Ptolemy as a Didactic Writer

In the final paragraph, let us conclude by adding some remarks on the author–reader relationship in the *Geography*.

Ptolemy states that his goal in his geographical handbook is 'to draft our *oikoumenē* as far as possible in proportionality with the real [earth]' (1, 2, 1).[19] All earlier attempts, especially that of Marinos, have been defective in method as well as in execution. Not only did he make a lot of mistakes, Marinos also ignored the principle of τὸ εὔχρηστον,[20] which I would not hasten to call the 'principle of practical applicability'. Since all earlier geographers have, to his mind, failed, Ptolemy feels obliged to do the job by himself.

He takes a stance like that of a teacher who has set exams, gets annoyed by the results and solves the problem presented all by himself. In my reading the *Geographikē hyphēgēsis* is neither a concise introduction to geography nor a full-fledged manual, nor is it a tutorial or instruction book for a reader on how to draw maps. I would compare it to a review of a masterly expert who criticizes what has been presented so far and who shows how the work should have been done properly. As a literary work, the *Geography* has more in common with Aristotle's treatises than with didactic works on geography like that of Dionysius of Alexandria or Pomponius Mela.

Also, what needs to be pointed out is the fact that Ptolemy is little, if it all, concerned with literary play (e.g. *imitatio/aemulatio*), which may set him apart from most other authors discussed in this volume. For sure, the *Geography* of Ptolemy has some didactic elements. But at no point in this lengthy introduction does Ptolemy talk directly to his reader or ask him to do his own research. Even when he uses the word 'we' (which he does quite rarely), it rings more of a *pluralis maiestatis* than of a sociative plural. This is quite understandable. As the above-mentioned passage with Kory and Kurula, where Ptolemy did not state the underlying mathematical principles, has shown too, the reader Ptolemy has in mind or is constructing is an *alter ego* of himself. The whole work makes it clear what Ptolemy thinks of the issue: one cannot do it better than he himself has already done.

If our reading of Ptolemy's work is sound, one may ask one last question: why did he write such a lengthy introduction in the first place, give practical advice and explain his methods at length, if all that his reader can do is

---

[19] Προκειμένου δ' ἐν τῷ παρόντι καταγράψαι τὴν καθ' ἡμᾶς οἰκουμένην σύμμετρον ὡς ἔνι μάλιστα τῇ κατ' ἀλήθειαν.

[20] Cf. *Ptol. geogr.* 1.6.2 (*supra*) and 1.6.3 καὶ δὴ τοῦτο ποιήσομεν ἀπερίττως, ὡς ἔνι μάλιστα, 'we will do this as concisely as possible'.

improve on only some minor details at best (e.g. if new discoveries or better observations are being made)? Or, to rephrase the question: why is Ptolemy concerned with the principle of practical applicability at all, when the reader simply has to use Ptolemy's work to get the best results? What is it good for?

I surmise that the reason Ptolemy is so concerned with this principle is probably the experience Ptolemy himself has had. To quote one final passage from his *Geography* (1.18.2–3) – it is also the passage where εὔχρηστον is used for the third and last time:

> Διπλῆς δὴ τῆς τοιαύτης οὔσης ἐπιβολῆς, καὶ πρώτης μὲν τῆς ἐν μέρει σφαιρικῆς ἐπιφανείας ποιουμένης τὴν τῆς οἰκουμένης διάθεσιν, δευτέρας δὲ τῆς ἐν ἐπιπέδῳ, κοινὸν μὲν ἐπ' ἀμφοτέρων ἐστὶ προκείμενον τὸ εὔχρηστον τὸ δεῖξαι, πῶς ἂν καὶ μὴ προυποκειμένης εἰκόνες ἀπὸ μόνης τῆς διὰ τῶν ὑπομνημάτων παραθέσεως εὐμεταχείριστον ὡς ἔνι μάλιστα ποιώμεθ τὴν καταγραφήν. Τό τε γὰρ ἀεὶ μεταφέρειν ἀπὸ τῶν προτέρων παραδειγμάτων ἐπὶ τὰ ὕστερα διὰ τῆς κατὰ μικρὸν παραλλαγῆς εἰς ἀξιόλογον εἴωθεν ἐξάγειν ἀνομοιότητα τὰς μεταβολάς. (3) Κἂν μὴ τὴν μέθοδον ταύτην τὴν ἐκ τῆς ὑπομνήσεως αὐτάρκη πρὸς ἔνδειξιν τῆς ἐκθέσεως εἶναι συμβαίνῃ, τοῖς οὐκ εὐποροῦσι τῆς εἰκόνος ἀμήχανον ἔσται τοῦ προκειμένου δεόντως τυχεῖν· ὃ συμβαίνει καὶ νῦν τοῖς πλείστοις ἐπὶ τοῦ κατὰ τὸν Μαρῖνον πίνακος, οὐκ ἐπιτυχοῦσι μὲν ἀπὸ τῆς ὑστάτης συντάξεως παραδείγματος, ἀποσχεδιάσασι δὲ ἐκ τῶν ὑπομνημάτων καὶ διαμαρτοῦσιν ἐν τοῖς πλείστοις τῆς ὁμολογουμένης συναγωγῆς διὰ τὸ δύσχρηστον[21] καὶ διεσπαρμένον τῆς ὑφηγήσεως, ὡς ἐξέσται παντὶ τῷ πειρωμένῳ σκοπεῖν.

This undertaking [i.e. drawing the map] can take two forms: the first sets out the *oikoumenē* on a part of a spherical surface, and the second on a plane. The object in both is the same, namely convenience; that is, to show how, without having a model already at hand, but merely by having the texts beside us, we can most conveniently make the map. After all, continually transferring [a map] from earlier exemplars to subsequent ones tends to bring about grave distortions in the transcriptions through gradual changes.

(3) If this method based on a text did not suffice to show how to set [the map] out, then it would be impossible for people without access to the picture to accomplish their object properly. And in fact this is what happens to most people [who try to draw] a map based on Marinos, since they do not possess a model based on his final compilation; instead they draw on his writings and err in most respects from the consensus of opinion, because his guide is so hard to use and so poorly arranged, as anyone who tries it can see. (trans. Berggren and Jones 2000, with slight modifications)

---

[21] Note the antonym τὸ δύσχρηστον for Marinos' work, as opposed to τὸ εὔχρηστον of his.

For sure, Ptolemy wants to save his work, especially his world map, from the rigour of time. Since he envisions it deteriorating in the process of being copied, he left to the future map-maker some practical advice on how to draft his – i.e. fundamentally Ptolemys – maps again.[22] The principle of practical applicability is expounded and propagated not least for his own benefit.[23]

REFERENCES

J. L. Berggren and A. Jones 2000, *Ptolemy's Geography: An Annotated Translation of the Theoretical Chapters* (Princeton)

H. Brunke 2011, 'Überlegungen zur babylonischen Kreisrechnung', *Zeitschrift für Assyriologie*, 101: 113–26

P. Gautier-Dalché 2009, *La Géographie de Ptolemée en Occident (IVe–XVIe siècles)* (Turnhout)

K. Geus 2002, *Eratosthenes von Kyrene: Studien zur hellenistischen Kultur- und Wissenschaftsgeschichte* (Munich)

  2007, 'Ptolemaios über die Schulter geschaut – zu seiner Arbeitsweise in der Geographike Hyphegesis' in M. Rathmann (ed.), *Wahrnehmung und Erfassung geographischer Räume in der Antike* (Mainz), 159–66

  2013, '". . . wie man auch beim Fehlen einer Kartenvorlage nur an Hand des Textes eine Weltkarte sehr leicht herstellen kann", Die Darstellung der Oikumenē; bei Klaudios Ptolemaios', *Periplus, Jahrbuch für außereuropäische Geschichte*, 23: 76–91

A. Herrmann 1930, 'Marinus von Tyrus' in H. Wagner, *Gedächtnisschrift: Ergebnisse und Aufgaben geographischer Forschung* (Gotha), 45–54

A. Jones 2010, *Ptolemy in Perspective: Use and Criticism of His Work from Antiquity to the Nineteenth Century* (New York, London, Heidelberg), Introduction, xi–xv

H. Mžik and F. Hopfner 1938, *Des Klaudios Ptolemaios Einführung in die darstellende Erdkunde. 1. Teil, Theorie und Grundlage der darstellenden Erdkunde (Γηωγραφική Ὑφήγεσις I, II, Vorwort)*, (Vienna)

K. Nielsen 1945, 'Remarques sur les noms grecs et latins des vents et des régions du ciel', *Classica et Medievalia*, 7: 1–113

A. Podossinov 1991, 'Himmelsrichtung (kultische)', *Reallexikon für Antike und Christentum*, 15: 233–86

Ptolemaeus 1838, *Claudii Ptolemaei geographiae libri octo. Graece et latine ad codicem manu scriptorum fidem edidit Dr. Frid. Guil. Wilberg* (Essendiae)

---

[22] Michael Rathmann 2013, 2014 has arrived at similar conclusions for the *Tabula Peutingeriana*.

[23] Lately, Gautier-Dalché 2009, in the first chapter of his book (*La Géographie de Ptolemée en Occident (IVe–XVIe siècles)*, widely discusses whether Ptolemy's *Geography* was made in order to make the drawing of maps possible at any given moment, and that the maps that accompany the text in the manuscripts are not ancient. Despite the fact that he found clear hints that there were such ancient maps, Gautier-Dalché deems it impossible to distinguish whether these maps where drawn by Ptolemy himself or served as models for the maps actually preserved.

M. Rathmann 2013, 'Die Tabula Peutingeriana und die antike Kartographie', *Periplūs, Jahrbuch für außereuropäische Geschichte*, 23: 92–120
2014, 'Orientierungshilfen für antike Reisende in Bild und Wort' in E. Olshausen and V. Sauer (eds.), *Geographica Historica 31, Mobilität in den Kulturen der antiken Mittelmeerwelt, Stuttgarter Kolloquium zur Historischen Geographie des Altertums 11, 2011* (Stuttgart), 411–23
A. Stückelberger and G. Graßhoff (eds.) 2006, *Klaudios Ptolemaios: Handbuch der Geographie*, vol. i, Einleitung und Buch 1–4 (Basel)

CHAPTER II

Living According to the Seasons
The Power of parapēgmata

Gerd Graßhoff

Among the writings of the great astronomer, mathematician and geographer Claudius Ptolemy, his *Phases of Fixed Stars and Collection of Weather Changes, Phaseis* in short, stands out for its brevity, yet it contains complex information on the knowledge of ancient authorities about star phases and related weather changes that had been handed down from earlier times.[*, 1] It is thus very characteristic of the Ptolemaic way of working. As in his other writings, Ptolemy evaluated traditional, partly heterogeneous sources and generated new data on the basis of these sources, providing descriptions that were far superior to those of his predecessors. By transforming traditional source material in this way, Ptolemy succeeded in improving the accuracy of statements derived from it, thus removing ambiguities and making the data accessible, even to non-professional users. The data were intended to be applicable to daily life, providing as they do practical information on agriculture, meteorology and medicine, as well as having other weather-related applications.

In his *Phaseis*, Ptolemy systematized an ancient scientific problem: how can astronomical phenomena that have been reliably observed be used as indicators of important or even extreme meteorological events? How can one predict potentially influential weather events (such as winds, changes in the weather, changes in temperature or precipitation)? The history of our earliest civilizations indicates that the inhabitants of even the smallest settlements invested much effort in carrying out weather forecasts so that they could organize their own economic activities accordingly.[2] As with making modern methodological predictions, the main difficulty ancient

---

[*] I am grateful to all the following: TOPOI and the Einstein Foundation generously supported the project, with many good suggestions from TOPOI-fellow Liba Taub. Kerstin Rumpeltes provided a first translation and Margarete Simmons the final draft. Elisabeth Rinner assisted with the building of the database and the Mathematica calculations.
[1] In the following I shall extensively use Alexander Jones' translation: Jones 2004.
[2] Systematic surveys of several hundreds of Kreisgrabenanlagen or circular ditch constructions are a very active field of current research, e.g. Neubauer and Melichar 2010.

astronomers faced was observing phenomena that correlated closely with the events to be predicted and that thus allowed them to make reliable predictions. To predict these events, ancient people tried to identify the causes of weather signs, the occurrence of which made a later event likely.

However, close correlations alone do not provide sufficiently valuable rules for making predictions. Correlations are more reliable if the weather signs and projected events have been brought about by common causes (*epiphenomena*). In order to find these, one needs to look for suitable *epiphenomena* that cause both observable signs and correspondingly significant events. The events that are used to make predictions are, in most cases, antecedent, as it is their prognostic power that is important. In antiquity, predictions were sought for political, social and economic occasions that were relevant to the ruling authority and influenced by the weather. The prognosis is much more accurate if the typical causes of both the changes in the weather and the events can be identified and can manifest themselves as signs. One of the main causes of weather change was identified early on in the history of the systematic observation and use of astronomical phenomena: the sun's changing position in the sky over the course of a year. This annual change determines the changes in the seasons, which are, of course, significant for agricultural cycles such as sowing, soil maintenance, irrigation and harvesting. But how could the position of the sun in the sky be determined?

As it is practically impossible to observe the sun directly owing to solar radiation, observing the horizon points of sunrises and sunsets is one of the simplest ways of positioning the sun in the sky. If you record these points on the horizon, they can then be used as signs for the weather events that occur over the course of the next few years. Despite its disadvantages (measuring equipment had to be protected from damage or alteration, and one was restricted to a fixed place of observation in order to observe the signs), this measuring method had been in common use since the Bronze Age. However, there is a second, historically more significant, method that can be used to locate the position of the sun – almost without the need for any auxiliary device: one can observe the heliacal risings and settings of the brightest fixed stars. Once a year, the sun reaches a position among the stars when it rises on the eastern horizon shortly after a bright star has also risen on that horizon. The star is briefly visible to an observer during the last moments of dawn – this first brief appearance of the star is known as its heliacal rising. Then it is masked by the sun's light, and it disappears from view, remaining hidden from sight both day and night for a period of some days because of its proximity to the sun. As the sun moves from west to east among the stars, on this one day of the year the star emerges and becomes visible. The star's setting is the moment when it is visible for the last time, just after sunset. Almost all ancient cultures had been able to identify twenty-five to thirty-five bright stars whose visibility during the year served as the easiest and most reliable indicator of the seasons. These indicators were later called star phases, or *phaseis*.

Using the phases of the fixed stars to correlate seasonal events was so entrenched in ancient culture that it became the basis for omen literature in Mesopotamia in the second millennium BCE. The observation of weather signs, their systematic documentation and their combinatorial variations were the main branch of study of early scholars, who tried to transform the signs into more complex ones by combining them with other prominent events and using them for increasingly sophisticated forecasts. Indeed, the interpretation of astronomical phenomena lies at the methodological core of almost every ancient science. Other phenomena related to the fields of theology, economics and medicine were interpreted in a similar fashion.

It was probably during the eleventh century BCE that the compendium MUL.APIN was created.[3] Widely copied and distributed, it represents the astronomical basis for the creation of omens according to the star phases. At this time, astronomers were trying to discover how to synchronize the seasonal data of the star phases with the corresponding changing calendar. Although a schematic rule for relating the star phases to the dates of the calendar was not introduced until much later (in the fifth century BCE), the basic scheme of using the star phases remained the same throughout this period: a certain seasonal date becomes observable from the specific heliacal rising or setting of a bright star, thus determining the position of the sun among the stars over the course of a year. Uncommon events are then correlated with these data as well as other additional conditions. It was believed that these frequent joint correlations were caused by the seasons. Changes in the weather over the course of a year are highly suitable for making correlations with star phases. From 500 to 60 BCE, the Babylonians systematically recorded their observations of the daily weather conditions and astronomical events, and thus of changes in the weather. Their observations have recently been published in three volumes as *Astronomical Diaries and Related Texts from Babylonia*.[4]

Like many other areas of knowledge, information on predicting the weather spread to Greece in the fourth century BCE, where, by correlating star phases with a selection of weather phenomena, rules for predicting

---

[3] Edited by Hunger and Pingree 1989.   [4] Sachs and Hunger 1988–96.

Fig. 11.1 Miletus *parapēgma*, Antikensammlung Berlin

the weather over the course of the year were made visible by means of *parapēgmata*, a popular genre.[5] However, in Greece, unlike elsewhere, these star calendars, which were made of stone, were set up in public places. The best-preserved examples of these calendars date from the second century BCE and were found in Miletus (see Fig. 11.1). Their striking physical form gave the genre its name: *parapēgma*. The calendar entries for the star phases and weather phenomena were cut into marble in tabular form. Each description had a small hole adjacent to it into which wooden pegs could be placed to indicate the actual day, thereby informing the public of the day of the year and its corresponding weather predictions. The public's main concern, however, was probably not the forthcoming weather *per se*, but the consequential events connected to the weather, that is, when it would be best to travel, to go to market and to order goods for storage.

The dates were recorded in a number of ways, the crudest and simplest of which was by giving the sun's position in the zodiac; the user then had to work out the exact date himself. Calendar dates always referred to the star phases occurring on these dates, because the user in Greece faced problems similar to those of the authors of the older MUL.APIN: the

---

[5] Rehm 1941; for a modern edition of these texts, see Sauter and Lahoux 2009. For the genre, cf. Taub 2003.

sequences of the calendar months occurred so irregularly that they could not be used to make an accurate schematic mapping of the seasonal dates of the star phases. In most regions with no systematic chronological recording method, observing star phases was, therefore, the only reliable means of determining the dates of the seasons.

The second-century BCE *parapēgma* of Geminos, one of the earliest preserved compilations of rules for predicting the weather, begins with a listing of the sun's path through the constellation of Cancer.[6] Occasionally, the heliacal rising and setting of individual stars or constellations of stars are described. However, most of the information collected by Geminos refers to the sun's path through the constellations. Yet the position of the sun in the constellations cannot be observed directly. Only the stellar risings can indirectly fix the date and, therefore, the daily position of the sun in the zodiacal sign:

> The sun passes through Cancer over 31 days.
>
> ... 1st day, according to Kallippos, Cancer begins to rise; summer solstice, and it signifies.
> ... 9th day, according to Eudoxus, the south wind blows.
> ... 11th day, according to Eudoxus, the whole of Orion rises in the morning.[7]

Parapēgmata attribute weather phenomena to certain dates, with an important qualification: the weather phenomena are always attributed to specific authorities. In Geminos' *parapēgma*, the star risings also appear as part of the statement referring the reader to the relevant authority. Yet this form of presentation had two disadvantages: it is not transparent to the user, who has to figure out the calendar date for himself; and the star phases attributed to an authority might not be relevant to the user's geographical latitude. It was Ptolemy who, in his *Phaseis*, succeeded in solving this ambiguity.

Ptolemy does not describe his *parapēgma* as a collection of epistemically equivalent opinions from various authorities about weather regularities. Indeed, the blank, absolute statements made by the authorities cited contradict each other. For example, one author states that the end of the first spring storm is on the seventh day; another author assigns it to the twelfth. Rather, the reader is meant to interpret the statements of the work's authorities as possible scenarios for different causal circumstances, and these must

---

[6] The authorship and dating are controversial, but not critical for the argument; Evans and Berggren 2006.
[7] Evans and Berggren 2006, 231–2.

Living According to the Seasons 205

be transferred to the respective local conditions for which an authority has extensive knowledge.

Particularly in comparison with Geminos' work, Ptolemy's compilation reveals his greater astronomical expertise and his more profound knowledge of the causal relations that combine astronomical and meteorological phenomena. Ptolemy adapts the seasonal data handed down from earlier sources to the Alexandrian calendar, which, like the Julian, approximates to the solar year quite well owing to the introduction of a leap year every four years. Ptolemy succeeds perfectly in making all the relevant information explicit and transparent. Unlike Geminos, Ptolemy lists the star phases for different regions. He organizes the days according to the calendar month. The star phases are inserted as markers for the calendar, as well as the statements on weather phenomena attributed to the authorities. These star phases are related to specific latitudes, which are expressed by the information on the climates given in units of the length of the longest day:

> Thoth.
> 1. 14½ hours: the one on the tail of Leo rises. Hipparchus: Etesians cease. Eudoxus: rains, thunders, Etesians cease.
> 2. 14 hours: the one on the tail of Leo rises, and Spica is hidden. Hipparchus: weather-change.
> 3. 13½ hours: the one on the tail of Leo rises. 15 hours: the one called Capella rises in the evening. Egyptians: Etesians cease. Eudoxus: shifting winds. Caesar: wind, rain, thunders. Hipparchus: Apeliotes blows.
> 4. 15 hours: the last one of Eridanus sets in the morning. Callippus: stormy and Etesians cease.
> 5. 13½ hours: Spica is hidden. 15½ hours: the bright one of Lyra sets in the morning. Metrodorus: bad condition of air. Conon: Etesians diminish.[8]

In the few introductory pages to his treatise, Ptolemy hints at the enormous computational effort that was required to compute the star phases from different geographical latitudes.

> Here for convenience we will set out those times of phases that have practical application, for which sake it was necessary to make a prior working out of the computations also of all those other [phases], and [this] only for the more noteworthy bright fixed stars, along with the weather-changes that have been observed by our predecessors at the phases.[9]

In Ptolemy's work, the rules for predicting the weather are attributed to a total of twelve authorities. Ptolemy presents all the information in a

[8] Jones 2004, 5.   [9] Jones 2004, 1.

single format, thereby creating a common frame of reference from which comparisons can be made. Much more clearly than in Geminos' earlier works, the statements attributed to specific authorities do not contradict each other. The focus here is not to 'play off' one authority against the other. Ptolemy's aim is to maximize the information about weather regularities that occur under various circumstances by reworking the older statements of his selected authorities. As in all his other texts, Ptolemy's concern is a systematic one. It is true that the traditional rules for making weather predictions are attributed to the authorities alone and not, for example, to special conditions occurring at the stated geographical latitudes. But Ptolemy explicitly says so, even though it is widely known that, even at the same latitudes, weather phenomena differ. So his insight that the main factors for weather regularities are, nonetheless, the geographical latitude and its conjunction with the position of the sun is remarkable.

Ptolemy makes it clear that he has decided not to consider the subjective differences of the observer or any historical circumstances in his rules. They are, after all, only secondary factors. From a causal viewpoint, the fact that even two major factors – the cycle of the sun during the course of the year and the respective geographical latitude – alone cannot strictly validate the respective rule for predicting the weather is a little tricky to deal with.

> And I have recorded the weather changes of these and set them down according to the Egyptians and Dositheus, Philippus, Callippus, Euctemon, Meton, Conon, Metrodorus, Eudoxus, Caesar, Democritus, [and] Hipparchus. Of these, the Egyptians observed here, Dositheus in Cos, Philippus in the Peloponnese and Locris and Phocis, Callippus on the Hellespont, Meton and Euctemon at Athens and the Cyclades and Macedonia and Thrace, Conon and Metrodorus in Italy and Sicily, Eudoxus in Asia and Sicily and Italy, Caesar in Italy, Hipparchus in Bithynia, Democritus in Macedonia and Thrace.[10]

Ptolemy finishes his introduction by advising his readers to relate the details concerning the weather conditions given by the authorities to specific geographical latitudes:

> Hence one would particularly fit the weather-changes of the Egyptians to the lands around this parallel, that is, [the parallel] On which the longest day is 14 equinoctial hours, those of the Dositheus and Philippus [to the parallel] On which longest day is 14½ [equinoctial] hours; those of Democritus and Caesar and Hipparchus [to the parallel] On which the longest day is 15 equinoctial hours, those of Callippus and Eudoxus and Meton and

[10] Jones 2004, 23.

Euctemon and Metrodorus and Conon generally [to the parallels] on which the magnitude of the longest days extends from 14½ hours to equinoctial 15th.[11]

In Ptolemy's eyes, the information given by the authorities is not irrefutable; it is only the most reliable information available. Therefore, the way he has arranged the reported rules for predicting the weather indicates that he believes the weather events referred to by the various authorities should be understood to be events that take place on a regular basis, although they do not always occur. None of the statements on star phases relative to geographical latitudes is attributed to any authority. Ptolemy makes his own calculations. Ptolemy clearly noticed that the details of the occurring star phases in his various sources did not always agree with his own calculations. In such cases, Ptolemy prioritizes his own mathematical derivation of the astronomical situations. In other places, though, Ptolemy stresses that the observations on the dates of the star phases are not free from fluctuations. Changing weather conditions and different visibility conditions are just as responsible for these variations as other fluctuating local circumstances. However, he cannot compensate for these variations by making geometrical calculations. Ptolemy is unspecific about the kind of circumstances that he would consider for rules:

> Perhaps it would not be out of place also to give a summary of the count of the fixed stars together with the [count] of the collected phases for the purpose of checking those that have been missed through copying errors, and also [a summary of] the men who marked the relevant circumstances, and the lands where each happened to have made his observations, so that we can perhaps more appropriately fit similar delineations to the [lands that lie] on the same parallel.[12]

Ptolemy thus considers his set of rules to be a dynamic schema with a secure astronomical core. This core depends only on the geometrical conditions of the places of observation and their relation to the astronomical facts. However, these astronomical facts are no longer dependent on additional empirical observations. They apply to any future rule for predicting the weather. For Ptolemy, the listed specifications of the authorities represent the accumulated knowledge of Greek literature. Thus, it is not important for him to cite their statements word for word or with philological accuracy. In particular, he corrects the information on star phases handed down from earlier times and attempts to transfer the specified weather events to the

---

[11] Jones 2004, 23.   [12] Jones 2004, 22–3.

most appropriate days in the Alexandrian calendar, which he considered the most user-friendly of the calendars then in use.

It is remarkable that Ptolemy did not add any of his own observations, at least not discernibly. True, the number of weather observations attributed to the Egyptians is disproportionally large, and it is quite possible that a simultaneous recording system of the weather conditions in Alexandria is hidden behind all this information. Ptolemy provides no information in his text about this peculiar feature. However, he does make it clear that the list can serve future users as a basis for choosing the appropriate geographical latitude and then adding additional rules solicited by further observations.

Ptolemy does not reveal how he thinks the project of the future addition of rules should be tackled methodologically. His list of authorities shows that no contemporary casual observer would have qualified as an authority. From this we may infer that Ptolemy was reluctant to state rules for predicting the weather on the basis of a few seasonal observations. However, the aforementioned quotation shows that Ptolemy assumed that the authorities themselves had made the observations. Yet, being an experienced astronomer, it was clear to him that such observations could not have been made simply on the basis of the weather reports from just one year. The complete coverage of rules for predicting the weather requires several years of observation in order to interpret the astronomical events and the regularly occurring annual changes in the weather. No out-of-the-ordinary weather events (such as hailstorms or a particularly severe thunderstorm, which occurred as one-off events over the village or town) would have been included in this list. In order for a weather event to be seen as a regular occurrence, it would be important that such events reoccurred around a particular date of the solar year, so that it could be identified as being the main cause over a longer period of time. Such methodology does not need to be particularly sophisticated. What it does need is time and relentless commitment. One could, for example, allow the inclusion of a rule if the corresponding weather event over a period of twenty years coincided at least two-thirds of the time with the corresponding reoccurrences of the sign. The qualification of weather predictions requires the systematic observation of the phenomena over the entire course of a year for a period of several years. It is doubtful, then, that all authorities would have been able to realize an observation programme that stretched over several years; hence, we are still in the dark as to the origin of their sources.

Rinner and I have broken down Ptolemy's sentences into their logical components and transferred these to a computerized database. The rules for predicting the weather are in a canonical form (which was what Ptolemy

distribution of bright stars in Ptolemy's Phaseis

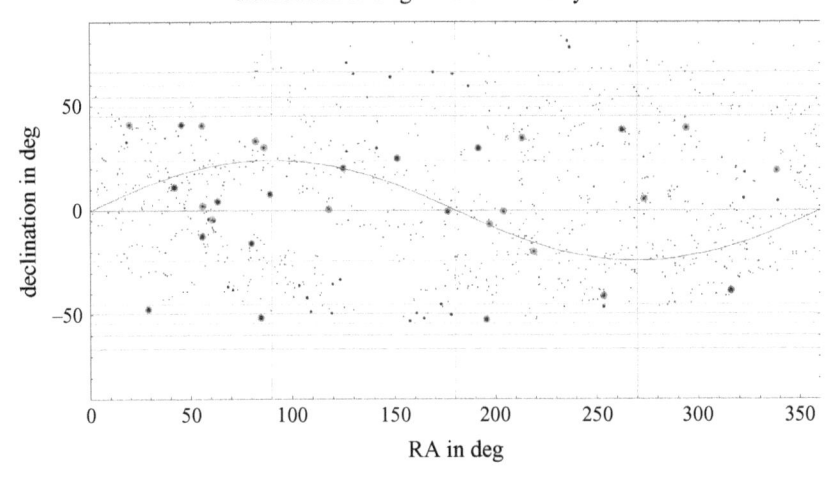

Fig. 11.2 Ptolemy's selection of 30 bright fixed stars using the equatorial coordinate system

intended), in that the heliacal risings and settings of the fixed stars were attributed to a specific day in the Egyptian calendar as a (complex) protasis of a rule with an apodosis to a specific authority followed by a weather description. The logic of this rule is typical for signs and categorically different from material conditionals. The whole rule depicts *epiphenomena* as signs. There is no weather event that is not attributed to a specific authority. In total, the text of Ptolemy's *parapēgma* records almost a thousand rules for predicting the weather. The internal relationships between these rules are so complex, and the resources used by Ptolemy were revised in such a sophisticated fashion, that a detailed investigation of the *Phaseis* will appear in a separate publication.[13] In this context, an overview of the findings will have to suffice.

To calculate the star phases, Ptolemy selected the thirty brightest fixed stars. According to his own words, he selected fifteen stars of the first magnitude and fifteen of the second. As shown in Fig. 11.2, the bright stars are evenly spread over the celestial sphere, occurring only slightly more predominantly in the northern hemisphere. Ptolemy did not have much room for manoeuvre in his choice of stars, as he had to include most stars cited by the older authorities. Historically, one could only observe the heliacal risings and settings of the brightest stars, which are frequently

---

[13] Cf. Jones 2009 concerning the relationship between Geminos' *parapēgma* to the *parapēgma* in Ptolemy's *Phaseis*.

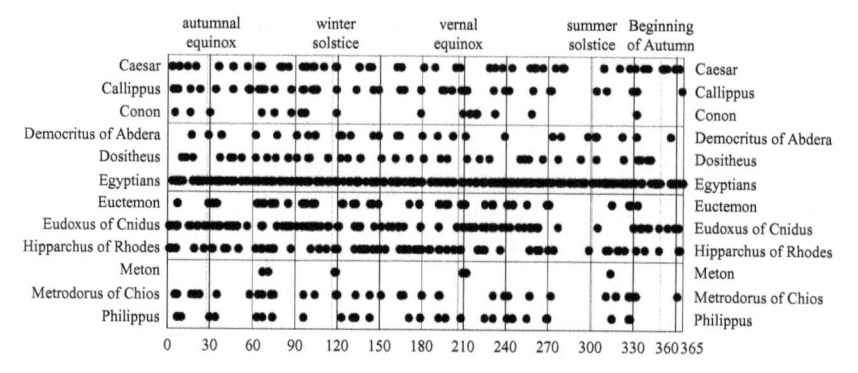

Fig. 11.3 Weather events as a function of the Alexandrian calendar (in days) and divided between the cited authorities

located close to the ecliptic (cf. the curve in Figure 11.2). The northern and southern boundaries are striking, as they incorporate the very bright southern stars, even though these were not visible from Athens. The fine horizontal lines in the diagram indicate the northern circles. North of the circles, the stars never set below the horizon, while the stars south of the limiting circles are not visible above the horizon. One gets the impression that Ptolemy did not use the extreme cases of the northern and southern boundaries only to indicate the dates of the seasons: they could also have illustrated the differences in geographical latitude. The distribution of the other bright stars along the ecliptic is sufficiently homogeneous to have enabled astronomers to use the star phases throughout the year in not too large intervals.

The Ptolemaic *Phaseis* that have been fully recorded in our database reveal a complex picture of the distribution and selection of the meteorological events recorded by the individual authorities according to the seasons. (The database can be consulted interactively with applets in the electronic document format CDF.) Fig. 11.3 shows the distribution of the recorded events attributed to the different authorities (along the y axis), with regard to the specific date of the year. The months of the Egyptian calendar are indicated on the x axis in intervals of thirty days (as Ptolemy calculated them in his text). The vertical grid lines indicate the date of the new month; the red lines mark the beginning of the seasons. The Ptolemaic year, which corresponds to the end of August in our calendar, begins just before the mark signifying the start of autumn in the first month of the Egyptian year. The diagram depicts all the recorded events as black dots. With the exceptions of Caesar (first century BCE) and Hipparchus (second century

Living According to the Seasons　　211

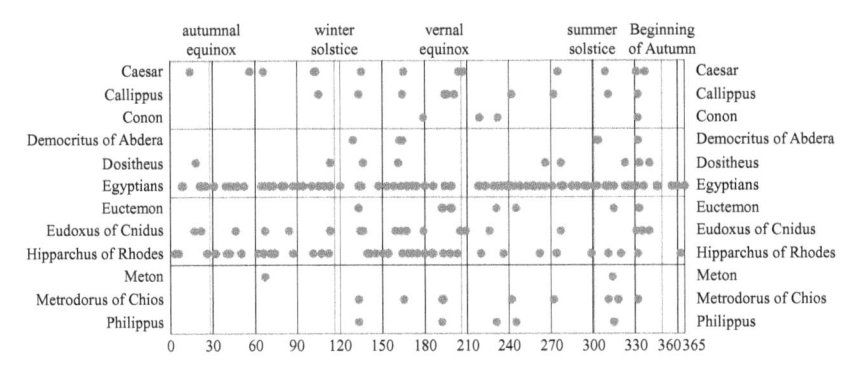

	autumnal equinox	winter solstice	vernal equinox	summer solstice	Beginning of Autumn	
Caesar	•	••	• • • •	•	• ••	Caesar
Callippus		• •	••	•	•	Callippus
Conon			• • •		•	Conon
Democritus of Abdera		• •		••	••	Democritus of Abdera
Dositheus	•	•• •	• •	•• •	• ••	Dositheus
Egyptians	• ••• •••••	• ••••••••••	••••••••••••••	•••••• ••• • ••	Egyptians
Euctemon		•	••	• •	•	Euctemon
Eudoxus of Cnidus	•• • • • •	• •••	•• •	•	••	Eudoxus of Cnidus
Hipparchus of Rhodes	• •••• ••	••• •••• •••	•• ••	•• •	•• •	Hipparchus of Rhodes
Meton		•			•	Meton
Metrodorus of Chios		• • •	•	•	••	Metrodorus of Chios
Philippus		•	• •	•	•	Philippus
	0　30　60　90　120　150　180　210　240　270　300　330　360 365					

Fig. 11.4 Distribution of named, directional winds

BCE), most of the authorities lived in the fourth or third centuries BCE. This diagram shows that there were nevertheless significant differences between the observations of the individual authorities on the seasonal distribution of the stars and the density of the described events.

The oldest quoted authority, Meton, lists very few events at the turn of the seasons. What is striking is the enormous relative frequency of meteorological events recorded by the so-called 'Egyptians'. Even taking into consideration the fact that Ptolemy worked in Alexandria and thus could easily access information on Egyptian sources at the city's library, the enormous density of their recorded weather events is impressive, particularly as the actual weather conditions were fairly stable. As yet, we have not been able to accurately determine the reason for this large number of recordings. However, it seems plausible to assume that these data included local records from Alexandria, which were selected according to an atypical set of criteria.

The various authorities did not always record the same meteorological events; their choice depended on the date and how it related to their geographical location. There are, for example, winds that blow from specific directions and that regularly occur during certain weather conditions (for example, the winds known as the *Apēliōtēs, Euros, Notos, Boreas* or *Aparktias*), which were referred to by many, but by no means all, of the authorities.

Amounting to 460 out of 946 total events, winds account for almost half of all the recorded weather events, and thus make up by far the most commonly occurring meteorological event. The relative proportion of wind data is strikingly high in the case of Hipparchus. As Hipparchus is assumed to have lived on the island of Rhodes, one might argue that the high proportion of wind data simply reflects his location. As Rhodes was a major trading centre, its inhabitants and especially its sailors would have been particularly dependent on forecasts for changing wind patterns. However, one should not overlook the fact that, besides having a direct impact (such as a change in strength or direction), changes in winds usually occur along with other important phenomena, such as temperature changes, precipitation or a period of dry weather. Wind changes are harbingers of general changes in weather configurations. So, given the frequency with which wind events are mentioned by all the authorities, it is much more likely that winds were generally considered to be important indicators of imminent weather changes in general. Furthermore, anyone can observe wind changes. Wind terminology is easy to understand, and wind patterns are predictable enough to be systematically recorded, and this must surely have been the basis on which weather predictions were formulated. A review of the weather observations listed in the Babylonian diaries confirms this interpretation; if the wind data did not indicate more general weather phenomena and their combined effects, observing the winds so extensively would have made little sense for the non-seafaring nation-state of Babylonia.

Fig. 11.5 plots the frequency of weather types in the form of a pie chart. The orange-toned right-hand side of the disc represents the proportion of wind events. The colours of the respective sectors correspond to the proportional numerical size of the basic types of weather: winds, precipitations such as thunder, air temperature, flora/fauna/the seasons, and so on. The sections themselves are divided into three or four groups of subclasses of events that stretch radially outwards, each of which expresses a classification hierarchy. In the inner zone, general terms for particular weather phenomena such as 'winds in general' or 'weather signs' are indicated; the more outward sections show the frequency of the respective subcategories. The catalogued expressions have been coded in lower-case letters, which are unfortunately illegible in the overview. However, the density of their distribution shows that the recorded weather phenomena occurred with a highly diversified number of other detailed phenomena. It also reveals that Ptolemy was not interested in making general weather predictions. His objective was to restate the specific weather descriptions of ancient authorities, even if their appearance was unique.

The distribution of meteorological events related to changes in the patterns of flora and fauna indicates that winds were used to forecast other seasonal events. In Fig. 11.6, bird-watching data, such as migrations, were recorded. It is clear that these data were used to indicate the beginning of

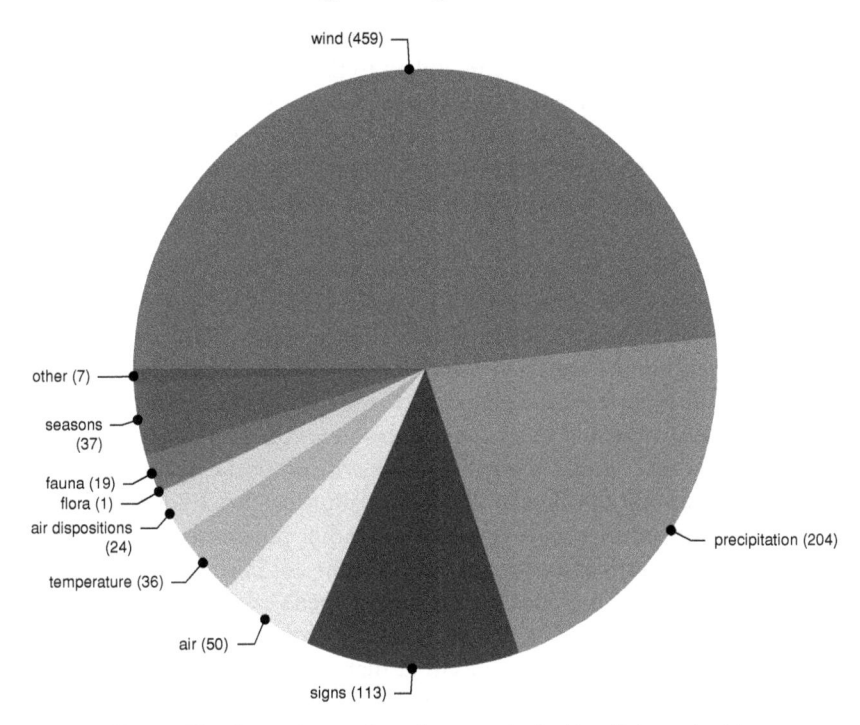

Fig. 11.5 Number and type of weather protasis cited by all the authorities

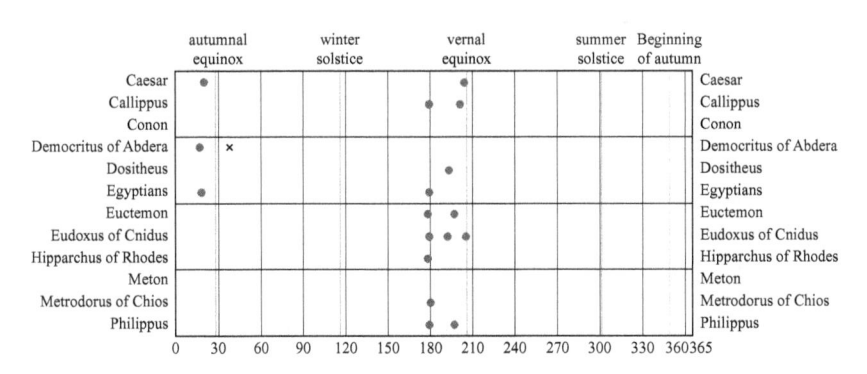

Fig. 11.6 Flora (×) and fauna (dots) events

spring and, to a lesser extent, the beginning of autumn. The occurrence of migrating birds in the corresponding regions is not an interesting event *per se*; rather, it marks the end of summer and thus the time when crops need to be harvested. Thus, Ptolemy's *parapēgma* is not just a list of the *epiphenomena* of signs indicating wind changes. The *epiphenomena* are themselves signs for events that are linked to the various winds.

In a recent controversy between Daryn Lehoux and Bowen and Goldstein over the empirical content of weather regularities, Lehoux questioned their direct empirical nature (Lehoux 2004, 235). According to Bowen and Goldstein 1988, who represent the majority opinion concerning the empirical content of *parapēgmata*, rules predicting the weather show that there is a connection between an observable star phase (P) and a simultaneously occurring, observable weather situation (W). The rule 'P -> W' is thus a generalization of empirically observed, simultaneous events of the same nature. Presenting the argument that, during periods of bad weather, star phases cannot be observed, Lehoux rejects the claim that such a form of simultaneously occurring event can ever be observed. As an alternative, he suggests that one consider the paradigm as a theoretical exercise in mathematical astronomy to calculate the path of the sun that leads to heliacal star risings and settings with co-existing weather phenomena ('sun -> P and W'). The marker of a *parapēgmata* had thus not been set according to the observations of star phases but had instead been calculated on the basis of astronomical data. Using the marker, the observable, simultaneously occurring weather events would occur at that date.

The analysis of Ptolemy's *Phaseis* has revealed a picture that differs from both the above accounts. The theoretical calculation and the empirical basis are indirectly related to each other. They are expressed in epiphenomenal fashion ('P <- sun -> W'). Ptolemy himself did not observe the weather regularities; he relied instead on the observations of the authorities and converted them into a common format. Ptolemy did not create a general theoretical model of weather forecasts by transforming the historical data to the identified positions of the sun. He used geometrical astronomy to calculate the possibility of the star phases being a direct function of the sun's position during the course of the year and used the horizon at the locations of these specific geographical latitudes, independently of the authorities. Therefore, he did not create new weather regularities in the form of 'A shows B'. Rather, he standardized traditional knowledge about the occurrence of significant weather events by representing the causal relationships of signs as A <- sun -> B. Ptolemy understands the relation '->' as a causal relationship. In the *epiphenomena* 'A <- sun -> B' there is no direct causal link between A and B, as both Geminos and Ptolemy emphatically reiterated.

The signs could be used for different purposes by applying additional knowledge about the effects or *epiphenomena* of B. It was not only professional astronomers who dealt with *parapēgmata*. Indeed, one of the main purposes of this genre was to convey knowledge about specific seasonal phenomena to the layman. Star phases had been used for this objective for centuries. They could be synchronized in remote areas of habitation through simple observation alone. By counting the days that had passed, one could establish the corresponding seasonal date by means of a plug-in calendar, without knowing that the star phases and weather had been observed simultaneously.

The importance of the weather phenomena as well as the significance of the more abstract recordings of significant dates underline the fact that the weather predictions were the users' primary concern. Naturally, this depended on the practical consequences of the events that were of particular interest to the user, for example, the most appropriate dates for seafaring and trading, astrological predictions, medically significant circumstances that announced a cure or an impending failure, or the management of dates relating to rural life that were important for agricultural activities.

Parapēgmata were not texts to be used for practical applications *per se*. They provided something that I would like to call 'hybrid practical knowledge': *parapēgmata* communicated knowledge about signs that occurred regularly. Other events linked to the weather in chains of signs were predicted but not listed in the *parapēgmata*. Clearly, information about such signs was either handed down orally or recorded in separate compendia. For example, sailors knew which winds would allow them to embark on the dangerous route from Alexandria to Rome. In essence, *parapēgmata* conveyed building-blocks of knowledge with which people could make decisions based on their practical navigational knowledge.

REFERENCES

C. Bowen and B. R. Goldstein 1988, *Meton of Athens and Astronomy in the Late Fifth Century* BC (Philadelphia)

J. Evans, J. L. Berggren 2006, *Geminos's Introduction to the "Phenomena": A Translation and Study of a Hellenistic Survey of Astronomy* (Princeton)

H. Hunger and D. Pingree 1989, *MUL.APIN: An Astronomical Compendium in Cuneiform*, Beiheft 24 (Horn, Austria)

A. Jones 2004, 'A Study of Babylonian Observations of Planets near Normal Stars', *Archive for History of Exact Sciences*, 58, 6: 475–536

(ed.) 2009, *Ptolemy in Perspective: Use and Criticism of His Work from Antiquity to the Nineteenth Century*, vol. XXIII (Dordrecht)

D. Lehoux 2004, 'Observation and Prediction in Ancient Astrology', *Studies in History and Philosophy of Science*, 35: 227–46

W. Neubauer and P. Melichar 2010, *Mittelneolithische Kreisgrabenanlagen in Niederösterreich: Geophysikalisch-Archäologische Prospektion – Ein interdisziplinäres Forschungsprojekt*, vol. LXXI. Mitteilungen der Prähistorischen Kommission (Vienna)

A. Rehm 1941, *Parapēgmastudien* (Munich)

A. Sachs and H. Hunger 1988–96, *Astronomical Diaries and Related Texts from Babylonia*, vol. CCIC. Denkschriften der philosophisch-historischen Klasse (Vienna)

J. Sauter and D. Lehoux 2009, 'Astronomy, Weather and Calendars in the Ancient World: *Parapēgmata* and Related Texts in Classical and Near-Eastern Societies', book review, *Journal of Astronomical History and Heritage*, 12: 84

L. Taub 2003, *Ancient Meteorology* (London)

CHAPTER 12

Auctoritas *in the Garden*
Columella's Poetic Strategy in De re rustica *10*

Christiane Reitz

L. Iunius Moderatus Columella's didactic poem is part of a voluminous work on agriculture.[1] In considering the practical applicability of such a text to contemporary Roman life, I shall make some assumptions about the intended readership of this work, and I shall try to integrate the poetical advice on gardening into the overall strategy of the *De re rustica*. I assume in the first place that the instructions the author offers for farming and for running a great estate are based both on expert knowledge and on his own experience. I also assume that these instructions are addressed to the Roman upper class.[2] The typical member of this class and reader of Columella's work would be a landowner with estates either in Italy or abroad. The ideal reader, so I argue, is not a character like the portrait the younger Pliny paints of himself, a cultured urban lawyer and intellectual seeking a luxurious hideaway and displaying his exquisite taste in the architectural qualities of his villas. Columella appeals to his reader as the one person responsible for the economic welfare of his estate and the people living on it and living from it. He is well aware of the multiple obligations that might prevent the landowner from looking after all the details of his farming business, but, in his advice, one of the ways to succeed is through the frequent presence and the well-founded and well-informed interest of the landlord. This becomes evident when he gives advice about building a house on the estate. It is important, so he argues, that comfort and beauty

---

[1] The date of the last two books – are they a later addition? – and the authorship of the *Liber de arboribus* will not be discussed in this chapter. For a short introduction to Columella with equally short bibliography, see Reitz 2013 and 2006. On Roman gardens the classical study is Grimal 1984. For the social and economic aspects, see also Frass 2006. On the archaeological evidence and spacial conception of the Roman garden, see von Stackelberg 2009.
[2] But agricultural texts were also intended to be read and were actually consulted by the *vilicus*, the person responsible for the smooth running of the farming business. This is convincingly argued by Christmann 2003.

be provided, as otherwise the landowner's wife may not be persuaded to visit the country as often as necessary.[3]

Whether *De re rustica* was actually used by contemporaries as a practical handbook is a question that we cannot answer with certainty, but that the text exudes practicality and applicability throughout is evident, and becomes even clearer when we look at the literary and stylistic devices the author uses in the prose part of *De re rustica*.[4] I shall therefore give some examples of this before I turn to the didactic poem.

In the proem and throughout the work Columella explains his procedure: while leaning on the established written authorities[5] in agriculture and giving attention to and valuing the traditional rules and prescriptions of his elders, it is important, so he argues, to test and introduce innovations on the basis of one's own experiments. In the proem and at other programmatic points of the work – mainly prompted by a change of topic – we find statements like the following:

> quippe eiusmodi scriptorum [Cato, Saserna, Hyginus, Celsus, Iulius Atticus etc.] monumenta magis instruunt quam faciunt artificem. Usus et experientia dominantur in artibus, neque est ulla disciplina, in qua non peccando discatur.
>
> for the treatises of such writers instruct rather than create the craftsman. It is practice and experience that hold supremacy in the crafts, and there is no branch of learning in which one is not taught by his own mistakes. (1.1.15–16)[6]

After discussing his main predecessors in agricultural writing, Columella warns his addressee Silvinus against leaning too heavily on tradition. This implies that Columella quotes the older literature where it seems useful for his argument, but that he stresses the importance of experience, and especially of personal inspection. In this context he also explicitly criticises the so-called authorities.

When discussing the fertility of the soil, Columella says:

> Plurimos antiquorum, qui de rusticis rebus scripserunt, memoria repeto quasi confessa nec dubia signa pinguis ac frumentorum fertilis agri prodidisse dulcedinem soli propriam... colorem nigrum vel cinereum. Nihil de ceteris ambigo, de colore satis admirari non possum cum alios tum etiam Cornelium Celsum, non solum agricolationis sed universae naturae prudentem virum, sic et sententia et visu deerrasse.

---

[3] See below, p. 220 (at quotation to 1.4.8.) [4] Columella, *Rust.* 1–9.
[5] On the interplay between authority and originality in technical writing, see Doody 2007, 181–2.
[6] All translations are from the edition listed under Columella 1941 in the References below.

Auctoritas *in the Garden* 219

> I recall that very many of the ancients who have written on agricultural topics have laid down as acknowledged and unquestioned evidence of fat and fertile grain-land the natural sweetness of the soil, its growth of herbage and trees, and its black or ashy colour. As to the other points I have no doubt; but in the matter of colour I cannot marvel enough, not only that other writers but especially that Cornelius Celsus, a man of discernment not merely in husbandry but also in nature as a whole, went so far astray, both in his thinking and in his observation. (2.2.14)

Inspection and observation are, in this case and generally, the basis for productive agriculture. Books are no alternative to actually touching and grasping the matter at hand:

> Nam perexigua conspargitur aqua glaeba manuque subigitur, ac si glutinosa est, quamvis levissimo tactu pressa inhaerescit et picis in morem ad digitos lentescit habendo ut ait Vergilius...

> For a clod is sprinkled with a little water and kneaded in the hand, and if it is viscous and cohesive when firmed with the slightest touch and, in the manner of pitch, is shaped to the fingers in handling, as Virgil says...[7] (2.2.18)

The praise of experiment gains more force through a quotation from Virgil. Does it not seem paradoxical that Columella first questions the authorities, but then cites Virgil? A similar procedure can be observed when a quotation from the *Georgics* is followed by the instruction never to neglect personal experience and innovative techniques. Columella speaks about the influence of climate:

> Haec autem consequemur, si verissimo vati velut oraculo crediderimus dicenti:
>
> ventos et proprium [*varium* codd. Verg.] caeli praediscere morem
> cura sit ac patrios cultus habitusque locorum
> et quid quaeque ferat regio et quid quaeque recuset;
>
> nec contenti tamen auctoritate vel priorum vel praesentium colonorum, nostra promiserimus exempla novaque temptaverimus experimenta.

> These results we shall attain, moreover, if we pay heed, as to an oracle, to the truest of poets, who says:
>
>> 'Be it our care to learn betimes the winds and moods of heaven,
>> To learn the tillage of our sires and nature of the place,
>> What fruits each district does produce and what it does refuse.'[8]

---

[7] Virg., *G.* 2.250.   [8] Virg., *G.* 1.51–3.

> And yet, not content with the authority of either former or present-day husbandmen, we just hand down our own experiences and set ourselves to experiments as yet untried. (1.4.4)

Virgil is omnipresent throughout the whole work, holding a special position compared with the other agricultural writers that Columella quotes, be they older or contemporary; yet even his authority does not suffice for the purposes of practical application, for which personal experience by trial and error is an essential guide.

There is another characteristic feature of Columella's way of thinking that is important for the question of applicability. Every so often the reader is informed of the importance of his personal presence in the country, on the estate. This starts with the choice of which land to purchase. The estate should not be too far from Rome, in order to make daily visits possible. Otherwise, the slaves (and the prospective heirs) may squander the property in no time. Continuous inspection, taking care of the servants, and attention to the seasons of the year and the duties that go with them are points that Columella stresses again and again. A nice example is the passage mentioned above, about the furnishing of the house. It should be equipped in such a way that the wife of the *pater familias* feels at home too:

> Pro portione etiam facultatum quam optime pater familiae debet habitare, ut et libentius rus veniat et degat in eo iucundius. Utique vero, si etiam matrona comitabitur, cuius ut sexus ita animus est delicatior, amoenitate aliqua demerenda erit, quo patientius moretur cum viro.

> Furthermore, the master should be housed as well as possible in proportion to his means, so that he may more willingly visit the country and find more pleasure in staying there. And especially, if his wife also accompanies him, since her disposition, like her sex, is daintier, she must be humoured by amenities of some sort to make her stay more contentedly with her husband. (1.4.8)

Keeping this concern with practical applicability in mind, we might now ask which role the garden poem plays in the work.[9] This work, as we have seen, is not devoid of a certain *ornatus*.[10] The poetic quotations and the author's self-confident stance should certainly appeal to an educated reader with a measure of *urbanitas*. In the prose *praefatio* to Book 10,

---

[9] On the interrelationship between technical prose and didactic poetry, generally and especially in late antiquity, see Formisano 2005. On the overall structure of the *De re rustica*, see Henderson 2002 and Pagán 2006.

[10] The most recent treatment of Columella under the aspect of self-fashioning is by Fögen 2009. On style, see especially 158–71, on the garden poem 182–5.

Columella explains to the addressee, a certain Silvinus,[11] his reasons for tackling the topic of gardening. The tone of the *praefatio* oscillates between modesty about the importance of the topic[12] and high ambition. The argument seems somewhat contradictory. First, it is stressed that gardening is nowadays much more important than in older times: *diligentius nobis, quam tradiderunt maiores, praecipiendus est* (10, praef. 3), 'the cultivation, therefore, of gardens, since their produce is now in greater demand, calls for more careful instruction from us than our forefathers have handed down.' This is important when we consider Columella's relation to his authoritative sources. Second, he explains that, strictly speaking, the theme of gardening should have been dealt with in prose. Yet two reasons, he points out, spoke for a poetic version of the topic, which the author of the *Georgics* left out of his agricultural poem: the request of the addressee (*frequens postulatio tua*) and, as a second reason, the summons of the *vates* himself: *neque enim aliter istud nobis fuerat audendum, quam ex voluntate vatis maxime venerandi* (10, praef. 3), 'for indeed I ought not to have ventured on the task, were it not in compliance with the wish of that greatly revered poet'.

While displaying such modesty as regards the size of the task, Columella as didactic poet speaks with self-assurance: not without hope of a prosperous result (*non sine spe prosperi successus*) did he tackle his subject matter (*tenuem admodum et paene viduatam corpore materiam*, 'material that is very meagre and almost devoid of substance' (10, praef. 4)); and the diligence he has applied (*elucubravimus*) would not be a disgrace to the preceding books (*si non sit dedecori prius editis a me scriptorum monumentis* (10, praef. 5)). If we look at the final verses of the tenth book, we find a further programmatic statement, which culminates in another quotation from Virgil:[13]

> hactenus hortorum cultus, Silvine, docebam
> Siderei vatis referens praecepta Maronis,
> qui primus veteres ausus recludere fontes
> Ascraeum cecinit Romana per oppida carmen.
>
> Thus far, Silvinus, I have sought to teach
> The cult of gardens and to call to mind
> The precepts taught by Maro, seer divine,

---

[11] The identity of Silvinus has been much discussed and must remain an open question. See Columella 1996, 95 and Fögen 2010, 157. That both the personality and the name may be fashioned after the addressees of Varro's *De re rustica* seems plausible.
[12] Praef. 10.4: *multa membra – tamen exigua*: 'many parts but smallish'.
[13] Virg., *G.* 2.175–6.

> Who first dared to unseal the ancient founts
> And sang through Roman towns the Ascraean lay.
> (10.433–6)

So why has Columella decided to write the final book in poetry, and how does this relate to the practical purposes of the work as a whole? In her detailed commentary, Francesca Boldrer deals with the botanic details and carefully lists the many parallels that illuminate Columella's intertextual technique. She also emphasises the 'virgilianesimo' of our author, but, on the other hand, she argues that there are several different poetic models for Columella. Columella, so she convincingly claims,[14] combines Virgilian elements with his specific goals of *variatio* and *abundantia* and with a sometimes sententious style, while also introducing elements of *sermo cottidianus*. Therefore, further poetic intentions may lie hidden behind the statement *vatis referens praecepta Maronis* (10.433).

In the following reading of the poem, I shall try to demonstrate that a specific persuasive strategy can be identified in our author's approach to his poetic models.

The structure of the tenth book parallels the overall arrangement of the whole work, and its concern with practical applicability. The first topic concerns the appropriate choice of terrain for gardening. At the beginning, the style is matter of fact. The list of indicative plants that show the ground's fertility is technical, but the catalogue is arranged in a rhetorically attractive way and is carefully structured (*principio – tum – quoque*). The fourth point introduces a new subject: the hopeful gardener should not decorate his garden with the statues of Greek masters, but with a traditional wooden Priapus.

> Nec tibi Daedaliae quaerantur munera dextrae,
> nec Polyclitea nec Phradmonis, aut Ageladae
> arte laboretur. Sed trunco forte dolato
> arboris antiquae numen venerare Priapi
> terribilis membro, medius qui semper in horto
> inguinibus puero, praedoni falce minetur.

> Seek not a statue wrought by Daedalus
> Or Polyclitus or by Phradmon carved
> Or Ageladas, but the rough-hewn trunk
> Of some old tree which you may venerate
> As god Priapus in your garden's midst

---

[14] Columella 1996, 22–6.

Who with his mighty member scares the boys
And with his reaping-hook the plunderer.
(10.29–34)

This instruction is witty both in itself and in its argument. Starting with Daedalus, there follows an elaborate enumeration, listing three Greek proper names of famous sculptors in one verse – *nec Polyclitea nec Phradmonis, aut Ageladae*. Priapus, on the other hand, is characterised as *terribilis membro, medio qui semper in horto inguinibus puero, praedoni falce minetur*. The end of the first precept ends with a striking effect: educated and learned versification stands in contrast to the verses on old-fashioned Priapus, which are equally elegantly composed with chiasmus and alliteration.

The following passage deals with the seasons and the appropriate moment to sow different flower seeds (*cultus et tempora*). In the introductory invocation of the Muses, the tone is much more solemn than in the preceding passage. The diction is close to Virgil's *Eclogues*, and contains poetological terminology too. In *Pierides tenui deducite carmine Musae* (40), *deducere* might be interpreted as a term for writing (refined) poetry, and the stress on *tenue carmen* as the stylistic model hints at Horace (*Epist.* 2.1.225) and at Ovid (*Met.* 1.3). This poetological discourse is taken up again in the central passage of the poem (215–29) where, in the middle of his praise of spring-time, the poet calls himself to order. There he turns against singing as if inspired by Apollo and Bacchus, and asks his Muse to bring him back, *parvo gyro* and *gracili filo*, in small compass and with slender thread, to his subject matter. We shall come back to this later.

After the introduction comes the gardener's calendar, starting in autumn after the vintage. The elaborate periphrasis of the season refers equally to the astronomical[15] circumstances and the details of wine-making by the use of a nice allegorical personification of Autumn, and sets the gardener his first task: digging, either with or without watering the soil, depending on weather conditions. Diction remains at a high level, without technical vocabulary apart from *pala* (45), but this technical term is ennobled by the juncture *robore palae*. From this rather elevated level of an astronomical and poetical description of the season, the text then moves, in a seemingly associative aside, to another topic. It is important, so Columella points out, not to miss the right time for digging (58). This is in fact pragmatic advice, though uttered in rather enigmatic phrasing, but the reason given leads to a mythological explanation. The earthy soil is not our mother who would

---

[15] On the treatment of time in Columella in general, see Pagán 2006, 27–8, who also gives a list of dates designated by constellations.

have a right to be treated with the care and reverence due to a mother. The earth-born race of Prometheus was destroyed by the deluge – so Columella in the manner of Ovid – and we, the digging gardeners of our time, are creatures of Deucalion. That this excursion into Ovidian mythology is at least partly ironic and not wholly serious is made evident in the way the author then abruptly turns from myth to the matter at hand:

> [nos] . . . Deucalioneae cautes peperere – sed ecce,
> durior aeternusque vocat labor: heia age segnis
> pellite nunc somnos, et curvi vomere dentis
> iam viridis lacerate comas, iam scindite amictus.
> Tu gravibus rastris cunctantia perfode terga
>
> [us] . . . rocks by Deucalion torn
> From mountain heights brought forth. But lo! a task
> Harder and endless calls. Come! drive away
> Dull sleep, and let the ploughshare's curving tooth
> Tear earth's green hair, and rend the robe she wears,
> With heavy rakes cleave her unyielding back[16]
> (10.67–71)

The author calls himself to order and gets back from Ovid to Virgil's *Georgics – durior aeternusque vocat labor*. But here, too, an undertone may be heard. As if encouraged by the theme of sleeping and rising, the act of digging and ploughing is described in erotic language: *iam viridis lacerate comas, iam scindite amictus.*[17] Neither metaphor, of hair and of garment for foliage and first flush, is unique, so a metaphorical reading is just possible, but the abundance of such metaphors (70), and the following *cunctantia (terga)*, makes the erotic subtext quite clear.

It is striking that the next passage, on manure, abounds in technical terminology and detailed precepts (76–93).

The reward for hard work and punctual sowing will be the rich flora in spring, which is listed in a long catalogue. The many names of flowers are ordered by being subsumed under their aesthetic values, such as colour, habitus – using comparisons with parts of the human body – and scent. The catalogue then shifts to edible plants, stressing the taste and effect of medicinal plants and aphrodisiaca. This links to a list of vegetables, where

---

[16] It should be noted that in the German translation by Will Richter (under Columella 1981–3 in the References below) the rendering obscures the evident sexual double entendre.
[17] For violence as part of the sexual act, see e.g. Ovid, *Am.* 1.5.13. For the (metaphorical) vocabulary, see Adams 1982, 145–55.

the cabbages form the climax. The many species of cabbage are enumerated in a geographically ordered catalogue, describing a circle south- and southeastwards around Rome (127–39). This reminds one of a conventional catalogue of troops in heroic epic, starting with the catalogue of ships in the Iliad.[18]

The next passage about the following sowing and care of the plants uses mainly medical terminology. This had already been prepared by the last entry in the vegetable catalogue *mater Aricia porri* (139). We hear of *gravida terra, concepto semine, partum, suboles, pullulat*. The earth is characterised both as child-bearing mother and as the object of care, in need of food and drink (140–54).

Medical language forms the basis for the next passage. There, after an introduction with a periphrasis for the time of the year in an astronomical-mythological tone, the act of sowing is described as part of a cult, as a ritual (155–65):

> Alma sinum tellus iam pandet adultaque poscens
> semina depositis cupiet se nubere plantis,
> invigilate, viri
>
> the bounteous Earth
> Will open her arms, claiming the adult seeds
> Wishing to wed the plants that have been set.
> Now, men, beware
>
> (10.157–9)

This is more than a personification or humanisation of the earth.[19] The text indubitably evokes the *Hieros gamos*, the ritual of mystic copulation between earth and seed. The strong appeal to the *viri* in the next verse adds to the moralising sense. All plants mentioned next have a mythical undertone, such as *myrrha*/*Myrrha* who is brought into connection with the story of her origin narrated in Ovid. The whole passage closes with a further metaphor for giving birth, in a figura etymologica (*semine sevit*, 177, looking back to 140 *credidimus semina*). The structure is at first sight apparently twofold. Therefore the insertion of even more species of cabbage seems irritating at first. Optical impressions (*fusco, candida*) and geographical details such as names prevail, likewise places of origin – among them Gades, the home of the poet, and, as a final item, Cyprus. From there

---

[18] Pagán 2006, 29 remains rather vague when she adduces this passage as an example of Columella's perception of space.
[19] Columella 1996 ad 157: 'terra umanizzata'.

the next passage moves on in calendar style, again on the topic of sowing, and here, too, the topic is developed associatively from the names of the months. Speaking about Mars and April, the month sacred to Venus, leads to the praise of spring-time as the natural time for the sexual act. The power of Eros is acclaimed in Lucretian language.[20] Columella presents himself as a didactic poet who is aware of love as the elementary force of creation and is able to describe it adequately.

But then this high-flown train of thought is abruptly abandoned. The poet admonishes himself to leave this path, as it is too sublime (*sublimi tramite*, 216). Such singing, so he claims in the following *recusatio*, belongs to other poets and other addressees, to poems and hymns addressed to Apollo and Bacchus (217–24). *sacra movere, orgia naturae* (218–19) is, so he argues, a task not in his power – while, at least for the moment, he is doing nothing less than singing the *orgia naturae* himself.[21] The author in the *recusatio* sheds light on the different stylistic and generic levels on which he is moving (*vagantem*, 225). While admonishing himself to turn back to the lighter genre in well-known terms (*leviore cura, parvo gyro, gracili filo* 225–7),[22] he introduces at the same time an original and new poetological metaphor: his task as a poet of gardens is to apply the 'secateurs' (*Musa modulante putator*). The last word of this self-admonition leads pointedly back to the main topic: it is the word *hortis*.

> Me mea Calliope cura leviore vagantem
> iam revocat parvoque iubet decurrere gyro
> et secum gracili conectere carmina filo
> quae canat inter opus Musa modulante putator
> pendulus arbustis, holitor viridantibus hortis.
>
> Me, Calliope, on a humbler quest
> Roaming, recalls and bids me to confine
> My course in narrow bounds and with her weave
> Verse of a slender thread, which tunefully
> The pruner perched amid the trees may sing
> Or gardener working in his verdant plot.
>
> (10.225–9)

True to this admonition, the topic now is the humble cress, growing in rows at small intervals (*parvo discrimini sulci*, 230). The cress and the following

---

[20] Cf. Lucretius, *De rer. nat.* 1.1 ff., especially 10, 13. See also the parallels in Columella 1996 ad 10.157.
[21] Pagán 2006, 35 just calls this a sign of 'excitement over the subject-matter' and calls the passage a 'temporary revery'. This does not, in my opinion, do justice to the multilayered meaning.
[22] For the poetological terminology, see above, p. 223.

plants are arranged according to the criteria of habitus, colour and scent. Scent leads to the use made of the scented rose, and the use of the rose in cultic contexts. So, after the seemingly modest, Callimachean start in 230, within twenty-five verses this modesty has been replaced by an entirely different mode of speaking. The rose leads to an invocation of the nymphs, the entourage of Proserpina; and Proserpina is associated with the myth narrating the abduction of Ceres' daughter while she was gathering flowers. Seemingly naively, the poet assures us that, in a cultivated garden, picking flowers carries no danger. The mythological digression is ironically tinted, but in a friendly and good-humoured way:

> Vos quoque iam posito luctu maestoque timore
> huc facili gressu teneras advertite plantas
> tellurisque comas sacris artate canistris.
> Hic nullae insidiae nymphis, non ulla rapina;
> casta Fides nobis colitur sanctique penates.
>
> come, lay aside
> Your mourning and sad fears and hither turn
> With gentle steps your tender feet and fill
> Your sacred baskets with earth's blossoming.
> Here are no snares for nymphs, no rapine here.
> Pure faith we worship here and household gods
> Inviolate.
>
> (10.276–9)

It seems all the more astonishing that, in the following praise of spring and its amenities, myth, which a moment ago had been characterised as dangerous and not to be tolerated in the garden, is now called upon again. Columella compares the advantages of this agreeable and well-tempered time of year (*ver egelidum*, 228) and its products to the beauty and fineness of heavenly bodies. All these stars, the moon and the rainbow receive elaborate mythological names, worthy of a poet like Manilius.

The last passage I want to discuss deals with a problem very much on the mind of any hobby gardener, and therefore especially interesting to a reader of didactic poetry and its literary and practical aims: the passage on rodents and the other enemies of the garden (322–36). A whole army of harmful creatures threatens the plants: the flea, the ant, the snail, the caterpillar, the worm. Later, diseases and pests like mildew are discussed. The wit lies in the contrast between the small size of the vermin and the great harm done. Here not only the list of parasites in Virgil, *G.* 1.181 ff.,[23]

[23] Columella 1996 ad 321–4.

but also the organisation of the bees as described by Virgil in *G.* 4.176 ff. – *si parva licet componere magnis* – forms the background. As well as the allusion to this high-flown passage from didactic poetry, there is also an epic background. The guileless gardener is often taken unawares by Jupiter's wrath. Rain and hail may ruin the work of men and animal. Why the work of animals, one is inclined to ask – hitherto only digging by hand has been discussed, never ploughing. After the attack of the different harmful forces, the gardener is left standing alone and looking sadly at his ravaged and murdered plantations:

> quae capitis viduata comas spoliataque nudo
> vertice trunca iacent tristi consumpta veneno
>
> Which, of their leaves bereft, with naked tops,
> consumed by baneful wastage, ruined lie.
>
> (10.335–6)

The reader versed in epic literature might be reminded of a well-known type of simile. Are not massacred soldiers often compared to plants, e.g. the earth-born fighters who rise up against Jason in Apollonius, *Argon.* 3.1399 ff.? Aeetes, the king of Colchis, looks at them in the same desperation as a wine-grower looks at his vineyards destroyed by a storm. And for epic characters like Jason and Aeetes, digging by hand somehow does not seem right – hence the more noble plough is introduced.

To conclude, let me recapitulate the different discourses that we have been able to identify in the poem and the question of how poetic language and imagery function in the context of a work primarily concerned with practical applicability and technical detail. The poem as a whole is arranged on the basis of chronological and thematic-pragmatic criteria. Epic is present in the allusion to an epic simile, but also in parts of the catalogues, like the organisation by geographic or biographical detail. Erotic-elegiac discourse is evident at different points: the earth must be treated like a recalcitrant girl; plants bear characteristics of the addressees of love poetry; they have a pretty mouth, beautiful eyes, a lovely scent. The didactic poet with philosophical and mythological expertise is present in the hymn about the powerful force of nature. An even higher level is sought by the poet of hymns who, even though in the form of a *recusatio*, sings a hymn to Bacchus and Apollo. The Callimachean poet of the small genre and of a seemingly modest subject matter has his say explicitly in the proem and in the invocation of the Muses, and implicitly in the many lists and learned names and aetiologies. Aetiology and allusions to mythographic material are part of the passage on pest control and magic as a last resource against harmful influences (337–68). Thus the well-meaning, sometimes naive advisor who might easily be identified with the *persona* who speaks to us in the prose parts of the *De re rustica*, and also in the praefatio of Book 10, changes his mask several times. Sometimes he does so abruptly, sometimes by way of association and with tongue in cheek. Contrary to his own statement, this is no faithful follower of the venerable *vates*, the poet of the *Georgics*, and his *Ascraeum carmen*.

For when we look back at the *praefatio* and the programmatic statements about the preferable methods in agriculture, we observe a striking interplay between the discourses of high poetry and practical, technical innovation. *Usus et experientia dominantur in artibus*, he said about agriculture (1.1.16), and the goal is *nec contenti tamen auctoritate vel priorum vel praesentium colonorum, nostra promiserimus exempla novaque temptaverimus experimenta* (1.4.4). If we replace *colonorum* with *poetarum*, this can be identified as Columella's poetic strategy: *auctoritate vel priorum vel praesentium poetarum, nostra promiserimus exempla novaque temptaverimus experimenta*. By knowing his predecessors well, and imitating what is pleasing and convincing in them, but also by trying to experiment with innovation, he has – in the shade of his lamp, because in daylight he is occupied in his garden (*elucubravimus*, 10, praef. 5) – created an oeuvre that fulfils its claim of shaming neither its author nor his reader (*non dedecori*).

REFERENCES

J. N. Adams 1982, *The Latin Sexual Vocabulary* (London)

E. Christmann 2003, 'Zum Verhältnis von Autor und Leser in der römischen Agrarliteratur: Bücher und Schriften für Herren und Sklaven' in M. Horster and C. Reitz, *Antike Fachschriftsteller, Literarischer Diskurs und sozialer Kontext* (Wiesbaden, Stuttgart), 21–152

Columella 1941, *Lucius Iunius Moderatus Columella, On Agriculture*, ed. H. B. Ash, vols. I–IV (Cambridge, MA), ed. E. S. Forster and E. H. Heffner, vols. V–XII (1941–55)

1972, *Über Landwirtschaft: Ein Lehr- und Handbuch der gesamten Acker- und Viehwirtschaft aus dem 1. Jahrhundert u. Z.*, ed. K. Ahrens (Berlin)

1981–3, *Lucius Iunius Moderatus Columella, Zwölf Bücher über Landwirtschaft, Buch eines Unbekannten über Baumzüchtung*, ed. W. Richter, 3 vols. (Munich, Zurich)

1996, *L. Iuni Moderati Columellae rei rusticae liber decimus*, ed. F. Boldrer (Pisa)

2010, *Columellae res rustica, incerti auctoris Liber de Arboribus*, ed. R. H. Rodgers (Oxford)

A. Doody 2007, 'Virgil the Farmer? Critiques of the *Georgics* in Columella and Pliny', *Classical Philology*, 102: 80–197

Th. Fögen 2009, *Wissen, Kommunikation und Selbstdarstellung: Zur Struktur und Charakteristik römischer Fachtexte der frühen Kaiserzeit*. Zetemata 134 (Munich)

M. Formisano 2005, 'Veredelte Bäume und kultivierte Texte: Lehrgedichte in technischen Prosawerken der Spätantike' in M. Horster and C. Reitz, *Antike Fachschriftsteller, Literarischer Diskurs und sozialer Kontext* (Wiesbaden, Stuttgart), 295–312

M. Frass 2006, *Antike römische Gärten: Soziale und wirtschaftliche Funktionen der Horti Romani*. Grazer Beiträge, Suppl. 10 (Horn)

P. Grimal 1984, *Les Jardins romains à la fin de la république et aux deux premiers siècles de l'empire*. 3rd edn. (Paris)

J. Henderson 2002, 'Columella's Living Hedge, The Roman Gardening Book', *Journal of Roman Studies*, 92: 110–33

M. Horster and C. Reitz 2003, *Antike Fachschriftsteller, Literarischer Diskurs und sozialer Kontext*. Palingenesia 80 (Wiesbaden, Stuttgart)

V. Pagán 2006, *Rome and the Literature of Gardens* (London)

C. Reitz 2006, *Die Literatur im Zeitalter Neros* (Darmstadt), 121–7

2013, 'Columella' in E. Buckley and M. T. Dinter (eds.), *A Companion to the Neronian Age: Blackwell Companions to the Ancient World* (Malden, MA)

K. von Stackelberg 2009, *The Roman Garden: Space, Sense, and Society* (London)

CHAPTER 13

The Generous Text
Animal Intuition, Human Knowledge and Written Transmission in Pliny's Books on Medicine

Brooke Holmes

What do we know without being taught? The question is at least as old as Psammetichus, the Egyptian king who, Herodotus reports, sequestered two small children away in an attempt to determine the language they would speak in isolation from other human beings (it turned out to be Phrygian).[1] Some seven hundred years later, Galen conducts a similar experiment to prove more conclusively that there is such a thing as untaught nature, not just in humans but in animals more generally. He takes a newborn goat away from its mother immediately after birth and waits a little while, then offers the kid a choice of honey, olive oil, wine, or milk. The milk wins out, to Galen's satisfaction. He repeats the test when the goat reaches the grazing age, going on to conclude triumphantly that animal natures are untaught (φύσιες ζώων ἀδίδακτοι).[2]

By animal natures, Galen means to include human nature, too. The excursus is, in fact, designed to explain the principle behind erections (for Galen, in strong contrast with the Church Fathers, these are a sign of providential teleology). Yet automatic and intuitive behaviors raise as many questions as they answer, especially in the most rational of creatures. How is it, Galen wonders, that, even as children, we move our arms and legs without the slightest clue as to the mechanisms involved, manipulating muscles and joints still largely unmapped by skilled anatomists?[3] Our ignorance becomes an obstacle only when things go wrong. Under these

---

[1] Herodotus II 2.
[2] Galen, *De loc. aff.* 6.6 (8.443 K.). For Galen on untaught nature (ἀδίδακτος φύσις), see also, e.g., *De caus. sympt.* 2.5 (7.178 K.); *De nat. fac.* 1.13 (2.38 K. =128–9 Helmreich); *De plac. Hipp. et Plat.* 9.1.21–2 (5.725 K. =544,13–17 De Lacy); *De sem.* 2.6.5 (4.643 K. =198,6–8 De Lacy); *De usu part.* 1.1, *bis* (3.6, 3.7 K. =1:4, 1:5 Helmreich); Galen is explicitly drawing on the well-known Hippocratic dictum regarding 'untaught nature' at *Epid.* VI 5.1 (5.314 L. =100,8–102,2 Manetti-Roselli; cf. *Alim.* 39 (9.112 L. =145,12 Joly).
[3] Galen, *De loc. aff.* 6.6 (8.445 K.):

ἐπεί τοι πῶς οὐ θαυμαστὸν τὸ τοὺς μὲν ἀνατομικωτάτους τῶν ἰατρῶν ζητεῖν, ὑπὸ τίνος μὲν μυὸς ἐκτείνεται τόδε τὸ ἄρθρον, οἷον φέρε τὸ κατ' ἰσχίον, ὑπὸ τίνος δὲ κάμπτεται, τίνες δ' εἰσὶν οἱ ἐφ' ἑκάτερα πρὸς τὸ πλάγιον ἀπάγοντες αὐτό, καὶ τίνες οἱ περιστρέφοντες

circumstances, we do indeed have recourse to a physician with anatomical knowledge, as Galen would be the first to admit. Our natures turn out, then, to be half untaught, half in need of teaching. The need to be taught is, in a basic sense, the void to be filled by medicine.

The need for teaching is the motor of the genres under consideration in this volume. Yet in seeking to understand the claims that technical or didactic texts make to be useful, we must also ask what these claims tell us about the need for teaching in the first place. What motivates the creation of these texts? That is, what are the gaps in human relations with the world that make such instruction manuals indispensable? These gaps are present from the very beginning of the Greek didactic tradition and remain an integral part of it. In Hesiod, advice about agriculture comes with a story about why the gods hid the means of life from men. Lucretius spends a good deal of energy in the *De rerum natura* explaining to his reader why he needs the poem in his hands. If Galen's experiment suggests that nature is the only teacher we require, the didactic scenario is by definition one in which untaught nature has failed or fallen short.

It is precisely on the basis of a failure that Pliny the Elder founds the largest didactic stretch of his *Natural History*, the books on medicine. In the opening lines of the first of these books, Pliny identifies this failure in neo-Hesiodic terms as ignorance about the means of life:

> Maximum hinc opus naturae ordiemur et cibos suos homini narrabimus faterique cogemus ignota esse per quae vivat.[4]

> From this point on, we will deal with a most important work of nature: we will tell man the foods proper to him and force him to admit that his means of living are unknown to him. (20.1)

This is not the first time Pliny has complained about the limits of untaught nature in human beings in the *Natural History*. In the preface to Book 7

ἑκατέρωσε· τὸ δ' ἐρίφιον ἢν ἂν ἐθελήσῃ κίνησιν ἑκάστης διαρθρώσεως εὐθέως ἐργάζευθαι, καθάπερ γε καὶ τοὺς ἀνθρώπους αὐτούς, καίτοι γ' ἀγνοοῦντας ὑπὸ τίνος μυὸς ἑκάστη γίνεται κίνησις;

So then, you see, is it not incredible that the most skilled men in anatomy are still trying to figure out by which muscle a joint is extended, let's say the hip joint, and by which one it's bent, and which are the muscles that lead it back on either side at a slant, and which are the ones that rotate it to either side; yet the kid straightaway performs the movement of whichever joint it wants to, just as, in fact, men do, too, and indeed being ignorant of which muscle it is that causes each movement?

See also *Foet. Form.* 6.8 (4.690 K. =94,15–18 Nickel). Galen calls the fact we use muscles without any knowledge of how they work one of the greatest enigmas in medicine and philosophy (4.692–3 K. =96,27–9 Nickel). He takes up the problem at length in his *On Problematical Movements*, where the example of the child returns (e.g., 4.5 [136, 13–16 Nutton]). See also the discussion at *In Hipp. Epid. VI comm.* 5.2 (259,6–261,14 Wenkebach-Pfaff).

[4] I have used Mayhoff's text. Translations are my own unless otherwise noted.

he writes that while all other animals have an instinctive awareness of their capabilities,⁵ 'man knows nothing, nothing but what he is taught: not speaking, nor walking, nor eating; in short, the only thing he does by natural instinct is weep' (*hominem nihil scire, nihil sine doctrina, non fari, non ingredi, non vesci, breviterque non aliud naturae sponte quam flere*, 7.4).⁶ Yet Pliny's concerns at the start of Book 20 about his compatriots' lack of knowledge about the means of life occupy a particular place in the *Natural History*. For one thing, they motivate one of the longest stretches of the work (thirteen books, nearly one-third of the total). Moreover, as we will see, these concerns are deeply entangled in value-rich assumptions about the position of human beings in the natural world in relation to other animals and about their relations with one another.⁷

In this chapter, I explore the nature of these assumptions in the interest of better understanding the stakes of Pliny's promise of a useful education in our "means of life," the subject of the books on *materia medica*. These assumptions are so important because Pliny's rhetoric of practical utility cannot be taken at face value. That is, the question of whether Pliny's claims on behalf of his work's practicality are plausible or valid – that is, whether anyone could use these chapters to gather herbs or apply medicinal plants in real life – is not as straightforward as it might appear.

The problem with this question has to do in part with the fact that Pliny himself is largely uninterested in the gritty realm of practical applicability. While he does dismiss the Greek practice of including illustrations to aid in the identification of medicinal plants as not especially useful (*HN* 25.8), he fails to offer a better solution. Instead, he seems to assume that anyone who wants to use his text will need a flesh-and-blood expert on hand to match the names on the page to actual plants, a role played in his own work by Antonius Castor (at least in cases where Pliny actually had firsthand knowledge of the plants discussed).⁸ It is true that the presence of an expert teacher to mediate the words of the text is an assumption made in some of our earliest technical texts, such as the more practical texts of the *Hippocratic Corpus*. Nevertheless, Pliny is remarkably unconcerned with how his reader will line his text up with the world of flora and fauna that

---

⁵ Pliny's reference to animals' innate awareness of their capacities recalls the Stoic concept of *oikeiōsis* (here suspended for human beings): see *SVF* 3.169–83.
⁶ For Pliny's pessimism about human life, see also *HN* 8.43–4; 8.167.
⁷ On the value and importance accorded to nature more generally in Pliny, particularly in relationship to human relations and society, see Wallace-Hadrill 1990, Hahn 1991 (with particular attention to the books on medicine), Beagon 1992.
⁸ For some of the problems created by Pliny's secondhand knowledge, see André 1955.

it describes.⁹ As Trevor Murphy has trenchantly observed, "'how to' is not a large part of the *Natural History*'s *raison d'être*."¹⁰

And yet, despite the text's apparent shortcomings as a practical manual, Pliny is convinced of the utility of the books on medicine. I therefore focus on what conditions this utility in Pliny's eyes, arguing that if we want to understand Pliny's claims on behalf of his text, we have to gain a sense of what motivates him to offer instruction to his fellow Romans in the means of life. One of the main drivers of his project, I suggest, is his understanding of human beings as uniquely ignorant of the natural resources around them, in comparison with all the other animal species. Non-human animals emerge in the books on *materia medica* as exemplars of untaught nature. Part of Pliny's project is to mediate between animal knowledge and the humans who need it to survive.

My strategy is twofold. I start by demonstrating that Pliny's vast compendium on *materia medica* responds to an implicit tension between humans and the natural world created by the shortcomings of our intuitive knowledge. What we lack, more specifically, is a grasp of how we fit into the web of connections structuring the natural world. Pliny refers to this web in sweeping terms in the preface to Book 20 (and again in Book 24):

> pax secum in his aut bellum naturae dicetur, odia amicitiaeque rerum surdarum ac sensu carentium et, quo magis miremur, omnia ea hominum causa, quod Graeci sympathiam et antipathiam appellavere, quibus cuncta constant.

> In what follows we will speak of the internal peace and war of nature, the hatreds and friendships of mute and insensate objects (and it is even more remarkable that all of this is for the sake of humans), what the Greeks call sympathy and antipathy, which form the basis of the whole system. (20.1; cf. 24.3)

Throughout these books, we see other animals navigating the web of sympathies and antipathies with what looks like innate knowledge. In the majority of those cases, animal behavior acts as a model for humans, implicitly confirming cross-species homologies.

Yet, at the same time, the reader of the *Natural History* is unlike an animal precisely because he lacks intuitive knowledge of his own regarding

---

⁹ Of course, it is also true that Pliny's work was appropriated as useful for medical purposes and enthusiastically excerpted into the early modern period: see Nauert 1979, 81, Mazzini 1986, 88–94, Doody 2010, 135–52. On challenges to the usefulness of Pliny's medical information in the Renaissance, see Nauert 1979, 81–5, French 1986. Modern scholars have not assessed the 'usefulness' of ancient guides to plants very kindly: see, e.g., Rackham 1996, 37–8.
¹⁰ Murphy 2004, 211.

natural sympathies and antipathies. The reader's need for instruction sits uncomfortably with Pliny's position that, as he insists in the preface to Book 20, everything in nature is 'for the sake of humans' (*hominum causa*), a position familiar to the reader by this point and reinforced two books later, when, at the start of Book 22, Pliny looks back on the awe-inspiring survey he has offered of all the kinds of plants 'created for the needs and pleasures of mankind [*utilitatibus hominum aut voluptatibus*]' (22.1). Ignorance is not only inherently problematic. More often than not, it results in harm. The very threats that the work aims to address undercut the idea, which Pliny introduces early in the *Natural History* and emphasizes repeatedly, that nature is not just anthropocentric, but also providentially oriented towards human beings.[11]

Nevertheless, Pliny does not represent his own efforts to supply the reader with a survival manual as a supplement to nature. Instead, as I show in the second part of the chapter, he recasts our ignorance as the outcome of a moral failing within human society, triggered by greed, opportunism and the desire for novelty, and exacerbated by the spread of professional, Hellenized medicine.[12] Pliny's approach succeeds in shifting attention away from the estrangement of human beings from their means of life – what I have elsewhere called the "negative exceptionalism" of the human in Greco-Roman antiquity – onto disruptions to the cultural transmission of natural knowledge.[13] Pliny elsewhere strongly upholds negative exceptionalism. Indeed, his remarks on human nature in the preface to Book 7 form a *locus classicus* for the idea of nature as a stepmother (*noverca*).[14] Yet in the books on medicine, he is working with a more subtle tension between animal intuitions about benefit and harm, represented as innate knowledge of sympathies and antipathies, and the human need for instruction in the underlying relationships structuring nature. By representing his own work as a mechanism for meeting that need, Pliny recuperates the providential

---

[11] On nature's orientation towards human flourishing, see, e.g., *HN* 2.154, 7.1, 18.1, 24.1.

[12] On Pliny's distrust of Greek physicians (and the decadence they foster in the Romans), see *HN* 29.1–28, with Nutton 1986, Hahn 1991, 231–9, von Staden 1996, 408–9.

[13] On the motif of negative exceptionalism in Lucretius, see Holmes 2013. On the exceptionalism of human beings more generally in Pliny, see Beagon 2005, 52–6, who argues that "overall, it seems that there is more evidence for stressing rather than underplaying the gap between man and other animals in the *HN*" (49); cf. Henderson 2011, 159–60. Renehan 1981, 246–52 provides a wealth of evidence for the belief in (positive) human exceptionalism in Greco-Roman and classically influenced Christian writers.

[14] See *HN* 7.1–6. On the idea of nature as a stepmother, see also Augustine, *Contra Iulianum Pelag.* IV 12.60 (paraphrasing Cicero, *Rep.* III 1); Philo, *Post. Caini* 46–7; and Goulon 1972, 8–11. Cf. Seneca, *Ben.* 2.29, *Ep.* 74.15–21; 90.18. For the motif of the helpless human infant, see Lucretius, *DRN* 5.222–7, with Goulon 1972, Rochette 1992, Sacré 1992.

orientation of nature and protects the place of human beings at its center. The didactic project of the books on *materia medica* in the *Natural History* is thus an exercise in using words to bridge the gap between human beings and the things in the world that sustain them.

Sympathy, Antipathy and Animal Knowledge

Pliny begins Book 20, as we have just seen, by referencing the loves and hates that organize nature: the Greeks, he tells us, call these sympathies and antipathies. The language of sympathy in an early imperial context is likely to trigger Stoic associations for modern scholars. Yet although Pliny's concept of nature is loosely consistent with the webbed, immanently intelligent cosmos of Stoicism, and although his philosophical proclivities are generally Stoic, it is unlikely that Stoicism is the most important source of the worldview articulated here.[15] For while the Stoics, at least by Pliny's time, embraced sympathy as a core commitment in their philosophy of cosmic unity, Stoic sympathy generally works on its own – that is, independent of antipathy – and on a grand scale.[16] It would be hasty, then, to conflate "the Greeks" to whom Pliny refers with the Stoics.

The idea of a natural world organized by what could be called local sympathies and antipathies between species is present, however, in the earliest examples of the natural history tradition, the writings of Aristotle and Theophrastus. Aristotle describes the friendships and wars (αἱ ... φιλίαι καὶ οἱ πόλεμοι, *Hist. an.* 8.1, 610a35) among wild animals, which usually arise from relationships of reciprocal benefit or the competition for resources: the snake is at war with the weasel and the pig, for example, but the raven and the fox are friends; in the sea, some animals form shoals and work together, while others are at war.[17] In Theophrastus, plants collaborate (συνεργεῖ)

---

[15] Gaillard-Seux 2003, 118–20 reaches a similar conclusion. See also Gaide 2003, 131. By contrast, Conte 1994, 92–9 approaches Plinian sympathy as a classic case of magical thinking.
[16] On Stoic sympathy, see, e.g., Chrysippus apud Alexander, *On Mixture* 3 (216,14–17 Bruns; see also 227,8 Bruns = SVF 2.473 [partial]): ἡνῶσθαι μὲν ὑποτίθεται τὴν σύμπασαν οὐσίαν, πνεύματός τινος διὰ πάσης αὐτῆς διήκοντος, ὑφ' οὗ συνέχεταί τε καὶ συμμένει καὶ συμπαθές ἐστιν αὑτῷ τὸ πᾶν ('[Chrysippus] holds that while the whole of substance is unified, because it is totally pervaded by a pneuma through which the whole is held together, is stable and is sympathetic with itself' [trans. Todd]), with White 2003, 128–33. See also the eclectic examples at Cicero, *Div.* 2.33, *Nat. D.* 2.120–6. The grand scale of Stoic sympathy and Stoic cosmology can be used to differentiate Seneca's (traditionally) Stoic perspective on nature from Pliny's more eclectic, local perspective: see Williams 2012, 37–48.
[17] Aristotle, *Hist. an.* 8.1, 609b28–9, 32–3, 8.2, 610b1–2. Richard Sorabji notes that Aristotle comes closest to attributing some kind of thought to animals in Books 8 and 9, but does not actually depart from his position that animals lack reason (Sorabji 1993, 13).

to preserve and propagate each other; they can also be antipathetic to each other, as the cabbage and the bay are to the vine.[18]

These friendships and hatreds in the plant and animal worlds were presumably developed in the occult study of what Pliny calls the more marvelous communities in nature, and it is likely that this area of study had a direct influence on Pliny. Pliny regularly makes reference to the Magi, and, while he finds reason to fault their methods, he nevertheless includes material credited to them.[19] The source that Pliny cites most frequently in his encyclopedia, "Democritus," is almost certainly the shadowy figure Bolus of Mendes, who worked at the crossroads of natural philosophy and learned magic in the fourth and third centuries BCE and wrote under the name of Democritus.[20] The title of Bolus' lost treatise, *On Sympathies and Antipathies*, bears witness to the development of an occult tradition of natural knowledge structured by the dynamics of sympathy and antipathy that developed alongside a tradition in natural history (with a fair amount of cross-pollination between the two).[21] Indeed, it is likely that sympathy and antipathy had become important vehicles for conceptualizing nature as an overarching category in the centuries between the Peripatetic naturalists and Pliny.

Whatever the provenance of Pliny's sympathies and antipathies, they thoroughly inform his understanding of the natural world in the chapters on *materia medica*. This is a world where the oak hates the olive 'with an inveterate hatred' (*pertinaci odio*), where oil alone mixes with lime, both being averse to water.[22] It is a world where, if you rub your hands with radish seed, you can handle a scorpion, where the blood of a goat can break the hardness of a diamond.[23] The web of sympathies and antipathies takes as its nodes specific natures, while at the same time cutting across boundaries between animals and plants and stones, as well as between human and non-human, to create an internally differentiated community coextensive with the entire natural world. It is a community bound together by criss-crossing "loves" and "hates" that shoot through even non-sentient beings (*sensu carentia*), governing their interactions with other natures.

The world of sympathies and antipathies matters deeply to Pliny because human beings, too, like all natures, are enveloped by it and subjected to its logic of benefit and harm. Yet people participate in this world unaware

---

[18] Theophrastus, *C.P.* 2.18.1, 2.18.4.
[19] Pliny of course at times takes his distance from information he reports: see Serbat 1973. But it seems clear that the tradition of learned magic shapes his perception of sympathy and antipathy.
[20] On Bolus, see Wellmann 1897, Kroll 1934, Gaillard-Seux 2010b.
[21] Gaillard-Seux 2003, 120–4. [22] *HN* 24.1, 24.3. [23] *HN* 20.25, 20.2.

of how it affects them, with the result that they are incapable of exerting control over it. Their ignorance is an anomaly. Indeed, Pliny regularly dwells on the ability of other animals to perceive benefit and harm and, thus, to navigate the web of sympathies and antipathies with great efficacy. I want to look at some examples of animal intuition in order to get a better sense of what it is that human beings lack – which is also, of course, what Pliny's text aims to provide.

The wondrous phenomenon of animals exercising therapeutic intuition appears at least as early as Aristotle's *Historia animalium*, where it is reported that Cretan goats struck by arrows seek out the herb dittany (which causes the arrows to fall out).[24] By the first centuries CE, such stories were circulating as part of a common repertoire. Plutarch, for example, offers plenty of examples to corroborate the claim that animals know how to care for themselves. In a dialogue on animal rationality, Gryllus, one of Circe's pigs, disabuses Odysseus of the assumption that the enchanted swine would prefer to return to human form in part by pointing to the special ability of animals to care for themselves: "each is his own specialist in medicine" (πρὸς ἴασιν αὐτότεχνόν ἐστιν).[25] He goes on to list a series of examples: pigs that go to the river and catch crabs when they are sick, tortoises that eat marjoram after swallowing snakes, and the Cretan goats using dittany to self-medicate. Similar examples appear in Philo's *De animalibus*, Aelian's *De natura animalium* and Cicero's *De natura deorum*.

Like Galen, Gryllus chalks these behaviors up to natural intelligence. Having rhetorically asked who taught each animal its particular medical expertise, he goes on:

> ἂν γὰρ εἴπῃς, ὅπερ ἀληθές ἐστι, τούτων διδάσκαλον εἶναι τὴν φύσιν, εἰς τὴν κυριωτάτην καὶ σοφωτάτην ἀρχὴν ἀναφέρεις τὴν τῶν θηρίων φρόνησιν· ἣν εἰ μὴ λόγον οἴεσθε δεῖν μηδὲ φρόνησιν καλεῖν, ὥρα σκοπεῖν ὄνομα κάλλιον αὐτῇ καὶ τιμιώτερον, ὥσπερ ἀμέλει καὶ δι' ἔργων ἀμείνονα καὶ θαυμασιωτέραν παρέχεται τὴν δύναμιν.

> For if you should say what is true, that Nature is the instructor in these things, you are elevating the intelligence of animals to the most authoritative and wisest first principle. If you think that it should not be called reason [*logos*] or intelligence [*phronēsis*],[26] it is time for you to seek out a fairer and even more honorable name for it, since it certainly presents a faculty in actions that is better and more wondrous. (Plutarch, *Mor.* 991F–992A)

[24] Aristotle, *Hist. an.* 9.6, 612a2–5. See also Cicero, *Nat. D.* 2.126, Pliny, *HN* 8.97, 25.92, Virgil, *Aen.* 12.423–4.
[25] Plutarch, *Mor.* 991E. See also 974B–C.
[26] In fact, some philosophers did allow *phronēsis* to animals: *nous* was what was withheld. See Aristotle, *EN* 6.7, 1141a26, *GA* 3.2, 753a11.

Gryllus is here taking the side of animals in a competitive, interspecies game – that is the premise of the dialogue. The real target of his attack is not so much Odysseus but presumably the Stoics, who were the most adamant in denying animals reason in this period (and who famously declared the purpose of the pig to be keeping the bacon fresh until it was ready to eat).[27] If merely instinctual action was not seen as a threat to the prerogative of reason in the debate about animal capabilities, Gryllus uses intuitive therapeutic knowledge to raise the stakes of the debate. He concludes that, whatever you want to call the animal intelligence behind such knowledge, it trumps reason.[28]

These larger philosophical questions appear out of place in the *Natural History*, which tends to avoid sustained reflection on the phenomena observed and reported. Nevertheless, Pliny is full of examples of animals capable of caring for and healing themselves. In some cases, animals gravitate toward plants that are beneficial to them. Snakes, for example, use fennel to aid in casting off their old skins and to improve their eyesight; wild goats eat rue to sharpen their vision.[29] Storks cure themselves of disease by eating marjoram, the goat with ivy and crabs thrown up from the sea.[30] In other cases, animals use plants or other natural substances as a prophylaxis against antipathies that they expect to encounter. Pliny tells us that weasels, going into battle against snakes, first fortify themselves with rue; deer reportedly eat wild parsnip for the same reason.[31] In a similar way, animals are said to use herbal remedies to heal injuries. Field mice, for example, eat the root of condrion to counteract the bite of a snake; swallows seek out the plant chelidonia to heal wounds to the eyes.[32]

---

[27] On the denial of reason to animals in the philosophical tradition, see Sorabji 1993, especially 7–61.

[28] Cf. Origen, *C. Cels.* 4.87 (=*SVF* 2.725), opposing such intuitions to *sophia* and *logos*. Other capacities used to prove animal intelligence included preparations for the future, virtues and vices, and the knowledge of how to use their body parts: see Sorabji 1993, 78–9. Note that these arguments do not always elevate animals at the expense of humans. On the strand of ancient thought that does elevate animals in this way, see Lovejoy and Boas 1935, 19–22, 389–420 (who coin the term "animalitarianism" to describe this line of thinking).

[29] On snakes and fennel, see *HN* 8.99, 20.254; see also Aelianus, *NA* 9.16. On wild goats: *HN* 20.132–3 20.134.

[30] *HN* 8.98.

[31] On weasels and rue, see *HN* 8.41 and 20.132–3. On deer: 22.79 (note the use of *fama*). Gaillard-Seux 2010a, 311–16 persuasively argues that poisonous animals loom large in these contexts in part because they enact the principles of antipathy so dramatically. These examples are also consistent with the principle that animals know not just what is advantageous for themselves but what is harmful for their enemies (*callent enim in hoc cuncta animalia sciuntque non sua modo commoda, verum et hostium adversa*, *HN* 8.91).

[32] On field mice, see *HN* 22.91. On swallows and chelidonia, see 8.98, 25.89; see also Aelianus, *NA* 3.25. For a long list of other examples of self-medicating animals, triggered by a digression on how hippopotamuses relieve plethora caused by overeating with bloodletting, see *HN* 8.96–101.

Most of the time, Pliny cites these examples because they are an aid to our own grasp of the attractions and repulsions in nature: by watching animals, we gain access to the knowledge they possess. The fact about snakes and fennel is used as the basis for the inference that fennel is also an aid to human vision. The legendary Melampus is said to have used hellebore to cure the madness of the daughters of Proetus after observing that his female goats were purged after eating it.[33] If a scorpion is touched by aconite, it grows numb and pale; white hellebore reverses the damage.[34] In these cases, animals are not so much sites of experimentation as they are vehicles to discovery. Indeed, Pliny observes that animals do as much good for the discovery of medicaments as they do in supplying them, casting animals as vehicles for knowledge much in the same way as they are in the world of portents and divination.[35]

In some cases, animals are used not only to back claims made by humans about plants but also to challenge those claims. Chrysippus, Pliny reports, thought that ocimum (basil) was bad for the stomach and capable of causing madness, lethargy and liver problems: "for this reason goats refuse to touch it, so that human beings also ought to avoid it [*fugiendum*]" (20.119). Chrysippus here takes the observation of goats as the springboard of a logical inference: because they avoid ocimum, human beings should, too, just as, in the example above, the recommendation to use fennel juice is derived (*inde*) from an observation of its benefit to snakes. But, Pliny goes on, after Chrysippus, ocimum found its defenders, who claimed that goats do indeed eat basil without it causing them any harm and, accordingly, drew the opposite conclusion – namely, that there is no need for us to avoid it, adding a range of benefits that it brings to human beings.

Underneath these observations, a subtle contrast is at work. On the one hand, conclusions about benefit and harm – and, by extension, what should be sought out and what should be avoided – are based, for humans, on rational inferences based on animal behavior. On the other hand, the animals themselves act on knowledge about harm and benefit that appears more intuitive and straightforward.

Yet such knowledge is, on close inspection, complex. If a bee avoids an olive blossom or a snake seeks out fennel, we can describe these behaviors as internal to relationships of antipathy and sympathy. By that I mean that the

---

[33] *HN* 25.47. On Melampus and the Proetides, see also Hesiod, frr. 37, 133 (M–W), Pausanias 2.7.8, [Plut.], *Fluv.* 21.4.

[34] *HN* 27.6.

[35] *HN* 28.3. See also 8.97, followed by a list of cases where animals help humans perceive danger (8.102–4).

nature of the animal is one node in an antipathetic or sympathetic nexus. The animal behaves as if it has a natural (assume innate) understanding of what is proper to it and what is hostile and responds accordingly. But in the case of animals seeking out prophylactics or antidotes, the situation is triangulated. These animals continue to pursue self-preservation, but they do so by acting on knowledge of natural antipathies (less often sympathies) between, say, rue and snake venom to protect or heal themselves. Gryllus calls such knowledge remarkable – indeed more remarkable (θαυμασιωτέραν) than the rational intelligence that humans claim as their own prerogative – and credits it to nature. Pliny tends to be more circumspect about animals' therapeutic intuitions. One case, however, shows him struggling to account for the complex but non-rational nature of this knowledge.

Early on in Book 27, Pliny mentions the fast-acting poison aconite, noting that it is especially toxic to leopards. Indeed, he reports that people in regions overrun with leopards take advantage of the natural antipathy between leopards and aconite to manage the leopard population, rubbing meat with the poison and setting it out for the animals. The leopards, however, have an antidote: if they happen to eat human excrement, they immediately recover from the poison.[36] But how do they know to do this? Pliny ends up with the following, rather muddled conclusion:

> quod certe casu repertum quis dubitet et, quotiens fiat, etiam nunc ut novum nasci, quoniam feris ratio et usus inter se tradi non possit? hic ergo casus, hic est ille qui pluruma in vita invenit deus – hoc habet nomen per quem intellegitur eadem et parens rerum omnium et magistra – utraque coniectura pari, sive ista cotidie feras invenire sive semper scire iudicemus.
>
> Who would doubt that [the remedy] has been discovered by chance and that, as often as it occurs, it is even now discovered as new, since reason and use do not allow wild animals to transmit it among themselves? Therefore, this chance, this is that god who discovers many things in life – this is the name by which is meant she who is the parent of all things and their master; either conjecture is likely, whether we determine that wild animals discover these things daily or whether they always know. (27.7–8)

Pliny rejects the idea that leopards learn from experience and from one another, on the widely accepted grounds that they, like all wild animals, lack the faculties – namely, reason – required to acquire knowledge and transmit it amongst themselves. He is left to conclude that they happen

---

[36] See also *HN* 8.100, Aelianus, *NA* 4.49 (where the leopard eats aconite "unwittingly" [ἀγνοοῦσα]), Aristotle, *Hist. an.* 9.6, 612a7–8, Cicero, *Nat. D.* 2.126.

across the remedy in question by chance. Yet it is hard to believe that every time a leopard cures itself in this way it is simply a lucky coincidence. Are we to imagine that every case of a self-medicating animal is a one-off?

Pliny, credulous as he sometimes is, seems to sense that the explanation is implausible and abruptly shifts into a paean to chance. By recasting randomness as the guiding hand of providence, Pliny evades the opposition between chance and reason and opens up another possibility – namely, that animals at some level "always know" (*semper scire*) the proper therapeutic response. By this, he seems to suggest that either the relevant knowledge is innate in the leopard or that it is immanent in the process by which the animal routinely discovers the cure. In other words, such knowledge either inheres in the individual nature or belongs to Nature as a transindividual principle.[37] Pliny's perhaps deliberate vagueness precludes a definite answer. Here, as elsewhere, he seems indifferent to the mechanisms of sympathy and antipathy. What matters is that nature is in some broad sense steering the animal toward self-preservation.[38]

In the end, then, Pliny leans on nature to explain even the most complicated cases of animals exhibiting therapeutic intuition. The primacy of nature allows him to collapse the difference between behaviors that other observers may wish to distinguish, that is, between animals acting on an intuitive sense of what is proper and what is foreign to them and animals who, like idiot-savant physicians, somehow grasp natural antipathies (e.g. between aconite and human ordure) and deploy them to their own advantage. For Pliny, both behaviors belong to the larger category of nature working to preserve the animal.

It is precisely nature understood in these terms that Gryllus had opposed to reason, concluding that it is in fact superior to rational intelligence. Pliny is working with an opposition of this kind when he makes animal behavior the basis for inferences about benefit and harm within the human community. Moreover, he, too, despite a general tendency to put human beings on top, is capable of privileging the alliance between animals and nature over the mediated relationship that humans have with the non-human world. The very need to observe animal behavior as a clue to the underlying attractions and repulsions in the natural world flags a fundamental ignorance in the human species. That ignorance lies at the heart of what I referred to earlier as Pliny's "negative exceptionalism."

---

[37] On *natura* as both individual and transindividual in Pliny, see French 1994, 196–202.
[38] Of course, one animal's benefit may be another's loss. The leopard that survives contact with aconite may attack a person. But Pliny does not here reflect on conflicts arising from the competition for survival, suggesting that by *natura* he means the specific nature of a given animal.

Although the excursus on the leopard begins with a nod to the animals' lack of reason and experience, it ends in a crescendo of self-critique that echoes the preface to Book 7: "it really is a disgrace that all animate beings should have knowledge of what is beneficial to them, with the exception of man" (*pudendumque rursus omnia animalia, quae sint salutaria ipsis, nosse praeter hominem*, 27.8–9). If animals are seamlessly integrated into the web of sympathies and antipathies, human beings, lacking an intuitive grasp of how to maximize benefit and minimize harm in their interactions with the natural world, suffer from an estrangement that Pliny pronounces shameful.[39]

The difference that Pliny stresses between humans and animals brings us to a central question of this chapter, namely, how written texts support – and, more important still, are *represented* as supporting – practices and acts. Pliny incorporates the intuitive knowledge of animals into a useful compendium of information for human beings. But insofar as such knowledge implicitly serves as a foil for our own epistemic helplessness in the face of nature's sympathies and antipathies, it must also be understood as part of Pliny's rhetoric of practicality. That is, animal intuition betrays our own need for teaching. And yet the motif of negative exceptionalism is difficult to reconcile with Pliny's conviction that we are the magnetic center of a providential nature. I turn now to examine how Pliny sidesteps the tension between these two ideas by blaming our ignorance not on a shortcoming of Nature but on persistent failures in the social and moral order, failures that his own work aims to correct.

Learning to Fit In

Pliny, as we have just seen, tackles the difficulty presented by the proposition that animals consistently happen upon cures by chance by conflating chance and nature and using the resulting hybrid as a counterweight to the signature skills of human beings, reason and experience. Yet chance also has a role to play in discoveries made by human beings themselves in a number of anecdotes that Pliny relates. In the time of Pompey, it was discovered that wild mint cures elephantiasis "by someone's chance experiment, the face being smeared with it from shame" (*fortuito cuiusdam experimento, propter pudorem facie inlita*, 20.144–5). In another example, a woman has a

---

[39] Cf. *HN* 18.3–4, where Pliny offers a variant on this theme: humans are aware of the poisons produced by nature, but they are not content with merely avoiding them, as other animals do; rather, they alone exploit poisons that belong to others to bring about the unnatural destruction of life (*nec ab ullo praeter hominem veneno pugnatur alieno*).

dream in which she sends her son, serving in the praetorian guard, the root of the wild rose – a plant she had happened to notice in her garden the day before – and asks him to drink an extract from it. Upon waking, the mother sends her son a letter describing the dream, which arrives just after he has been bitten by a rabid dog. He complies with her request and so, Pliny concludes, "he was saved unexpectedly [*ex insperato*], and afterward each person who tried a similar remedy [was saved]" (25.17).[40]

The discovery of a cure for rabies, with its suspiciously adventitious dream, suggests that humans are not entirely excluded from the providential dynamics of nature even as the very narrative of discovery emphasizes that the routes that exist for the transmission of knowledge eliminate the need for nature to engineer the discovery serially.[41] Such is, in fact, the theme in the preface to Book 27, where Pliny traces the entire body of human knowledge about plants to divine inspiration and nature, the common parent of all.[42]

Yet chance is insufficient on its own. More often, as in Hesiod, we have to work for our knowledge. It is true that labor is not incompatible with the idea of nature's beneficence. But Pliny's argument to redeem labor shows some strain at its extremes, as when he tackles the anti-providential argument that the earth is full of things hostile to human beings by countering that even what is most hateful has been devised for the benefit of humans (*causa hominum*). Those plants that are ugly to look at and dangerous to touch have been made that way, he argues, to ward off other animals and preserve their benefits for humans.[43] Pliny suggests at another point that nature sometimes makes the honey crop poisonous as a way of keeping men on their toes and keeping their greed in check.[44] Moralizing aside, though, Pliny admits that the labor we undertake to compensate for our lack of intuitive knowledge is not easy. More seriously still for the claim that nature does everything for the sake of humans is the problem that we encountered earlier – namely, that humans are the only living beings untutored by nature in what is beneficial to them.

---

[40] For other chance discoveries, see *HN* 22.120, 23.56–7.
[41] For dream cures, see also *HN* 22.44. On the peculiarly human capacity to transmit knowledge, see *HN* 11.271, 27.7. The rudiments in Pliny of a "signatures" theory of plants, later one of the central elements of sympathy, also point to nature's providence: see, e.g., *HN* 22.57–60, 27.98–9, 36.55–6. See further Stannard 1982, 14–15, Amigues 1995, Gaillard-Seux 2010a, 313 and Hahn 1991, 225–6 on the indications of benefit that nature gives to humans in Pliny.
[42] On nature as a parent, see also *HN* 24.1 and Formisano 2004, 135.
[43] *HN* 22.17. For some of these arguments, see Cicero, *Acad.* 2.120; Lucretius, *DRN* 2.177–82, 5.195–221; Philo, *Prov.* II 56–65.
[44] *HN* 21.78.

In formulating human nature as ignorant, however, Pliny plants the seed of the strategy he adopts in the books on *materia medica* to neutralize the problem that negative exceptionalism presents to his dream of anthropocentrism. In the discussion of aconite, he had characterized ignorance in strongly moral terms. The choice of language testifies to Pliny's decision to recast the problems that hinder our understanding of benefit and harm as social and cultural, rather than as inherent in nature, our nature or Nature as a whole. The strategy works at two levels: first, it displaces attention from the fact that humans are naturally in the dark about the sympathies and antipathies that govern their flourishing; second, it generates problems that Pliny himself is in a better position to solve. By shifting blame away from nature onto people themselves for their ignorance, Pliny establishes the utility as well as the broader cultural value of his own book.

In specifying the problems that lie behind the gap between human beings and their "means of life," Pliny focuses not on the generation of knowledge but failures in its transmission.[45] To start with, the conventional hierarchy of erudition is upended when it comes to the knowledge of plants: those most learned in their medicinal properties are the unlearned peasants who live among them.[46] It is often the case, Pliny says, that simples in widespread use have not even received proper names. The trouble here is not that country-dwellers will not share their knowledge. Rather, no one bothers to consult them, because everyone is seduced by the tribes of medical "experts" hawking remedies for their own profit.[47]

It is nothing other than greedy miserliness that poses the second threat to the transmission of plant knowledge. Pliny takes it as an indictment of modern times that elites who do have expert knowledge refuse to share it freely.[48] In the early days of learned botany, the situation was different: "Nothing was left untried or unattempted by them, and nothing was kept secret, nothing they wanted to be of no benefit for posterity" (*nihil ergo intemptatum inexpertumque illis fuit, nihil deinde occultatum quodque non prodesse posteris vellent*). The moderns, by contrast, want "to defraud life of good things won by others" (*fraudare vitam etiam alienis bonis*, 25.1–2). On the one hand, then, the experts jealously guard what they have learned, feeding the illusions that the knowledge of medicinal plants is scarce and specialized, and that healing requires exotic and expensive remedies – the

---

[45] See also Hahn 1991, especially 225–31.   [46] *HN* 22.94.
[47] Galen makes a similar complaint: see *Antid.* 1 (14.30 K.), cited by Nutton 1986, 54 n. 46.
[48] On the financial metaphors that Pliny uses here, see Lao 2011, especially 37–44; see also Murphy 2004, 203–9. Galen's physician-centered polemics against pharmacological predecessors and rivals a century later, on which see Totelin 2012, especially 310–12, offer a fascinating foil here.

very illusions that elevate the prestige of ostensibly esoteric knowledge. Only dogs are as stingy with their therapeutic knowledge (25.91). On the other hand, there is no prestige attached to the knowledge of simples, with the result that people do not seek it out from the people who have it. Such knowledge remains local and uncatalogued, largely inaccessible to urban elites.

The perverted chain of knowledge leads, as we have seen, to a situation that Pliny views as entirely unnatural: not just ignorance, but, still worse, indifference among elite, urban Romans to the very plants that preserve health, allay the pains of the body and ward off death. The blame, in other words, lies not only with the broken links in the transmission of medical knowledge, but also with the shortcomings of the consumers of that knowledge, who fail to take responsibility for their own education. The culprit is the Roman moralist's preferred vice: luxury. The very care of our life, Pliny complains, echoing Plato in Book 3 of the *Republic*, is looked upon as the duty of others.[49] Moreover, distracted by opulence, people are deceived into seeking salvation in unnatural remedies, antidotes created not by nature but by humans, who mask the essential sympathies and antipathies of things, and who are driven above all by the profit motive.[50] The major advantage that reason gives humans over animals – namely, the ability to share knowledge – is corrupted by the evils of commercialization, a taste for the exotic and conspicuous consumption.

Andrew Wallace-Hadrill has shown that the tension between luxury and nature closely identified with proper social order is a persistent theme of the *Natural History* and especially the books on medicine.[51] Here I am interested more specifically in the way that luxury intervenes in these books to mask the more muted, but also more insidious problem of the estrangement of humans from nature and the natural web of sympathies and antipathies. By recasting the lack of intuitive knowledge about medicinal plants as a vacuum filled by luxury, Pliny succeeds in redefining our ignorance of nature in cultural terms. He is, as a result, in the position of offering his own text not as the supplement to correct a fundamental flaw in nature, but, rather, as the remedy for a crisis of our own making, one implicated in the larger crisis of Roman decadence and decline. The books on *materia medica* in the *Natural History* are addressed to a *cultural* failure.

---

[49] *HN* 22.14–15; see also 29.19–20. For Plato, see *Rep.* 3, 405a–b.
[50] *HN* 22.106 (on masking sympathies and antipathies), 22.117, 24.4.
[51] Wallace-Hadrill 1990, especially 86–92. For Wallace-Hadrill, the opposition forms part of a strategy that enables Pliny to "sell" Greek natural history to a Roman audience focused on utility. For the argument that Pliny's views on luxury are not purely negative, see Lao 2011. On luxury in Pliny, see also Beagon 1992, 17–18, 75–9, 190–4, Carey 2003, 76–9, 91–9.

One of the ways Pliny orients his work is by trying to carve out a place for the authorial self that avoids – or at least neutralizes – rusticity, on the one hand, and self-serving secrecy, on the other. Beginning with the second aspect of his self-presentation, it is obvious Pliny is not stingy with what he knows – indeed, for most modern readers, he is overly generous. More specifically, Pliny lays claim to a long tradition of philanthropic botanists that extends from a handful of pioneering Greeks to a group of Romans: Cato, Gaius Valgius, and Pompeius Lenaeus (25.4–5). His own role in this tradition is to make the vast store of knowledge accumulated by the ancients publicly available. The generous spirit that the ancients showed in transmitting the results of their research is thus kept alive in Pliny's vast compendium, which counters the meanness of his own age.[52]

Pliny is touchier about the rusticity of the knowledge he has to offer. On more than one occasion, he assures the reader that what looks trivial is, in fact, of great significance: it is the fault of luxury, he insists, to see the things conducive to health as paltry (25.22). He makes it part of his mission, in the preface to Book 22, to assign the lowliest plants the *auctoritas* they deserve. These contexts are where we most frequently find Pliny explicitly justifying his work in terms of its larger utility and significance, at the same time begging the reader's pardon for dealing with apparently lowly subjects. Both these devices are on display, for example, in the prelude to his discussion of the "kitchen garden" in Book 20. Two books later, he reports that he has been accused of busying himself with trifles.[53]

What gives Pliny comfort in the face of this derision is knowing that *nature herself* incurs such contempt. It is precisely the idea that he is allied with nature in his opposition to a corrupt culture that endows his claims in the books on *materia medica* on behalf of his work's practical utility with *ethical* significance. Once he has identified the unique helplessness of human beings vis-à-vis the web of sympathies and antipathies as a cultural, rather than a natural, problem, he can present his work as the way out of the quandary. The future that Pliny envisions is, in one sense, a return to the collective cultural knowledge of a utopian past: he counters the greed and luxury that have warped his own culture with the willingness and the courage to teach, traits grounded in a fundamental generosity of spirit. If such generosity is the signature virtue of the early botanists, it is also what defines nature. Indeed, in the preface to Book 27, Pliny explicitly compares the great early botanists and nature, claiming that their

---

[52] For praise of the open transmission of knowledge among the ancients, compare the discussion of Vitruvius in Long 2001, 31–3.
[53] *HN* 22.15; see also 20.1, 25.22.

"munificent bounty" nearly surpasses "the munificent disposition of nature herself" (*naturae ipsius munificentia*, 27.1). The hint of hybris is immediately countered: Pliny hastens to add, as we saw earlier, that even discoveries are gifts from nature. Nevertheless, the very fact that these findings have to be transmitted to others grants considerable responsibility and prestige to figures like Pliny himself, who become vehicles for nature's munificence.

Pliny's drive toward inclusiveness in the books on medicine, persuasively interpreted in recent years through the lens of Rome's imperial ambitions and encompassing claims on the world,[54] is thus also rooted in a belief that only sheer magnitude can replicate the abundance of nature. The efficacy of the text under these conditions is tethered not to the success or failure of individual recipes or recommendations. Rather, it belongs to its capacity for mimesis on a grand scale.[55] In other words, what we might think of as practical utility is integrated into – if not subordinated to – the creation of a text as generous as nature itself: utility functions at both the micro- and the macro-level. Pliny, in turn, occupies a position analogous to nature in the books on medicine insofar as he makes available to his readers a rich supply of resources tailored to their flourishing.[56]

Yet Pliny is not simply an analogue to but also a surrogate for nature. The fact that one of the most significant functions of the books on medicine is to close the gap between a natural world saturated with things useful to humankind and the targeted beneficiaries of this bounty is a reminder that the gap exists – that is, a reminder that the human condition must be defined by a lack of intuitive knowledge about where the human belongs in the network of sympathies and antipathies underlying the natural world. By aligning his own labor with nature's proper ends and targeting

---

[54] The book is "patterned after the vast empire that has made the universe available for knowing" (Murphy 2004, 2); it "follows a principle of accumulation which mirrors the rule of Roman power" (Naas 2011, 70); "like the 'heaps' which his work contains... Pliny's text is also an *acervus* which demonstrates the magnitude of Rome and her empire" (Carey 2003, 99). On the specific relationship between imperialist expansion and pharmacology, see Flemming 2003, focusing on the Hellenistic world. Cf. Doody 2010, 23, who sees the accumulation of facts part of Pliny's "revolutionary vision" of the nature of things, a vision that takes nature as "exactly the sum of its parts." See also Doody 2009, especially 17–21 and Doody 2010, 64–75, emphasizing the role of the reception of Pliny's text in shaping perceptions – especially anti-imperialist and post-colonial perceptions – of the ideological commitments of Pliny's encyclopedia.

[55] For the idea of the text as a mirror, see Carey 2003, 17–40, 99–101. For Carey, however, the world to be represented is the vast Roman Empire, more than nature or its benefits. See also Henderson 2011, 146–7.

[56] Several centuries later, Theodorus Priscianus, self-consciously modeling himself on Pliny, adopts a similar role, mediating between Nature's bounty and an ignorant public (*Euporista*. 1.3–4): see Formisano 2004, especially 135–40 (on the relationship of Theodorus to Pliny). I thank Marco Formisano for bringing this source to my attention.

contemporary culture as corrupt, Pliny manages not to dwell too much on our exceptional exclusion from natural knowledge. Nevertheless, the existence of the text – and, in fact, our very ability to read it in the first place – testifies to that exclusion. It also transforms the intuitive knowledge of animals into a moral question. The *Natural History* is there. Whether we use it is up to us.

REFERENCES

S. Amigues 1995, 'La Signature des plantes, source de croyances ou de savoir dans l'antiquité gréco-romaine?' in M.-Cl. Amouretti and G. Comet (eds.), *La Transmission des connaissances techniques* (Aix-en-Provence), 127–38

J. André 1955, 'Pline L'Ancien botaniste', *REL*, 33: 297–318

M. Beagon 1992, *Roman Nature: The Thought of Pliny the Elder* (Oxford)
  2005, *The Elder Pliny on the Human Animal: Natural History Book 7* (Oxford)

I. Bruns 1887, *Alexandri Aphrodisiensis praeter commentaria scripta minora*. CAG suppl. 2.1. (Berlin)

S. Carey 2003, *Pliny's Catalogue of Culture: Art and Empire in the Natural History* (Oxford)

G. B. Conte 1994, 'The Inventory of the World: Form of Nature and Encyclopedic Project in the Work of Pliny the Elder' in *Genres and Readers: Lucretius; Love Elegy; Pliny's Encyclopedia*. Trans. Glenn W. Most. (Baltimore), 67–104

Ph. De Lacy (ed. and trans.) 1992, *Galeni De semine*. CMG V 3,1 (Berlin)
  2005, *Galeni De Placitis Hippocratis et Platonis*, 3 vols. 3rd edn. CMG V 4,1,2 (Berlin)

A. Doody 2009, 'Pliny's *Natural History*: Enkuklios Paideia and the Ancient Encyclopedia', *JHI*, 70: 1–21
  2010, *Pliny's Encyclopedia: The Reception of the Natural History* (Cambridge University Press)

R. Flemming 2003, 'Empires of Knowledge: Medicine and Health in the Hellenistic World' in A. Erskine (ed.), *A Companion to the Hellenistic World* (Oxford), 449–63

M. Formisano 2004, 'The "Natural" Medicine of Theodorus Priscianus', *Philologus*, 148: 126–42

R. French 1986, 'Pliny and Renaissance Medicine' in R. French and F. Greenaway (eds.), *Science in the Early Roman Empire: Pliny the Elder, His Sources and Influence* (London), 252–81
  1994, *Ancient Natural History, Histories of Nature* (London)

R. French and F. Greenaway (eds.) 1986, *Science in the Early Roman Empire: Pliny the Elder, His Sources and Influence* (London)

F. Gaide 2003, 'Aspects divers des principles de sympathie et d'antipathie dans les textes thérapeutiques latins' in Palmieri (ed.), 129–44

P. Gaillard-Seux 2003, 'Sympathie et antipathie dans l'*Histoire Naturelle* de Pline l'Ancien' in Palmieri (ed.), 113–28

2010a, 'Morsures, piqûres et empoisonnements dans *l'Histoire Naturelle* de Pline l'Ancien' in D. Langslow and B. Maire (eds.), *Body, Disease and Treatment in a Changing World: Latin Texts and Contexts in Ancient and Medieval Medicine* (Lausanne), 305–17

2010b, 'Un pseudo-Démocrite énigmatique: Bolos de Mendès' in F. Le Blay (ed.), *Transmettre les savoirs dans les mondes hellénistique et romain* (Rennes), 223–44

R. K. Gibson and R. Morello (eds.) 2011, *Pliny the Elder: Themes and Contexts* (Leiden)

A. Goulon 1972, 'Le Malheur de l'homme à la naissance: Un thème antique chez quelques pères de l'Église', *Revue des études augustiniennes*, 18: 3–26

J. Hahn 1991, 'Plinius und die griechischen Ärzte in Rom: Naturkonzeption und Medizinkritik in der *Naturalis Historia*', *Sudhoffs Archiv*, 75: 209–39

G. Helmreich 1893, *Galeni Pergameni Scripta Minora*, vol. III (Leipzig)

1907–9, *Galeni De usu partium libri XVII*, 2 vols. (Leipzig)

J. Henderson 2011, 'The Nature of Man: Pliny, *Historia Naturalis* as Cosmogram', *MD*, 66: 139–71

B. Holmes 2013, 'The Poetic Logic of Negative Exceptionalism in Lucretius, Book Five' in D. Lehoux, A. D. Morrison and A. Sharrock (eds.), *Lucretius: Poetry, Philosophy, Science* (Oxford), 153–91

W. Kroll 1934, 'Bolos und Demokritos', *Hermes*, 69: 228–32

C. G. Kühn (ed. and trans.) 1821–33, *Medicorum Graecorum Opera quae exstant: Claudius Galenus*, 20 vols. (Leipzig, repr. Hildesheim 1964–5)

E. Lao 2011, 'Luxury and the Creation of a Good Consumer' in R. K. Gibson and R. Morello (eds.), *Pliny the Elder: Themes and Contexts* (Leiden), 35–56

É. Littré (ed. and trans) 1839–61, *Œuvres complètes d'Hippocrate*, 10 vols. (Paris)

P. O. Long 2001, *Openness, Secrecy, Authorship: Technical Arts and the Culture of Knowledge from Antiquity to the Renaissance* (Baltimore)

A. O. Lovejoy and G. Boas 1935, *Primitivism and Related Ideas in Antiquity* (Baltimore)

D. Manetti and A. Roselli (eds. and trans.) 1982, *Ippocrate: Epidemie, Libro Sesto* (Florence)

K. F. T. Mayhoff 1906–9, *C. Plini Secundi Naturalis historiae libri XXXVII* (Leipzig, repr. 1967–70)

I. Mazzini 1986, 'Présence de Pline dans les herbiers de l'Antiquité et du haut Moyen-Age', *Helmantica*, 37: 83–94

T. Murphy 2004, *Pliny the Elder's Natural History: The Empire in the Encyclopedia* (Oxford)

V. Naas 2011, 'Imperialism, Mirabilia, and Knowledge: Some Paradoxes in the *Naturalis Historia*' in R. K. Gibson and R. Morello (eds.), *Pliny the Elder: Themes and Contexts* (Leiden), 57–70

Ch. Nauert 1979, 'Humanists, Scientists, and Pliny: Changing Approaches to a Classical Author', *American Historical Review*, 84: 72–85

D. Nickel (ed. and trans.) 2001, *Galeni De foetuum formatione*. CMG V 3,3 (Berlin)

V. Nutton 1986, 'The Perils of Patriotism: Pliny and Roman Medicine' in R. French and F. Greenaway (eds.), *Science in the Early Roman Empire: Pliny the Elder, His Sources and Influence* (London), 30–58

V. Nutton and G. Bos 2011, *Galen, On Problematical Movements* (Cambridge University Press)

N. Palmieri (ed.) 2003, *Rationnel et irrationnel dans la médecine ancienne et médiévale: Aspects historiques, scientifiques et culturels* (Saint-Étienne)

O. Rackham 1996, 'Ecology and Pseudo-Ecology: The Example of Ancient Greece' in J. Salmon and G. Shipley (eds.), *Human Landscapes in Classical Antiquity: Environment and Culture* (London), 16–43

R. Renehan 1981, 'The Greek Anthropocentric View of Man', *HSCP*, 85: 239–59

B. Rochette 1992, 'Nudus... infans...: À propos de Lucrèce, V, 222–227', *LEC*, 60: 63–73

D. Sacré 1992, 'Nudus... infans (Lucrèce, V, 222–227): La survie d'un littéraire τόπος dans la poésie néo-latine', *LEC*, 60: 243–52

G. Serbat 1973, 'La Référence comme indice de distance dans l'énoncé de Pline l'Ancien', *Revue de philologie, de littérature et d'histoire anciennes*, ser. 347: 38–49

R. Sorabji 1993, *Animal Minds and Human Morals: The Origins of the Western Debate* (Ithac a, NY)

J. Stannard 1982, 'Medicinal Plants and Folk Remedies in Pliny, *Historia Naturalis*', *History and Philosophy of the Life Sciences*, 4: 3–23

R. B. Todd 1976, *Alexander of Aphrodisias on Stoic Physics* (Leiden)

L. Totelin 2012, 'And to End on a Poetic Note: Galen's Authorial Strategies in the Pharmacological Books', *Studies in History and Philosophy of Science*, 43: 307–15

H. von Staden 1996, 'Liminal Perils: Early Roman Receptions of Greek Medicine' in F. J. Ragep and S. P. Ragep (eds.), *Tradition, Transmission, Transformation* (Leiden), 369–418

A. Wallace-Hadrill 1990, 'Pliny the Elder and Man's Unnatural History', *Greece and Rome*, 37: 80–96

M. Wellmann 1897, 'Bolos [3]', *RE* 3, 1: coll. 676–7

E. Wenkebach and F. Pfaff 1956, *Galeni In Hippocratis Epidemiarum librum VI commentaria*. CMG V 10,2,2 (Berlin)

M. J. White 2003, 'Stoic Natural Philosophy (Physics and Cosmology)' in B. Inwood (ed.), *The Cambridge Companion to the Stoics* (Cambridge University Press), 124–52

G. Williams 2012, *The Cosmic Viewpoint: A Study of Seneca's "Natural Questions"* (Oxford)

CHAPTER 14

From Descriptions to Acts
The Paradoxical Animals of the Ancients from a Cognitive Perspective

Pietro Li Causi

This chapter takes a somewhat different move from others in this volume, since it considers the question of applicability not so much in practical but rather in cognitive terms. In fact, the zoological sections of the *naturales historiae* of the ancients cannot be regarded, *stricto sensu*, as technical texts. For example, while consulting an *animalium historia*, it is hardly likely that the reader faces the problem of transition from written instructions to their practical application in the context of extra-linguistic reality, as is the case with agricultural manuals, cookbooks, or cosmetic and architectural treatises.[1] However, even for the texts that I am going to focus on in this chapter, it is possible to speak of a problem of applicability in cognitive terms: in fact, even though the reader of *naturales historiae* does not have the problem of translating into practical action what he reads, there is always a process of negotiation going on in his mind between the written text and the external world. To put it simply, when we read of the fabulous animals we find in Pliny or in Solinus, it is easy to formulate questions such as the following: 'How is it possible to "know" objects like the manticore, the *corocotta*, the unicorn ass, the griffin, or the giant ants of India?'[2] We tend to classify these beasts as fantastic or imaginary animals; yet from an experience-oriented perspective, we must consider that the accounts

---

[1] As for the question of the passage from words to acts in Vitruvius' *De architechtura* (*De arch.*) or in ancient cosmetic texts, see respectively the contributions of E. Romano and L. M. V. Totelin in this volume (respectively Chapters 4 and 8).
[2] See Li Causi 2003, 23 ff. For the manticore, see e.g. Ctesias, *FGrHist* III C 688, F 45, 15 (=Phot. *Bibl.* 45b 31–46 a 12 Henry), Aristotle, *Hist. an.* II 1, 501a24–b1, Philostratus, *VA* III 45, Aelianus, *NA* IV 21, Pausanias IX 21, 4, Solinus 52, 37 f., Eusebius, *Hierocl.* 22, Pliny, *HN* VIII 75 and 107. For the griffin, see Aeschylus, *Pr.* 804 ff., Herodotus III 116, 1, Ctesias, *FGrHist* III C 688 F 45, 6, Lucianus, *Ver. hist.* I 11, Aelianus, *NA* IV 27, Philostratus, *VA* VI 1, Pliny, *HN* VII 10, Solinus 15, 22, Mayor 2000, 15 ff. For the giant ants of India, see Herodotus III 98–105, Strabo XV 1, 44, Pliny, *HN* XI 111, Propertius III 13, 1–5, Arrian, *Ind.* XV 4–9, Philostratus, *VA* VI 1, Lucianus, *Gall.* 16, Harpocration, *Lex.* s.v. χρυσωχοεῖν, pp. 307–9 Dindorf, Dio Chrysostomus 35, 23–5, Heliodorus Eroticus X 26, 2, Li Causi and Pomelli 2001–2, 177 ff. For the unicorn ass, see below; for a complete dossier of texts ranging from ancient times up to the nineteenth century, see Shephard 1984.

of ancient travellers and historians depicted them as realistic or even as *realia*.

Of course, we do not have any direct experience of these 'somethings' the ancients considered actual beings, but we do possess descriptions of them from a number of texts. The problem is, therefore, to understand how these descriptions may lead us to the recognition of beings of which the only visual representations available to us are dependent on verbal accounts fixed in writing rather than on autoptic sightings. In other words, the question is to understand whether the descriptions transmitted by the *historiae* are 'applicable' or not. Before answering this question, however, a brief theoretical digression may be appropriate.

Cognitive Types and *camelopardaleis*

Umberto Eco uses the term Cognitive Type (henceforward CT) to refer to the mental procedure for constructing the three-dimensional and multimedial image of a known 'something', e.g. a known animal.[3] But what happens when we run into a being we have never encountered before and for which we have no CT to assimilate it to?

As Eco points out, in such cases 'when faced with an unknown phenomenon, we react by approximation: we seek that scrap of content, already present in our encyclopaedia, which, for better or worse, seems to account for the new fact'.[4] In other words, the unknown is reduced to the known. This is why, for example, the descriptions of many exotic creatures that are indexed by the *naturales historiae* of the ancients make large use of comparisons, similes and analogies.

Such rhetorical artifices operate as *hypotyposeis* aimed at creating an illusion of ostensive evidence that, through a reality effect, can replace the autoptic perception. We can see an example of this in the following passage from Pliny's *Naturalis Historia* (Pliny, *HN* VIII 69): 'The Ethiopians give the name of *nabus* to one [i.e. animal] that has a neck like a horse, feet and legs like an ox, and a head like a camel, and is of a ruddy colour picked out with white spots, owing to which it is called the camelopard [*camelopardalis*].'[5] The similes of the passage suggest that those who saw the *camelopardalis* (or *nabus*) for the very first time reacted to a bundle of traits that activated the proper domains of a series of known animals.

[3] See Eco 2000, 130 ff.   [4] See Eco 2000, 57.
[5] English trans. Rackham 1947. Horace, *Ep.* II 1, 195 refers to this beast, when speaking of an object calculated to excite the vulgar gaze, as *diversum confusa genus panthera camelo* ('the different race of the panther mingled with the camel').

The CT of the *camelopardalis*, in other words, was built initially on the basis of metaphorical impressions that resulted from the joint experience of agglutinated false positives. The items belonging to the actual domain of the previously unknown animal (the *camelopardalis*) did not fit the proper domain of all the known animals that it resembled: the neck looked like that of a horse, but the animal in question was not a horse. Its head reminded one of the proper domain of a camel, but it was not a camel. Its coat was like that of a leopard, but it was not a leopard. Thus we can say that the unknown is reduced to the known, but the agglutination of analogical traits creates the instructions for a cognitive act, i.e. the recognition of a new 'generic species' whose CT can be culturally transmitted by the accounts of the ancients and whose proper domain can be activated in our brain.[6] These cognitive instructions are fairly simple to follow, and, once confronted with a *camelopardalis*, we would not hesitate to apply them and recognize the animal. We would not call it a *camelopardalis*, however, but a 'giraffe', or, if we wanted to use the Linnaean name, *Giraffa camelopardalis*.

An adventure similar to that of the giraffe for the Romans, after all, is what must have happened when the first horse set foot in the Americas. In this regard, Umberto Eco (2000, 127 ff.) imaginatively reconstructed the cognitive process of the discovery of this animal by the Aztecs before the destruction of their civilization. As the sightings of the strange four-legged beast with long hair on its neck increased, the Aztecs began to understand even more of the features of the animal that they had started calling 'kawayo'. They were, in fact, building what Eco calls Nuclear Content (henceforward NC), i.e. the set of *interpretants* and their publicly transmitted representations of the animal in question, thus circumscribing the CT of the strange mammal introduced by the Spaniards. 'In other words, the Aztecs gradually interpreted the features of their CT in order to homologate it as much as possible. While their CT (or CTs) might have been private, these interpretations were *public*.'[7]

In a similar way, the first CT of a *camelopardalis* might have been built around a private mental representation constructed by the first Roman (or Greek) who sighted it, but then, little by little, a series of public descriptions and interpretations (including the one in Pliny's *Naturalis Historia*) began to circulate, thus homologating the image of the animal and creating the modules responsible for its recognition.

---

[6] The notion of 'generic species' has been developed by Atran 1996, 5 ff. and 2006, 6 ff., who notices that the notions of genus and species are usually co-extensive in folk taxonomies.

[7] See Eco 2000, 136 f. As for the notions of 'proper domain' and 'actual domain' in cognitive sciences, see Sperber and Hirschfeld 2004, 40.

The point is, however, that often, for many of the paradoxical and unusual creatures described in the natural histories of the ancients, it is not easy to get to the level of approval and homologation that had been reached for the horse or the *giraffa camelopardalis*. For many of these beings, there are only written descriptions and definitions, and the *percepienda in se* are completely lacking. We know that the giraffe was shown to the Romans for the first time by Caesar during a circus,[8] but no one – maybe with the exception of Ctesias – had ever been able to claim having seen, for instance, a manticore.[9]

The descriptions of some strange creatures that circulated, give, as we have seen, the ostensive illusion of the vision by means of analogies and similes. The point is, however, that though vivid, such illusions are either always deprived of actual domains or their actual domains are destined to create necessarily false positives. Such descriptions, in other words, cannot become 'factual beliefs' – i.e. beliefs based on perception – or acts of recognition, and remain 'representational beliefs' that are culturally transmitted only by means of an epidemiology of ideas (and, of course, words).[10]

Umberto Eco would say, in this regard, that the CTs the ancients had of many exotic animals did not arise from perceptual experience but were transmitted through the formation of public NCs (natural histories, travellers' reports) that implicitly refer to perceptual experiences to come. These perceptual experiences, however, almost never materialized, and consequently the descriptions that circulated could only rarely be applied. Indeed, it is a matter of fact that the ancients, though their brains probably worked exactly like ours, were equipped with boundaries of reality that were less extensive than ours. For example, at the time of the Greeks and the Romans, journeys to far-off lands were much more complicated, so the direct verification of the data transmitted by the written tradition was extremely problematic.

The Desire of Cognitive Experience and the Survival of Monsters

Through the centuries, news of the fantastic animals of the *eschatiai* fed an enormous curiosity. Their descriptions created expectations, so that those who visited the distant regions of the known world wanted to see

---

[8] Pliny, *HN* VIII 69: *dictatoris Caesaris circensibus ludis primum visa Romae* ('it was first seen in Rome in the Circensian games held by Cæsar, the Dictator').
[9] We are told by Aelian (*NA* IV 21) that Ctesias has professed that he had personally seen a manticore in India, at the court of the king of Persia. For Aelian's manticore, see Li Causi 2003, 83 ff. and 251 ff.
[10] As for the notions of 'factual belief' (dealing with easily graspable external entities or events) and 'representational belief' (whose object is a semipropositional representation of an incomplete conceptual content), see Sperber 1982, 169ff.

the strange beasts. To give just one example, Philostratus tells us that Apollonius of Tyana, the charismatic philosopher and healer of the first century AD, after going to India, wanted to find out if all the biological and geographical oddities described by Ctesias' *Indika* really existed. He needed, in other words, to enforce and 'apply' the CTs that were part of his cultural encyclopaedia. In particular, he needed to know whether there were indeed sources of liquid gold and whether the manticore, that fabulous beast with the face of a man, the body of a lion and the tail of a dart-throwing scorpion, existed. However, the answer given by his guide and informant Iarcas was disappointing (Philostratus, *VA* III 45):

> What do I have to tell you about animals or plants or fountains which you have seen yourself on coming here? For by this time you are as competent to describe these to other people as I am; but I have never yet heard in this country of an animal that shoots arrows or of springs of golden water [θηρίον δὲ τοξεῦον ἢ χρυσοῦ πηγὰς ὕδατος οὔπω ἐνταῦθα ἤκουσα].[11]

However, perceptual experiences are not always so disappointing; or, at least, not entirely. Pliny the Elder, for example, says that 'the horns of an Indian ant, suspended in the temple of Hercules, at Erythræ, have been looked upon as quite miraculous because of their size' (*HN* XI 111: *Indicae formicae cornua Erythris in aede Herculis fixa miraculo fuere*). We could say that we are dealing with a partial application of a CT transmitted by the zoological encyclopaedia of the ancients. News of the giant ants of India was reported for the first time by Herodotus, and then by Nearchus and Megasthenes, who told the story of a fight between the animal and gold prospectors living in the northern part of India, near Caspatyrus.[12] In this regard, the Plinian account refers to an animal relic, a single organic part that was used semiotically as an argument to demonstrate the existence of the whole strange creature: what has happened in this case is that the presence of a 'something'– whatever it was – that had no resemblance to the horns of the commonly known horned animals has supported a transfer from the words of a culturally transmitted belief to an act of recognition.

Of course, we do not know much about these horns that the Roman encyclopaedist refers to. We can imagine, however, the creation and, consequently, the circulation of a local story that told how Hercules had broken the horns of a monstrous exemplar of the gigantic animal near Erythrae.

---

[11] English trans. Conybeare 1948. For this passage, see Li Causi 2003, 290 ff.
[12] See Herodotus III, 102 ff. The works of Nearchus, admiral of the fleet of Alexander, and Megasthenes, ambassador of Seleucus I, have been lost. For their experiences with the giant ants, see Strabo XV 1, 44, Arrian, *Ind.* XV 4 ff., and Li Causi and Pomelli 2001–2, 177 ff.

The tale, associated with the exhibition of the relic, could have created a powerful reality effect, thus feeding hopes and encouraging wishful thinking: sooner or later, the perceptual experience of the animal in its entirety would be realized, and the felicitous recognition would become a reality. The monster could have survived, and, consequently, its description was still thought to be applicable.

Applying a description: the unicorn is a rhinoceros (or not?)

In the natural histories of the ancients, there are cases even more complex than these. To give just one example, the vicissitudes of the CT of the unicorn, of whose epidemiological chain I shall quickly cover just a few links, are emblematic.

After a brief mention in Herodotus (IV 191, 4), the first exhaustive attestation of the existence of the beast is in Ctesias (*FGrHist* F 45q =Aelianus, *NA* IV 52):

> I have heard that there are wild asses in India no smaller than horses, which have a white body, a head which is almost crimson, and dark blue eyes. They have a horn on their brow one and a half cubits in length. The lower portion of the horn is white, the upper part is crimson, and the middle is very dark. I hear that the Indians drink from these multicolored horns, but not all the Indians, only the most powerful, and they pour gold around them at intervals as if they were adorning the beautiful arm of a statue with bracelets. They say that the one who drinks from this horn will never experience terminal illnesses. No longer would he suffer seizures or the so-called sacred disease nor could he be killed with poison. If he drank the poison first, he would vomit it up and return to health. It is believed that the other asses throughout the world, both tame and wild, and the rest of the other solid-hoofed animals do not have an astragalus in their ankle nor do they have bile in their liver. According to Ctesias, however, the one-horned Indian asses have astragaloi and are not lacking bile. They say their astragaloi are black and if someone should grind them up they would be the same on the inside. These creatures are not only faster than other asses, but horses and deer as well. They begin to run lightly, but gradually they run harder and to pursue one is, to put it poetically, to chase the intangible. When the female gives birth and guides her newborns about, the sires join them in the pasture and watch over their young. These asses are found on the most desolate plains in India. When the Indians set out to hunt them, the asses allow those that are still young and tender to graze behind them while they fight and charge the horsemen at close quarters and strike them with their horn. Such is their strength that nothing can endure their impact. Everything succumbs to them and gets pierced; however, if by chance it

is crushed to pieces, it is rendered useless. They have attacked the sides of horses and ripped them open disemboweling them. For that reason, the horsemen are too afraid to go near them because the price for getting too close is a horrible death for both themselves and the horses. The asses also have a deadly kick and their bite reaches such a depth that whatever is caught in its grip is completely torn away. You could not capture a full grown ass alive, but they are killed with javelins and arrows and when it is dead, the Indians remove the much revered horn from the animal. The flesh of the Indian ass is inedible because it is so bitter. (English trans. Nichols 2008)

Almost all modern commentators agree that the beast described by the physician of Cnidus is actually the rhinoceros.[13] In fact, as Karttunen asserts, this animal was actually present in the region of the Indus at the time in which the *Indika* was written. It is also worth noting that, according to Karttunen, some traits attributed to the Indian ass – its way of running and, of course, the characteristic bulge on its forehead – seem to coincide with those of the rhinoceros.[14] Moreover, it is important to point out that the rhinoceros has always been hunted in India for the healing properties attributed to its various bodily parts.[15] The similarities, however, end here. The rhinoceros' horn – which is not polychrome – is on the animal's nose, not its forehead. Furthermore, an arrow cannot dent its armour and its meat – it seems – is edible.

Details such as these perhaps should not be overlooked, especially if we consider that when the Romans first began to see real rhinoceroses at the circus, they never thought of applying the description kindly furnished by Ctesias. Pliny (*HN* VIII 71), for one, speaks of the rhinoceros shown at the circus by Pompey as a unicorn beast, but its features do not coincide with those described by Ctesias, whereas Martial (*Spect.* 22) notes that the animal in question is *gemino cornu*, i.e. having two bumps on its head.

Thus, for the encyclopaedia of the Romans, the descriptions (and the CTs) of the two beings are incompatible, and the rhinoceros and the unicorn continue to be considered different generic species. The same happens with the *kartazonus*, a one-horned being – allegedly a rhinoceros – described for the first time by Megasthenes. Aelian speaks of this animal in his *De natura animalium*. Although some of its traits appear similar to those of the rhinoceros, he does not suspect that the animal described in *NA* XVI 20 is the same as the one described afterwards in Book XVII,

---

[13] See Lenfant 2004, 315 n. 871, Nichols 2008, ad l. In the first chapter of his book, Shepard 1984 sees in Ctesias' unicorn the result of the agglutination of three different actual animals: the rhinoceros, the Tibetan antelope (i.e. the so called 'chiru') and the onager. Lavers 2010, 15 ff. proposes the substitution of the onager with the 'kiang'.
[14] See Karttunen 1989, 170.   [15] See Ylla 1958, 73, Prater 1971, 192.

of which creature he says that 'there are many Greeks and Romans who know it from having seen it' (ἴσασι γὰρ καὶ Ἑλλήνων πολλοὶ καὶ Ῥωμαίων τεθεαμένοι: *NA* XVII 44):

> In certain regions of India (I mean in the very heart of the country) they say that there are impassable mountains full of wild life ... in these same regions there is said to exist a one-horned beast, which they call *Cartazonus*. It is the size of a full-grown horse, has reddish hair, and is very swift of foot. Its feet are, like those of the elephant, not articulated and it has the tail of a pig. Between its eyebrows it has a horn growing out; it is not smooth but has spirals of quite natural growth, and is black in colour. This horn is also said to be exceedingly sharp. And I am told that the creature has the most discordant and powerful voice of all animals. When other animals approach, it does not object but is gentle; with its own kind, however, it is inclined to be quarrelsome. And they say that not only do the males instinctively butt and fight one another, but that they display the same temper towards the females, and carry their contentiousness to such length that it ends only in the death of their defeated rival. The fact is that strength resides in every part of the animal's body, and the power of its horn is invincible. It likes lonely grazing-grounds where it roams in solitude, but at the mating season, when it associates with the female, it becomes gentle and the two even graze side by side. Later when the season has passed and the female is pregnant, the male Cartazonus of India reverts to its savage and solitary state. They say that the foals quite young are taken to the King of the Prasii and exhibit their strength one against another in the public shows, but nobody remembers a full-grown animal having been captured.[16] (Aelian, *NA* XVI 20)

We find ourselves faced with a proliferation of CTs. Cores of *pensée sauvage* concerning unicorn animals are put in quotes, stored and accumulated in the hypothetical inventory of the world that was the zoological encyclopaedia of the ancients. The image we have is that of a huge mass of data waiting to be processed, but which – due to the objective difficulty of travelling in the ancient world – can only be transmitted culturally and is difficult to apply to that 'something' which is reality.[17] Thus the possibility of empirical verification, a necessity for the modern sciences, was a matter of chance in the ancient world, and it was normal, in problematic situations, to rely on the principle of authority. The result was that different traditions resulted in different CTs (and NCs) of animal species that were perhaps the same.

One reason for the proliferation of CTs was not simply the application of the descriptions, but the enumeration and the catalogue itself. As some recent studies have pointed out, the geographical explorations and

[16] English trans. Scholfield 1972.   [17] The definition of reality as *something* is in Eco 2000, 12 ff.

discoveries of new species that took place after the expansion of the Roman Empire did not arise from the need to check the stock of knowledge accumulated previously, but rather were clearly born of the urge to multiply exponentially the files of the zoological encyclopaedia of the time.[18] Therefore, it is no coincidence that this mechanism of proliferation appears to stop in the Middle Ages, when the perception of the world faced a process of restriction and when animals were no longer as important for their physical characteristics as for their moral and theological affordances.[19] As we shall see, such new affordances will allow the CT of the unicorn to survive throughout the Middle Ages and for part of the modern age with the complicity of two translations: one from Greek, the other from Latin.

The Unicorn Exists, Guarantees the Bible

In the King James Bible there are several passages in which the unicorn is mentioned. Here are just two examples:

> His glory is like the firstling of his bullock, and his horns are like the horns of unicorns: with them he shall push the people together to the ends of the earth. (*Dt.* 33:17)

> Save me from the lion's mouth; for thou hast heard me from the horns of unicorns. (*Ps.* 21:22)

As noted by Odell Shepard, 'one thing is evident in these passages: they refer to some actual animal'.[20] The point is, however, that the term used in the original texts – *Re'em* – does not necessarily refer to an animal with a single horn on its forehead. From the attestations, we understand that the mysterious animal described in the Old Testament was certainly wild and as fierce as the lion, the bull and the unicorn asses of the classical tradition. Moreover, it was apparently provided with bony bulges. Its final metamorphosis into the unicorn, however, is due to the *Septuagint*, which translates *Re'em* with μονόκερως, and to Jerome's Vulgate, which sometimes uses *rhinoceros* and sometimes *unicornis*.[21]

---

[18] See Beagon 1992, 9 ff. and Li Causi 2003, 202 ff. and bibliography.
[19] See e.g. Zambon 1993, 11 ff., Morini 1996, vii ff., Shepard 1984, 33 ff. An *affordance* is a quality that an object has and that *affords* a limited set of actions. The concept has been developed by the psychologist James J. Gibson (1979) and afterwards applied to the field of cultural studies by Bettini (1998, 202 ff.). In Bettini's view, the affordances correspond substantially to the possibilities of symbolic re-use which the individual features of an object, or an organism, quite naturally possess.
[20] See Shepard 1984, 34. But see even Lavers 2010, 44 ff., who proposes an original interpretation of *Dt.* 33:17.
[21] For the term *rhinoceros*, see *Nm.* 23:22; *Dt.* 33:17; *Job* 29:9–12. The term *unicornis* is used in *Ps.* 21:22; 28:6; 77:69; 91:11; *Is.* 34:7 instead. It would be interesting to analyze the reason for this variation.

What happens is that the CT transmitted by the tradition of the ancient *naturales historiae* interferes with the process of translation from one language to another. By using a cognitivist metaphor, we could say that the bundle of traits implicitly attributed by the biblical texts to the *Re'em* functions as an actual domain, which creates a false positive, and refers, in the minds of the translators, either to the proper domain of the rhinoceros or to the one of the unicorn ass, i.e. to an animal of which there are hardly any sightings but which has always been thought to exist.

It goes without saying that since the Bible speaks of the unicorn as if it were real, this is proof enough that the unicorn described by Ctesias – or something similar – exists. The translators, while waiting to apply the CT of the Indian animal to a real occurrence, are content to cognitively overlay it with an enigmatic animal – the *Re'em* – whose ethological contours and Gestalt appear blurred and ambiguous. The mental image that the Septuagint and Jerome have of the unicorn is applied to make an unknown and mysterious object less opaque and elusive and, in a certain way, to clarify it.

To complete the work, then, there is the *Physiologus*, which finally consigns the CT of the unicorn to the mediaeval bestiaries, thus helping to create a new tradition destined to survive for centuries:[22]

> There is an animal which is called *monoceros* by the Greeks, and *unicornis* in Latin. The *Physiologus* says that the unicorn has this nature: it is a small animal and is similar to a goat, very fierce, having one horn in the middle of its forehead. And no hunter is able to capture it. But by this trick it is captured: a virgin girl is led to the place where it lives, and is left there alone in the woods. And as soon as the unicorn sees her, it leaps into her lap and embraces her, and thus, it is seized by those who are spying on it and is put on display in the palace of the King. And thus our Lord Jesus Christ, spiritual unicorn, descending into the uterus of the virgin, by the means of the flesh assumed from her was captured by the Jews and was sentenced to die on the cross, he who, up until then, had been invisible to us with his father. (*Physiologus latinus*, versio bis, 16)

From this moment on, Ctesias and Aelian, whose descriptions of the Indian ass and the rhinoceros were applied in the translations of the Septuagint and Jerome, are no longer needed. The nature of the unicorn has changed: it no longer has an equine form, but looks like a goat and, above all, becomes a symbol of Christ.[23] Moreover, a limit is placed on its legendary

However, it should be noted that the Latin version of the *Physiologus* (16) explains that the term *rinoceros* and the term *monoceros* (translated into Latin by *unicornis*) have the same meaning.

[22] See Shepard 1984, 40 ff. Ambrose and Tertullian, however, co-operate to form and consolidate the tradition inaugurated by the *Physiologus*: see Lavers 2010, 59 ff. in this regard.

[23] As for the Christianization of the unicorn, see Lavers 2010, 44 ff. and 63 ff.

ferocity with the invention of the expedient of hunting with a virgin. The transformations and adaptations, however, do not end there.

Once Again, the Unicorn is a Rhinoceros

When Marco Polo travels to India and Java, he sees a strange animal with a single horn on its forehead, and he does not believe his eyes. Once the Roman circuses began to decline in Europe, the rhinoceros disappeared. The term *rinoceros* still existed, but the etymology given by the *Physiologus* explained that it was a synonym for *monoceros*, the Greek word for *unicornis*:

> The Greeks call it *rhinoceros*, which in Latin means 'horn on the nose'; it is the same as *monoceros*, i.e., the unicorn, because it has, on its forehead, one horn four feet long, so sharp that it can puncture whatever has attacked it or has brandished a weapon in the air. Indeed, it often fights with elephants and kills them by wounding them in the stomach. (*Physiologus latinus*, versio bis, 16)

As is clear from the reference to the fight with the elephant, some traits of the rhinoceros have been applied to the CT of the unicorn, and, ironically, the legendary being has survived, while the rhinoceros has disappeared in the current zoological encyclopaedia of the Middle Ages. For this reason, when Marco Polo sees a specimen of the beast, he has no doubts. It is, for sure, a unicorn:

> There are wild elephants in the country, and numerous unicorns, which are very nearly as big. They have hair like that of a buffalo, feet like those of an elephant, and a horn in the middle of the forehead, which is black and very thick. They do no mischief, however, with the horn, but with the tongue alone; for this is covered all over with long and strong prickles [and when savage with anyone, they crush him under their knees and then rasp him with their tongue]. The head resembles that of a wild boar, and they carry it ever bent towards the ground. They delight much to abide in mire and mud. 'Tis a passing ugly beast to look upon, and is not in the least like that which our stories tell of as being caught in the lap of a virgin; in fact, 'tis altogether different from what we fancied. (Marco Polo, *Il Milione*, 162, 14–17)[24]

This is Umberto Eco's comment on the cognitive experience of the Venetian merchant:

> Marco Polo seems to have made a decision: rather than rearrange the content by adding a new animal to the universe of the living, he has corrected the

[24] English trans. Yule 1993. See G. R. Cardona's commentary on this passage in Bertolucci Pizzorusso 1975, ad l.

contemporary description of unicorns, so that, if they existed, they would be as he saw them and not as the legend described them. He has modified the intension and left the extension unchanged. Or at least that is what it seems he wanted to do, or in fact did, without bothering his head overmuch regarding taxonomy.[25]

In other words, by a simple mechanism of cognitive economy, Marco Polo decides to 'apply' a known CT to something that otherwise would have run the risk of sounding too new and unusual to be true. To do this, it is, therefore, necessary to change a few specific traits of the animal. It is important, however, to understand which ones.

In this regard, Umberto Eco, in *Kant and the Platypus*, distinguishes between the 'cancellable properties' and the 'indelible properties' that every object has. More specifically, 'cancellable properties are *sufficient* conditions for recognition (such as striking a match to produce combustion), while indelible properties are seen as *necessary* conditions (there can be no combustion in the absence of oxygen)'.[26] Eco then specifies that 'the recognition of a property as indelible depends on the history of our perceptual experiences. The zebra's stripes strike us as indelible properties, but it would be sufficient if evolution had produced breeds of horse or ass with striped coats; the stripes would become all too cancellable because we would have shifted our attention to some other characterizing feature.'[27] It follows that the contexts and the frontiers of knowledge in which we are immersed are the variables that determine the erasability or non-erasability of certain properties. Thus for Marco Polo – who probably inherited the CT of the unicorn from the mediaeval bestiary (rather than from Ctesias) – the uniqueness of the horn on the forehead suddenly becomes the only indelible property that can lead to recognition and to the full applicability of a description, while all other traits can be easily modified.

The change of these traits, however, determines the change of the entire NC of the unicorn and eventually the modification of part of the whole categorial system of the tradition:[28] the unicorn cannot be captured with virgins, it does not have a slender, agile body, but it is similar in size to an elephant, and rather than hoofs it has elephant feet as well. The description of the beast – and its CT – can finally be applied to what we clearly recognize as a rhinoceros. In other words, the known can now be applied to the unknown. Unfortunately (or fortunately, depending on one's point of view), this does not mean that truth has won out over fantastic theories, or that facts have won out over words.

[25] Eco 2000, 58.   [26] Eco 2000, 239.   [27] Eco 2000, 240.   [28] See Eco 2000, 248 f.

The Unicorn Survives Marco Polo

It might seem that the history of the CT of the unicorn ends here, and that the sightings by Marco Polo finally make a clean sweep of a misunderstanding that had lasted for centuries. However, the stories of cultural representations are often more complex than we might imagine. In fact, after Polo's journey to Java, rhinoceroses still continue to be invisible in the West for at least another two centuries; this lasts until 1498, when an African exemplar of the animal doubles Cape Horn and is brought to Lisbon, to the palace of the king of Portugal, and is later portrayed in a famous engraving by Albrecht Dürer.[29]

The long desuetude of the Western world regarding the class of the *rhinocerotidae* leads back to a doubling of the CTs (that of the unicorn and that of the rhinoceros) that Marco Polo had implicitly proposed to unify. In addition, from the fourteenth century on, in all the courts of Europe, strange objects called 'alicorns' (allegedly the horns of the unicorn) begin to circulate, and kings, dukes, popes and physicians begin to attribute miraculous curative properties to these objects, which, when crushed, dissolved in a liquid and drunk can immunize anyone against poisoning.[30] In many cases they are the teeth of a narwhal, or, sometimes, pieces of worked ivory.[31] However, for the notables and noblemen who buy them, or obtain them in mysterious ways, there is no doubt that these are, indeed, the horns of unicorns. As with the horns of the Indian ant, these precious objects become semiotic arguments ready to prove the existence of the whole fantastic creature. The belief in the animal described for the first time by Ctesias and – albeit with some changes – by the bestiaries, is further reinforced by biblical testimonies, and thus becomes an article of faith.

In 1556, in his *Discorso contra la falsa opinione dell'Alicorno* (*Speech against the False Opinion of the Alicorn*), the physician Andrea Marini tries to show the falsity of the belief in the beast, and especially in the curative properties of its horn; but ironically his work does not have the expected success. Not only do the natural histories of Gesner, Aldovrandi and Topsell continue to file the unicorn among the things worth talking about, but, most importantly, in the wake of what was written by Johann Homilius in his dissertation *De monocerote*, discussed in Leipzig in 1667, scholars who doubt the existence of an animal spoken about in the Scriptures begin to be accused of blasphemy.[32]

---

[29] See Shepard 1984, 261.
[30] See Shepard 1984, 113 ff. and Lavers 2010, 94 ff.
[31] See Shepard 1984, 309 ff.
[32] For the Modern Age's debate on the existence of the unicorn, see Shepard 1984, 183 ff.

The CT of the beast described for the first time by Ctesias is constantly being renegotiated, and still survives. Or at least some indelible traits continue to have more chance of survival than others. In this regard, on the basis of what Odell Shepard points out in his *The Lore of the Unicorn*, it is understandable how, in a Europe whose dukes and kings are nothing short of obsessed with the risk of poisoning, the property of an animal that becomes indelible is linked with the horn's hyperbolic capacity to operate as an antidote.[33]

Updates and Conclusions: Unicorns, Roe-Deer and TV News

In conclusion, from the stories I have traced here one could say that the CT of the unicorn works as a sort of 'chewing-gum notion', i.e. as an always negotiable mental image, which assumes 'configurations that vary according to circumstances and cultures'.[34] We are, therefore, faced with a very complex mechanism, which concerns the cultural encyclopaedia of the Western world itself, whose zoological knowledge tends, economically, to safeguard over the centuries some of its files – i.e. its own CTs and NCs – always believing in their cognitive applicability.[35]

This belief, however, does not result – as usually happens with the transmission of technological knowledge – in mechanisms of mimetic repetition,[36] since on each occurrence of the description of the unicorn – whether it is generated by perceptual experience or simply by textual transmission – the enabled concepts and categories modify their indelible traits by means of what we could call a *cognitive contract*. It is through this contract that the words of the descriptions are each time adapted to the facts or their contexts.

In this way, I think it is possible to confirm Dan Sperber's perspective on the epidemiology of beliefs, according to which at each public transmission of a cultural representation, a mechanism of change is activated rather than a 'memetic' copy of the same.[37] That happens because, although

---

[33] See Shepard 1984, 113 ff., especially 138 ff.   [34] See Eco 2000, 271.

[35] It is worth noting that whereas the CT of the unicorn tends to be preserved and applied to actual animals, the CT of the rhinoceros is often removed and forgotten in mediaeval and modern zoological encyclopaedias.

[36] E. Romano has just pointed out, in this volume (p. 54, n. 4), that in the field of technological knowledge practical action is regulated by what Assmann 1997, xv f. calls 'mimetic memory'.

[37] In Dawkins' view (Dawkins 1976), a *meme* is composed of a recognizable amount of information about the human culture. Every *meme* is replicable by a mind or a symbolic support of memory (such as a book, or another mind). In more specific terms, a meme is a 'self-propagating unit' of cultural evolution, analogous to what the gene is for genetics. On several occasions, Dan Sperber has taken a position against memetic theories: e.g. Sperber 1996, Sperber 2005, Sperber 2009. For an explanation of memetic theories, however, see Acerbi 2009.

the culturally provided descriptions act as ostensive instructions for the recognition of instances, the cognitive actions directed by them cannot be completely guided (or at least not always).

In his novel *A Caverna* (*The Cavern*), Jose Saramago, taking a cue from the unhappy experiences of the potter Cipriano Algor, says that there is a small brain in each of the phalanges of the hand, which autonomously and automatically interprets the instructions it receives from the mind and, above all, from technical manuals. As the writer suggests, the brain is made of 'models', of 'Platonic ideas' of what the hand is required to do, but the fingers, in practice, shape the clay on their own.[38] Similarly, one could paraphrase Saramago's text and say that the eyes have an autonomous brain. In the end, what we see – what Marco Polo saw – even in the face of the objective evidence of the rhinoceros is somehow an approximation and a negotiation of the CTs communicated to us by the public representations and descriptions with which our encyclopaedia is equipped.

There is, however, another lesson that can be learnt from the history of the intertwining of the CT of the unicorn and the CT of the rhinoceros. We understand that, in the absence of a *percipiendum*, it is the desire for possible perceptual experience – i.e. for the acts of recognition – that allows beliefs to survive. It is what happens in the case of the horns of the Indian ants, or in the case of the alicorns. In other words, the survival of fanciful beliefs is sometimes linked to the dynamics of potential applicability. The more things are described as mysterious – because of their being distant or difficult to experience – the more one has the desire to test the information that has been received by means of textual transmission and that shows us the unknown through analogical mechanisms. In such situations, the whole world is potentially an actual domain of inputs and stimuli that activate proper domains constructed on the basis of the principle of authority rather than on actual experience.

But we must say that the whole story is not just about Marco Polo or the ancients. The Venetian merchant is, in this sense, in good company. If we type 'unicorn' on the *YouTube* search engine, the first video that appears shows us a roe-deer with a single horn peacefully walking in the Tuscan countryside. It is a freak of nature, of course, but the voiceover of the local news commentator agrees only to a certain point, since the commentator himself claims that the specimen sighted clearly shows that 'le iconografie e le leggende antiche non rappresentavano solo fantasie ma un animale realmente esistito' ('the iconographies and ancient legends were not just fantasies, but represented an animal that actually existed').[39]

---

[38] I am quoting from the Italian edition: Saramago 2000, 73.
[39] See www.youtube.com/watch?v=No2jYgls_8U

Once again, it's the same old story. The cancellable properties of the beast are modified, and the CT that the interpreter (in this case the TV commentator) desires to be true is applied to the 'something' he is facing. Once again, a word – an evocative word: 'unicorn' – triggers an act of recognition. Under these conditions, unicorns can never die, and it is always possible to know and apply – albeit with some modifications – their descriptions.

REFERENCES

A. Acerbi 2009, 'Modelli evoluzionistici della cultura', *Rivista Italiana di Filosofia del Linguaggio* 1, 2–19: https://acerbialberto.files.wordpress.com/2013/03/evocult.pdf

J. Assmann 1997, *Das kulturelle Gedächtnis: Schrift, Erinnerung und politische Identität in frühen Hochkulturen* (Munich) (Italian trans. Francesco de Angelis, *La memoria culturale: Scrittura, ricordo e identità politica nelle grandi civiltà antiche*, Turin 1997)

S. Atran 1996, *Cognitive Foundations of Natural History: Towards an Anthropology of Science* (Cambridge University Press)

2006, 'Folk Biology and the Anthropology of Science: Cognitive Universals and Cultural Particulars', *Behavioral and Brain Sciences Online*, 1–41 (previously in *Behavioral and Brain Sciences*, 21, 4 (1998): 547–69): https://halshs.archives-ouvertes.fr/file/index/docid/53281/filename/index.html

M. Beagon 1992, *Roman Nature: The Thought of Pliny the Elder* (Oxford)

V. Bertolucci Pizzorusso (ed.) 1975, *Marco Polo: Milione* (Milan)

M. Bettini 1998, *Nascere: Storie di donne, donnole, madri ed eroi* (Turin)

F. C. Conybeare (ed.) 1948, *Philostratus: The Life of Apollonius of Tyana*, vol. 1 (London, Cambridge, MA)

R. Dawkins 1976, *The Selfish Gene* (Oxford)

U. Eco 2000, *Kant and the Platypus: Essays on Language and Cognition* (New York) (orig. title *Kant e l'ornitorinco*, Milan, 1997)

J. J. Gibson 1979, *The Ecological Approach to Visual Perception* (Boston)

K. Karttunen 1989, *India in Early Greek Literature* (Helsinki)

C. Lavers 2010, *The Natural History of Unicorns* (New York)

D. Lenfant (ed.) 2004, *Ctésias de Cnide: La Perse. L'Inde. Autres Fragments* (Paris)

P. Li Causi 2003, *Sulle tracce del manticora: La zoologia dei confini del mondo in Grecia e a Roma* (Palermo)

P. Li Causi and R. Pomelli 2001–2, 'L'India, l'oro, le formiche: Storia di una rappresentazione culturale da Erodoto a Dione di Prusa', *Hormos: Rivista di storia antica*, 3–4: 177–246

A. Mayor 2000, *The First Fossil Hunters: Paleontology in Greek and Roman Times* (Princeton)

L. Morini (ed.) 1996, *Bestiari medievali* (Turin)

A. Nichols 2008, '*The Complete Fragments of Ctesias of Cnidus*', PhD diss., University of Florida

S. H. Prater 1971, *The Book of Indian Animals* (Bombay)
H. Rackham (ed.) 1947, *Pliny: Natural History*, vol. III (London, Cambridge, MA)
J. Saramago 2000, *La caverna* (Turin) (orig. title *A caverna*, Lisbon 2000)
A. F. Scholfield (ed.) 1972, *Aelian: On Animals*, vol. III (London, Cambridge, MA)
O. Shepard 1984, *La leggenda dell'unicorno* (Florence) (orig. title *The Lore of the Unicorn*, New York, Boston 1930)
D. Sperber 1982, *Le savoir des anthropologues* (Paris)
   1996, *Explaining Culture: A Naturalistic Approach* (Oxford) (orig. title *La Contagion des idées*, Paris, 1996)
   2005, *Cultura e modularità* (Florence)
   2009, 'Culturally Transmitted Misbeliefs', *Behavioral and Brain Sciences*, 32: 493–561
D. Sperber and L. A. Hirschfeld 2004, 'The Cognitive Foundations of Cultural Stability and Diversity', *Trends in Cognitive Sciences* 8, 1: 40–6
Ylla 1958, *Animaux des Indes* (Lausanne)
H. Yule (ed.) 1993, *The Travels of Marco Polo* (Mineola, NY)
F. Zambon (ed.) 1993, *Il Fisiologo* (Milan)

Index Locorum

Aelianus, Claudius
De natura animalium, 238
III 25, 239
IV 21, 252, 255
IV 27, 252
IV 49, 241
IV 52, 257
IX 16, 239
XVI 20, 258, 259
XVII 44, 258, 259
Aelianus Tacticus
Tactica
III 2.266 Köchly-Rüstow, 186
Aeneas Tacticus
Poliorcetica
1, 44
5, 44
10.3, 44
13, 44
Aeschylus
Prometheus vinctus (Prometheus Bound)
804 ff., 252
Aetius Amidenus
Tetrabiblos
VI 56 (CMG VIII 2, 205 Olivieri), 157
VIII 6 (CMG VIII 2, 408 Olivieri), 144, 157
Alexander Aphrodisiensis
De Mixtione (On Mixture)
III (216, 14–17 Bruns), 236
III (227, 8 Bruns), 236
Alexander Trallianus
Therapeutica
II 375.11–15, 117
Anonymus
Physiologus latinus, versio bis
16, 261, 262
Apollonius Citiensis
In Hippocratis De articulis commentarius, 5
I 1 (Kollesch-Kudlien CMG XI 1, 1, 14.7), 103

II 10 (Kollesch-Kudlien CMG XI 1, 1, 38.7), 104
II 10 (Kollesch-Kudlien CMG XI 1, 1, 38.14), 104
Apollonius Rhodius
Argonautica
III 228–31, 43
III 1399 ff., 228
Archimedes
De planorum aequilibris (Equilibrium of Planes)
I 6 (I 152.11–13 Heiberg), 47
Aristophanes
Aves (Birds), 114
Aristoteles
Analytica posteriora
I 9 (76a23–5), 47
I 13 (78b35–9), 47
Categoriae (Categories), 48
10 (11b17–19), 48
De generatione animalium
III 2 (753a11), 238
Ethica Nicomachea (Nicomachean Ethics), 6
II 1–2, 6
VI 7 (1141a26), 238
VII 4–5 (1146b35–1147b19), 29
X 9 (1181b2–6), 6
Historia Animalium
II 1 (501a24–b1), 252
VIII, 236
VIII 1 (609b28–9), 236
VIII 1 (609b32–3), 236
VIII 1 (610a35), 236
VIII 2 (610b1–2), 236
IX, 236
IX 6 (612a2–5), 238
IX 6 (612a7–8), 241
Metaphysica (Metaphysics)
I (981a5–b17), 3
Physica (Physics), 48
IV 10 (217b29–224a17), 48

Aristoteles (cont.)
  Politica (Politics)
    III 15 (1287a35), 6
  Topica (Topics), 48
(Pseudo?-)Aristoteles
  Mechanica, 7, 27
    1, 848b3–5, 46
    1, 849b19–21, 46
    3, 850a39–b2, 47
    17, 853a19–32, 45
Arrianus
  Indica
    XV 4 ff., 256
    XV 4–9, 252
Athenaeus Mechanicus
  De machinis (On Machines)
    4–5 Wescher = 44 Whitehead-Blyth, 27, 28
    15 Wescher = 50 Whitehead-Blyth, 28
    27 f. Wescher = 56 Whitehead-Blyth, 40
    31 Wescher = 58 Whitehead-Blyth, 41
    39 Wescher = 60 Whitehead-Blyth, 34
Athenaeus Naucraticus
  Deipnosophistae
    322a, 150
Augustinus
  Contra Iulianum Pelagium
    IV 12.60, 235

Biton Pergamenus
  Instructiones bellicorum instrumentorum et tormentorum, 27
    70 Marsden = 52 Wescher, 41
    75 Marsden = 61 f. Wescher, 35
    76 Marsden = 65 Wescher, 32
    76 Marsden = 67 Wescher, 32
    76 f. Marsden = 67 f. Wescher, 33

Caelius Aurelianus
  Tardae Passiones
    V 1.20 (Bendz and Pape, CML VI 1.2.866), 99
    V 1.24 (Bendz and Pape, CML VI 1.2.868), 99
Caesar
  Commentarii de bello civile, 76
    II 9.2, 83
  Commentarii de bello Gallico, 76–88
    IV 16, 70
    IV 16.1, 70
    IV 16.2–4, 70
    IV 16.4, 70
    IV 17, 68, 70
    IV 17.1, 71, 72
    IV 17.2–10, 71
    IV 17.6, 80
    IV 18, 70
    IV 19, 70
Callimachus
  Epigrammata
    46, 120
  Hymni
    2, 1–4, 43
    5, 2 f., 43
Cassiodorus
  Institutiones
    I 25, 195
Cato Maior
  De agricultura
    18.1–2, 30
Catullus
  Carmina
    16, 131
Celsus
  De medicina
    III 21.2, 117
    VI 4, 139
    VI 5, 139, 140
    VI 6.25, 139
Cicero
  Academici libri
    II 120, 244
  Brutus
    185–200, 115
  De divinatione
    II 33, 236
  De natura deorum
    II 120–6, 236
    II 126, 238, 241
  De oratore
    I 6, 125
    I 92–3, 126
    I 109–10, 126
    I 130, 115
    I 145–6, 126
    II 18.75, 47
    II 28–33, 126
    II 121, 115
    II 144, 115
    II 232, 126
  De re publica
    III 1, 235
Chrysippus
  Fragmenta
    SVF II 473, 236
    SVF III 169–83, 233
Clemens Alexandrinus
  Paedagogus
    III 2.7.3, 144
Columella
  De arboribus, 217
  De re rustica, 10, 15, 217

I–IX, 218
I 1.15–16, 218
I 1.16, 229
I 4.4, 220, 229
I 4.8, 220
II 2.14, 219
II 2.18, 219
X praef., 229
X praef. 3, 221
X praef. 4, 221
X praef. 5, 221, 229
X 29–34, 222, 223
X 40, 223
X 45, 223
X 58, 223
X 67–71, 224
X 70, 224
X 76–93, 224
X 127–39, 225
X 139, 225
X 140, 225
X 140–54, 225
X 155–65, 225
X 157, 226
X 157–9, 225
X 177, 225
X 215–29, 223
X 216, 226
X 217–24, 226
X 218–19, 226
X 225, 226
X 225–7, 226
X 225–9, 226
X 228, 227
X 230, 226, 227
X 276–9, 227
X 322–36, 227
X 335–6, 228
X 337–68, 229
X 433, 222
X 433.6, 221, 222
Ctesias
  Fragmenta
    FGrHist III C 688 F 45, 6, 252
    FGrHist III C 688 F 45.15 (= Phot. Bibl. 45b31–46a12 Henry), 252
    FGrHist III C 688 F 45q, 257

Dio Chrysostomus
  Orationes
    35.23–5, 252
Diocles Carystinus
  fr. 6 van der Eijk (2000), 5
Diodorus Siculus
  Bibliotheca Historica
    XVII 17.1–3, 71

Diogenes Apolloniates
  fr. 64 B 1 D.-K., 37

Ennius
  Annales, 130
Epictetus
  Dissertationes (Discourses)
    I 4.23, 171
    I 4.24, 165
    I 11.31, 171
    I 18.18, 172
    I 22.10, 170
    I 29.18, 165
    II 2.2, 171
    II 4.10–11, 165
    II 7, 169, 172, 173
    II 7.3, 172
    II 7.9, 170
    II 7.10–14, 173
    II 7.11, 169
    II 7.12, 169, 172
    II 7.14, 172
    II 16.38, 172
    II 19.21–4, 165
    II 25, 163
    III 2.1–5, 164
    III 2.5–8, 175, 182
    III 4.9–10, 171
    III 6.5, 173
    III 12.2, 171
    III 12.10, 171
    III 12.17, 171
    III 14.4, 171
    III 14.4–6, 171
    III 21.23, 172
    III 22.95, 165
    III 22.102, 173
    III 24.28, 172
    III 24.84, 172
    III 24.84–8, 172
    III 24.85, 172
    III 24.86, 172
    III 26.26, 172
    IV 1, 163
    IV 1.67, 172
    IV 1.100, 172
    IV 1.111, 172
    IV 4.4, 172
    IV 4.21, 165
    IV 4.29, 177
    IV 5.14, 171
    IV 10.13, 175
  Enchiridium, 6, 9, 10
    1–28, 175
    1–29, 174
    1.1, 170

Epictetus (*cont.*)
    1.10, 180
    1.18, 169
    2.6–7, 180
    2.7, 178
    2.7–8, 180
    2.10–11, 180
    3, 174, 178
    3–6, 175
    4, 175, 178
    4.6, 171
    4.9–10, 171
    5a, 177, 178
    5a.5, 169
    5b, 172, 178, 180
    5b.2, 180
    6, 178
    7, 175, 176, 178
    8, 172, 178
    9, 178
    10, 175, 178
    11, 176, 178
    12, 175, 178
    12–13, 175, 182
    13, 175, 179
    13.1–2, 180
    13.2–3, 180
    14a, 178
    14b, 179
    15, 175, 178
    16, 175, 178, 179
    16.3–4, 169
    17, 175, 176, 178
    18, 178, 179
    19a, 173, 174, 178, 179
    19b, 178
    20, 178
    21, 178
    22, 175, 176, 178, 179, 180, 182
    22–5, 175
    23, 179
    24, 175, 178, 179
    25, 175, 178, 179
    26, 178, 179
    26.7–8, 171
    27, 172, 174, 178
    28, 169, 175, 178, 179
    30, 175, 178
    30–51, 174
    31, 175, 176, 178, 179
    32, 169, 172–176, 178, 179
    32.5–6, 169
    32.6, 170, 172, 173
    32.10, 172
    32.15, 172
    33, 174, 178, 179
    33^2, 180
    33^9, 180
    33.12, 177
    34, 175, 178
    35, 175, 179
    36, 178
    37, 178
    39–41, 178
    42, 175, 178, 179, 180
    42–5, 175
    42.8, 171
    43, 178
    44, 178
    45, 175, 180
    46, 171, 175, 179, 180
    46–53, 175
    46.7–13, 180
    47, 171, 174, 178, 179
    47.4–5, 171
    48a, 172, 178
    48a–48b, 180
    48b, 175, 178, 183
    48b.1, 180
    48b.2, 180
    48b2–3, 180
    48b.3, 180
    48b.3–4, 180
    48b.4, 180
    48b.5–6, 180
    48b.6–7, 180
    48b.7–8, 180
    48b.8, 180
    48b.8–9, 180
    49, 175, 180, 182, 183
    49–50, 175
    49–53, 180
    48a, 172
    48a–53, 181
    49, 175, 180, 183
    50, 175, 178–180
    51, 180, 183
    52, 169, 174, 175, 180, 182, 183
    53, 181
Eusebius
    Contra Hieroclem
        22, 252

Galen
    De alimentorum facultatibus
        I 1.47 (VI 480 K. = 217 Helmreich), 5
    De antidotis
        I (XIV 30 K.), 245
    De compositione medicamentorum per genera, 142
        I 1 (XIII 367 K.), 142

Index Locorum 273

De compositione medicamentorum secundum
    locos, 139, 141
  I 1 (XII 402–3 K.), 143
  I 1 (XII 403 K.), 157
  I 1 (XII 408 K.), 144
  I 1 (XII 416 K.), 144, 158
  I 2 (XII 432 K.), 157
  I 2 (XII 434–5 K.), 140
  I 3 (XII 442 K.), 141
  I 3 (XII 444.6–9 K.), 153–157
  I 3 (XII 445 K.), 141
  I 3 (XII 445–6 K.), 143
  I 4 (XII 454 K.), 147
  I 8 (XII 492 K.), 157
  VI 1 (XII 894 K.), 5
De facultatibus naturalibus
  I 13 (II 38 K. = 128–9 Helmreich), 231
De foetuum formatione
  6.8 (IV 690 K. = 94.15–18 Nickel), 232
  6.14 (IV 692–3 K. = 96.27–9 Nickel), 232
De locis affectis
  VI 6 (VIII 443 K.), 231
  VI 6 (VIII 445 K.), 231
De methodo medendi
  I 7 (X 53.17 K. = I 84 Johnston and Horsley), 103
De motis dubiis
  4.5 (136,13–16 Nutton), 232
De placitis Hippocratis et Platonis
  IX 1.21–2 (V 725 K. = 544.13–17 De Lacy), 231
De remediis parabilibus
  II 1–2 (XIV 390–3 K.), 139
De semine
  II 6.5 (IV 643 K. = 198.6–8 De Lacy), 231
De simplicium medicamentorum
    temperamentis ac facultatibus
  VI praef. (XI 971 K.), 5
  VIII 56 (XII 47–8 K.), 144
  X 1 (XII 249 K.), 144
  X 19 (XII 291 K.), 145
De symptomatum causis
  II 5 (VII 178 K.), 231
De usu partium
  I 1 (III 6 K. = I 4 Helmreich), 231
  I 1 (III 7 K. = I 5 Helmreich), 231
In Hippocratis Epidemiarum librum VI
    commentaria
  5.2 (259.6–261.14 Wenkebach-Pfaff = XVIIB 233–237 K.), 232
Gellius
  Noctes Atticae
  VI 12, 147

Geminus Rhodius
  Elementa astronomiae, 204, 209

Harpocration
  Lex. s.v. χρυσωχοεῖν, pp. 307–9 Dindorf, 252
Heliodorus Eroticus
  Aethiopica
  X 26.2, 252
Hero Alexandrinus
  Automata, 42
  I 1–3, 38
  Belopoeica
  73. 6–11 Wescher = 18 Marsden, 49
  73.6–74.4 Wescher = 18 Marsden, 119
  96 Wescher = 30 Marsden, 35
  Dioptra
  190.14–21 Schöne, 41
  Mechanica, 27, 36, 45
  I 1 (II 4.13–15 Nix-Schmidt), 39
  I 12 f. (II 27 ff. Nix-Schmidt), 39
  I 23 (II 61 ff. Nix-Schmidt), 39
  II 7 (II 111–14 Nix-Schmidt), 39
  II 8 ff. (II 113 ff. Nix-Schmidt), 39
  II 17 (II 139 ff. Nix-Schmidt), 39
  II 21 ff. (II 147 ff. Nix-Schmidt), 39
  II 32 (II 171 Nix-Schmidt), 39
  III 13 (II 226 f. Nix-Schmidt), 31
  Pneumatica/Spiritalia
  I praef. (I 2 Schmidt), 41
  I 38 (I 174 Schmidt), 42
Herodotus
  Historiae
  I 29.1, 49
  II 2, 231
  III 98–105, 252
  III 102 ff., 256
  III 116.1, 252
  IV 85, 71
Hesiodus
  fr. 37 (M-W), 240
  fr. 133 (M-W), 240
  Opera et dies
  289–90, 118
Hippocrates
  Aphorismi
  VII 87 (IV 608.1 L. = 476.11 Magdelaine = IV 216.13 Jones), 95
  De aere, aquis, locis, 107
  De alimento
  39 (IX 112 L. = 145.12 Joly), 231
  De arte, 107
  De articulis, 104
  4 (IV 82 L. = II 114 Kühlewein = III 205 Withington), 105

Hippocrates (*cont.*)
   9 (IV 100 L. = II 125 Kühlewein = III 220 Withington), 108
   11 (IV 106 L. = II 128 Kühlewein = III 224 Withington), 98
   11 (IV 108 L. = II 129 Kühlewein = III 226 Withington), 108
   30 (IV 142 L. = II 146 Kühlewein = III 252 Withington), 108
   33 (IV 148 L. = II 151 Kühlewein = III 258 Withington), 105
   40 (IV 174 L. = II 163 Kühlewein = III 278 Withington), 108
   41 (IV 182 L. = II 167 Kühlewein = III 282 Withington), 108
   42 (IV 182 L. = II 167 Kühlewein = III 282 Withington), 106
   45 (IV 190 L. = II 171 Kühlewein = III 288 Withington), 108
   47 (IV 212 L. = II 182 Kühlewein = III 302 Withington), 106
   48 (IV 212 L. = II 182 Kühlewein = III 302 Withington), 106
   57 (IV 246 L. = II 202 Kühlewein = III 332 Withington), 108
  De affectionibus, 5, 93–5, 101
   5 [[7,7]] (VI 214 L. = V 246 Potter), 102
   20 (VI 228 L. = V 34 Potter), 96
   20 (VI 230 L. = V 36 Potter), 97
   22 (VI 232 L. = V 38 Potter), 96
  De capitis vulneribus, 104
  De fracturis, 104, 107, 108
   1–2 (IV 414 L. = II 46 Kühlewein = III 94 Withington), 106
   1 (IV 414 L. = II 47 Kühlewein = III 96 Withington), 107
   4 (IV 430 L. = II 52 Kühlewein = III 102 Withington), 106
   25 (IV 500 L. = II 83 Kühlewein = III 152 Withington), 107
   30 (IV 518 L. = II 90 Kühlewein = III 165 Withington), 106
  De genitura, 107
  De internis affectionibus, 95, 98, 99
   9 (VII 212 L. = VI 132 Potter), 95
   19 (VII 214 L. = VI 134 Potter), 98
   25 (VII 230 L. = VI 156 Potter), 96, 97
   32 (VII 250 L. = VI 184 Potter), 97
  De locis in homine, 95, 99
   24 (VI 314 L. = 64 Craik), 97
   24–5 (VI 314 L. = 64 Craik), 96
  De medico
   7 (IX 212 L. = I 22 Heiberg = VIII 306 Potter), 100

  De morbis I, 95, 97
   1 (VI 140 L. = 2 Wittern = V 98 Potter), 107
   6 (VI 152 L. = 16 Wittern = V 112 Potter), 97
   10 (VI 158 L. = 26 Wittern = V 120 Potter), 95
  De morbis II, 95, 102
   33 (VI 50 L. = V 246 Potter = X 2.167 Jouanna), 100, 101
   34–7 (VI 50 L. = V 248 Potter = X 2.168 Jouanna), 101
   57 (VI 90 L. = V 300 Potter = X 2.197 Jouanna), 95
   60 (VI 94 L. = V 304 Potter = X 2.199 Jouanna), 95
  De morbis III, 95
   16 (VII 154 L. = 94 Potter CMG = 54 Potter Loeb) 95
  De morbis IV
   57 (VII 610 L. = X 180 Potter), 96
  De morbis popularibus
   VI 5.1 (V 314 L. = 100.8–102.2 Manetti-Roselli), 231
  De morbo sacro, 107
  De mulierum affectibus I
   89 (VIII 214.8–13 L.), 146
  De mulierum affectibus II
   127 (VIII 272–4 L.), 148
   185–91 (VIII 366–70 L.), 139
  De mulierum affectibus III/*De sterilibus*
   214 (VIII 416.2–5 L. = X 338 Potter), 148
  De natura ossium, 104
  De natura pueri, 107
  De officina medici, 104
  De ventis, 107
  De vetere medicina, 107
  De victu in acutis, 107
  Vectiarius, 104
Homilius, Johann
  De monocerote, 264
Horatius
  Ars poetica
   343, 115
   344, 115
  Epistulae
   II 1.195, 253
   II 1.225, 223
  Epodes
   12.11, 144

Iustinianus
  Iustiani digesta
   XLVIII 8.3.3, 145

Index Locorum

Iuvenalis
  Satirae
    II 99–109, 148
    VI 462–3, 149

Kallixeinos
  FGrHist III C 627 F 2, 43

Lucianus
  Dialogi meretricum
    36, 138
  Gallus
    16, 252
  Verae Historiae
    I 11, 252

Lucretius
  De rerum natura, 120, 130, 232
    I 1 ff., 226
    I 10, 226
    I 13, 226
    II 177–82, 244
    V 195–221, 244
    V 222–27, 235

Marcellus
  De medicamentis
    7 (CML V 1.100–10 Liechtenhan), 139

Marco Polo
  Il Milione
    162.14–17, 262

Martialis
  Epigrammata
    XII 43.4, 150
  Liber spectaculorum
    22, 258

Onasander
  Strategicus, 21–4

Oribasius
  Eclogae medicamentorum
    5 (CMG VI 2.2.187 Reader), 139

Origenes
  Contra Celsum
    4.87, 239

Ovidius
  Amores
    I 5.13, 224
    II 1.5–6, 121, 122
    II 10.27–8, 114
    III 12.41–4, 128
  Ars amatoria, 6, 9, 15
    I 1–2, 112, 113
    I 3–6, 125
    I 3–8, 121
    I 8, 120
    I 31–4, 122
    I 33–4, 129
    I 347–50, 123
    I 611–12, 135, 136
    I 615–18, 136
    II 99, 119
    II 99–108, 119
    II 155–8, 122
    II 285, 119
    II 293, 119
    II 331, 119
    II 415–417, 119
    II 415–25, 119
    II 448, 123
    II 479–80, 127
    II 537–40, 118
    II 615–6, 131
    II 709–10, 115
    II 731–2, 115
    II 735–8, 121
    III 79–80, 124
    III 89–92, 124
    III 270, 144
    III 353, 119
    III 611–16, 122
  Fasti, 18
  Heroides
    7, 114
    12, 114
  Medicamina faciei femineae, 139, 144
  Metamorphoses, 114, 128
    I 3, 223
  Remedia amoris, 123, 133, 134
    17–22, 113
    51–4, 113
    55–68, 114
    70–1, 117
    225–6, 117
    249–50, 119
    249–90, 119
    423–4, 118
    487–8, 115
    489–90, 115
    495–6, 117
    715, 119
    801, 120
  Tristia
    I 1.112, 113
    II 63–4, 128
    II 105–10, 131
    II 221–4, 128
    II 239–40, 129
    II 343–50, 131

Ovidius (cont.)
    II 250, 129
    II 251–4, 129
    II 255–6, 129, 130
    II 259, 130
    II 261–2, 130
    II 265–8, 130
    II 311–14, 130
    II 353–6, 131, 132

Pappus Alexandrinus
  Collectio
    VIII, 36, 48
    VIII 11 (III 1060.2 Hultsch), 39
    VIII 11 (III 1060.7 Hultsch), 39
Paulus Aegineta
  De re medica libri septem
    III 2.1 (Heiberg CMG IX 1, 132), 157
    VI 25 (Heiberg CMG IX 2, 64.28), 101
Pausanias
  Graeciae descriptio
    II 7.8, 240
    IX 21.4, 252
Petronius
  Satyrica, 152
Philo Alexandrinus
  De animalibus, 238
  De posteritate Caini
    46–7, 235
  De providentia
    II 56–65, 244
Philo Byzantinus
  Belopoeica, 27
    50 Thévenot = 106 Marsden, 33, 34
    51 Thévenot = 108 f. Marsden, 34
    62.15 Thévenot = 126 Marsden, 34
    76.21–77.6 Thévenot = 150 f. Marsden, 119
  Mechanica syntaxis, 36
Philostratus Athenaeus
  Vita Apollonii
    III 45, 252, 256
    VI 1, 252
Plato
  Crito
    43d, 165
  Phaedrus, 30
    275A ff., 4
  Res publica
    III 405a–b, 246
Plinius Maior
  Naturalis historia, 11
    II 154, 235
    VII praef., 235, 243
    VII 1, 235

VII 4, 233
VII 1–6, 235
VII 10, 252
VIII 43–4, 233
VIII 41, 239
VIII 69, 253, 255
VIII 71, 258
VIII 75, 252
VIII 91, 239
VIII 96–101, 239
VIII 97, 238, 240
VIII 98, 239
VIII 99, 239
VIII 100, 241
VIII 102–4, 240
VIII 107, 252
VIII 167, 233
XI 111, 252, 256
XI 238, 149
XI 271, 244
XVIII 1, 235
XVIII 3–4, 243
XVIII 317, 31
XX praef., 233–6
XX 1, 232, 234, 247
XX 2, 237
XX 25, 237
XX 47, 240
XX 119, 240
XX 132–3, 239
XX 134, 239
XX 144–5, 243
XX 254, 239
XXI 78, 244
XXII praef., 247
XXII 1, 235
XXII 14–15, 246
XXII 15, 247
XXII 17, 244
XXII 44, 244
XXII 57–60, 244
XXII 79, 239
XXII 91, 239
XXII 94, 245
XXII 106, 246
XXII 117, 246
XXII 120, 244
XXIII 56–7, 244
XXIV 1, 235, 237, 244
XXIV 3, 234, 237
XXIV 4, 246
XXV 1–2, 245
XXV 4–5, 247
XXV 8, 233

Index Locorum 277

XXV 17, 244
XXV 22, 247
XXV 89, 239
XXV 91, 246
XXV 92, 238
XXVII praef., 247
XXVII 1, 248
XXVII 6, 240
XXVII 7, 244
XXVII 7–8, 241
XXVII 8–9, 243
XXVII 98–99, 244
XXVIII 3, 240
XXVIII 38, 158
XXVIII 66, 158
XXVIII 81, 158
XXVIII 82, 158
XXVIII 108, 144
XXVIII 183, 149
XXVIII 184, 140, 144
XXVIII 262, 158
XXIX 1–28, 235
XXIX 19–20, 246
XXIX 107, 145
XXXII 67–8, 139
XXXII 84, 139
XXXII 135, 139, 158
XXXII 140, 158
XXXVI 55–6, 244
Medicina Plinii
  1.5 (CML 3.12 Önnerfors), 142
Plutarchus
  Moralia
    974B–C, 238
    991E, 238
    991F–992A, 238, 239
  Vita Marcelli
    17.307C–D, 44
Pseudo-Plutarchus
  De Fluviis
    21.4, 240
Polybius
  Historiae
    V 97.5 f., 41
    IX 12–19, 41
Propertius
  Elegiae
    III 13.1–5, 252
Ptolemaeus mathematicus
  Geographia, 10, 186
    I 2.1, 196
    I 4.2, 195
    I 6.2, 187, 196
    I 6.3, 196

I 10.2, 191
I 13.1, 192
I 13.1–4, 188
I 18.2–3, 197
II 1.4–5, 193
Phaseis, 6, 10, 200, 204, 209, 210, 213, 214
Pythagoras
  Carmen aureum, 166

Quintilianus
  Institutio oratoria
    II 15.11, 126
    II 15.33, 126
    II 15.38, 126
    II 16.1, 126
    II 16.5–6, 127
    II 17.9–11, 127

Scribonius Largus
  Compositiones
    59–60, 139, 149
Seneca Maior
  Controversiae
    I 8–9, 147
    I 9.10, 125
Seneca Minor
  De beneficiis
    II 29, 235
  Epistulae
    I 8.1–2, 118
    XX 74.15–21, 235
    XX 90.18, 235
Simplicius
  In Epicteti Enchiridium commentarius
    P 2–3, 167
    P 4–11, 167
    P 58–60, 182
    13, 178
Solinus
  Collectanea rerum memorabilium
    15.22, 252
    52.37 f., 252
Strabo
  Geographia
    XV 1.44, 252, 256
Suda
  s.v. Ἀστυάνασσα, 150
Suetonius
  Divus Julius
    45.2, 148
  Domitianus
    18, 148
  Tiberius
    43.2, 150

Tatianus
  Oratio ad Graecos
    34.3, 150
Tertullianus
  De cultu feminarum, 141
Theodorus Priscianus
  Euporista, 142
    I 3–4, 248
    I, 142
Theophrastus
  De causis plantarum
    II 18.1, 237
    II 18.4, 237
Vergilius
  Aeneis
    I 32, 128
    IV, 114
    XII 423–4, 238
  Georgica, 10, 15, 125, 224, 229
    I 51–3, 219
    I 181 ff., 227
    II 175–6, 221
    II 250, 219
    IV 176 ff., 228

Vetus Testamentum
  Deuteronomium
    33:17, 260
  Isaias
    34:7, 260
  Iob
    29:9–12, 260
  Numeri
    23:22, 260
  Psalmi
    21:22, 260
    28:6, 260
    77:69, 260
    91:11, 260
Vitruvius
  De architectura, 8, 15, 17
    I 1.1–2, 54, 55, 57
    I praef. 3, 57
    I 1.4–5, 57
    I 1.7, 57
    I 1.15, 54
    I 1.16, 57
    I 1.18, 116
    I 2.2, 58
    I 4.12, 62
    I 5.3, 85
    I 6.1, 62
    I 6.6, 60
    II 1.2–3, 53, 54
    II praef. 3, 17, 63

II 1.8–9, 53
II 8.4, 61, 62
II 8.8, 58
II 8.16–17, 63, 64
II 8.20, 64
II 10.3, 58
III 3.3, 65
III 3.6, 63
III 3.7, 60
III 3.8, 63
III 3.9, 55
III 3.13, 60
III 4.3, 61
III 4.5, 60
III 5.8, 60
IV 3.1–3, 60
IV 3.1, 59
IV 3.3, 58, 59, 63
IV 3.5, 59
IV 5.1–2, 65
IV 8.6, 65
V 1.4–5, 56
V 1.6, 56
V 3.1–2, 65
V 4.1, 60
V 5.6, 60
V 5.7, 56
V 6.2, 65
V 6.7, 65
V 9.1, 65
V 11.1, 63
V 12.4, 83
VI praef. 6, 58, 119
VIII 5.3, 60
IX, 36
IX praef. 5, 60
IX 8, 60
X 1, 37
X 2.1, 84
X 2.3, 83, 84
X 3, 38
X 3.9, 38
X 4–8, 38
X 6.1–4, 48
X 9, 38
X 10–15, 38
X 15.2–7, 48
X 13.1–3, 38
X 13.3, 49
X 16.3–8, 38, 49
X 16.5, 38

Xenophon
  Memorabilia
    I 1.7–9, 173

General Index

adaptability, 8, 66
addressee, 8, 28, 31, 44, 58, 59, 60, 61, 178, 179, 218, 221, 226, 228
adultery, 130, 149, 165
Aelian, 186, 238, 255, 258, 259, 261
Aetius of Amida, 139, 144, 157
Agennius Urbicus, 14
Aineias the Tactician, 23, 44
Aldovrandi, 264
Alexander the Great, 62, 71, 195, 256
Alexandria, 32, 49, 208, 211, 215
Alfonso II of Naples, 74
allelopoiesis, 73
allotria/ἀλλότρια, 164, 178
anatomical illustrations, 5
animal, 11, 144, 145, 228, 231–66
Antikythera device, 27
Antonius Castor, 233
Aparktias, 192, 211
Apeliotes, 189, 192, 205, 211
Apollo, 173, 223, 226, 228
Apollonius of Citium, 5, 103, 104
Apollonius of Rhodes, 228
Apollonius of Tyana, 256
apprentice, 3, 33
Archimedes, 39, 41, 44, 193
Aristophanes, 114
Aristotle, 3, 5–7, 27, 29, 37, 151, 153, 196, 236, 238
Arrian, 9, 166–74, 181–3, 252, 256
ars amandi, 6
ars vivendi, 6
artillery, 32–4, 39, 44, 47, 49, 119
askēsis/ἄσκησις, 177, 178, 181
Athenaeus mechanicus, 7
auctoritas, 247
Augustan moral legislation, 9, 112, 121
automata, 41, 43, 47

Babylonian Astronomical Diaries, 10
Bacchus, 223, 226, 228

bad practice, 40, 41, 107
Bakhtin, Michail, 13, 19, 24
Barthes, Roland, 20, 21, 24
belopoiika, 43
Biton, 7, 27, 32–5, 40, 41, 44
Boreas, 188, 189, 191, 192, 211
Bramante, 75
bridge-building, 8, 68, 70, 88, 236

Caelius Aurelianus, 99
Caesar, 8, 44, 68, 69, 70–88, 205, 206, 210, 255
calendar, 202–10, 223, 226
Callixeinus, 43
camelopardalis, 253–5
capital letters, 78, 79
cartography, 4, 10, 188
Cato maior, 7
cauterization, 8, 97–101
Celsus, 14, 58, 139–41, 218, 219
Charon of Magnesia, 50
chrēsimon, to, 1
Chrysippus, 175, 183, 240
Cicero, 77, 115–17, 121, 126, 238, 241, 244
Clausewitz, 22, 23
Clement of Alexandria, 144
Cleopatra, 143, 144, 148–50, 152, 157
code words, 145
cognitive contract, 265
cognitive texts, 1
cognitive type, 11, 253–67
Columella, 10, 15, 75, 217–29
commentarii, 58, 68, 69, 72, 77, 87
contemplation, 14, 24, 33
cosmetics, 6, 9, 138–57
Crito, writer on cosmetics, 142, 146–9, 153, 157
Croesus, 49, 63
Ctesias, 11, 252, 255–67
Ctesibius, 36, 41
cupping vessels, 8, 99, 100

D'Alembert, 20
Damius of Colophon, 50
Darius I, 71
decision-making, 7, 29, 44
Democritus, 206, 237
diagram, 20, 27, 30–6, 42, 46, 47, 60, 61, 210, 211
didactic poem, 217, 218
didactic prose, 19, 61
didagma/δίδαγμα, 107
Diderot, 20
Diocles of Carystus, 4, 5
Diodorus Siculus, 71
Dionysius of Alexandria, 119, 196
docilitas, 53, 54
drug, 3, 8, 94, 95, 97, 102, 141, 145, 149
dulcis, 117
dunameis/δυνάμεις, 39
Dürer, Albrecht, 264

ekklisis/ἔκκλισις, 164, 174–7
Elephantis, 149–51, 157, 158
encheirēsis/ἐγχείρησις, 5
Epictetus, 6, 9, 163–77, 182, 183
epiphenomena, 201, 209, 214
epistemic texts, 1
ergon/ἔργον, 178
Eros, 226
erotodidaxis, 9, 112–25, 135
ethics, 6, 29, 181, 182
etiquette, 6
eucheiriē/εὐχειρίη, 95
euchrēston, to, 4, 10, 186–8, 193, 197
Eudoxus, 204–6
Euros, 192, 211
exempla, 22, 23, 219, 229
extra-textual world, 2, 6

fabrica, 8, 54, 55, 66, 119
Fachtexte, 1, 16
fertilisers, 146
fertilising effect, 145
fertility, 146, 148, 218, 222
fibulae, 79, 84, 85, 86
fomes fomentarius, 99

Galen, 4, 5, 6, 18, 44, 49, 95, 103, 116, 139–47, 150, 153, 157, 158, 231, 232, 238, 245
garden, 10, 217, 220–9, 244
Gellius, Lucius, 166, 168, 182
Geminos, 204–6, 209, 214, 215
gender, 139, 142, 148
genre, 1, 5, 12, 13, 22–4, 28, 45, 48, 86, 106, 203, 215, 226, 228
geography, 1, 3, 4, 10, 77, 186–98, 225, 259
Gesner, 264

Giuliano de' Medici II, 76
Gryllus, 238, 239, 241, 242
gynaecology, 9, 94, 139, 142, 146–53

happiness, 10
Hegetor's turtle, 48
helepolis, 41
Heracleides of Tarentum, 143
Hermogenes, 55, 59
Hero, 7, 27, 31, 34, 52
Herodotus, 187, 231, 252, 256, 257
Hipparchus, 187, 205, 206, 210, 211, 213
Hippocratic writings, 4, 5, 8, 93–109, 139, 148–50, 231, 233
historiae, 11, 57, 252, 253, 261
Homilius, Johann, 264
Horace, 115, 143, 144
hybrid practical knowledge, 6, 215
hypomochlion, 6
hypotypōseis, 40

idiōtes, 5
illustrations, 5, 20, 32, 35, 77, 81, 84, 102–4, 233
imitatio/aemulatio, 196
incision, 8, 95–7, 100
instruction, 2–5, 15, 18–22, 28, 30, 58, 60, 61, 68, 87, 88, 93, 94, 97, 99, 105, 108, 109, 117, 122, 124–6, 182, 196, 217–23, 232–5, 252, 254, 266
instruments, 9, 100, 101, 109
Isocrates, 121

Jerome, 260, 261
Julian (Methodist physician), 103
Justinian, 145
Juvenal, 147

kakotechnia, 9, 138, 141, 153
Kallippos, 204
kommōtikē technē, 140, 141
kosmētikē technē, 140, 141

literary, 1, 3, 6, 8, 12–24, 35, 49, 58, 72, 73, 87, 112, 126, 128, 132, 166, 172, 187, 196, 218, 227
literature of knowledge, 16, 19
Lorenzo Magnifico de' Medici, 75
Louis XII, 75
Lucian, 138, 139, 141, 153
Lucretius, 117, 118, 120, 130, 226, 232, 235
luxury, 142, 217, 246, 247

Machiavelli, 22
machine, 7, 20, 27–50, 84, 85, 152
magic, 119, 120, 145, 229, 237
Manilius, 227

General Index

manual, 5, 7, 14, 22, 24, 34, 44, 109, 120, 125, 142, 150, 151, 153, 163, 166, 196, 232, 234, 235, 252, 266
maps, 4, 10, 76, 194, 196, 198
Marcellus (Roman general), 41
Marcellus of Bordeaux, 139
Marcellus, Claudius, 28
Marco Polo, 262–4, 266
Marini, Andrea, 264
Marinos of Tyre, 187
Martial, 258
mathematics, 16, 28, 29, 33–6, 42, 45–7, 49, 187, 190–3, 200, 207, 214
means of life, 11, 232–5, 245
mechanical writings, 5, 7, 21, 27–49
medicinal ingredients, 9, 142–6, 151, 153
medicinal plants, 224, 233, 245, 246
medicine, 1, 3–6, 8, 11, 14, 16, 49, 94, 95, 97, 99, 101–3, 107, 117, 118, 120, 127, 134, 140, 146, 200, 202, 232–5, 238, 245, 246, 248
Melampus, 240
Meliteia, 40
Mesopotamia, 202
Meton, 206, 210, 211, 213
Middle Ages, (medieval, mediaeval) 20, 102, 103, 124, 132–6, 148, 150, 166, 260–3, 265
Miletus, 35, 203
military treatise, 8, 14, 22–4, 41, 43, 44, 47, 69, 85, 87, 186
MUL.APIN, 202, 203

nabus, 253
natural history, 11, 232–7, 246, 252, 253, 258, 261
nature, 10, 11, 21, 47, 53, 118, 140, 157, 165, 171, 228, 231–48
non-literary, 1, 12, 14
Notos, 192, 211

Odysseus, 238, 239
olive presses, 30–2, 38, 39
Onasander, 14, 21–4
opus, 54, 57, 58, 60, 63, 64, 79, 80, 82–4, 88, 128, 226, 232
opus reticulatum, 60
orality, 4, 5, 6, 30, 101, 106, 107, 161, 215
Orestinus, 143
orexis/ὄρεξις, 164, 169, 170, 174, 175, 176, 177, 180
Oribasius, 139
ornatus, 10, 220
Ovid, 6, 9, 15, 112–36, 139, 144, 223–5

Palladio, 76
Pappus, 36, 39, 48
papyri
   magical, 145

parapēgmata, 4, 6, 200, 203, 204, 214–15
Paul of Aegina, 101, 139
pensée sauvage, 259
Pergamum, 32, 141
Petronius, 152
pharmacology, 1, 141, 245, 248
Philippus (writer on weather prediction), 201, 206, 210, 211
Philo of Alexandria, 235, 238, 244
Philo of Byzantium, 7, 27–44, 119
Philostratus, 252, 256
phōnē zōousa, 5
physical mechanisms, 7
Plato, 4, 23, 30, 35, 36, 120, 165, 246, 266
Pliny the Elder, 11, 31, 139–50, 158, 231–68
Pliny the Younger, 75, 217
Plutarch, 41, 44, 238
pneumatika, 34, 41
poetry as a medium of transfer, 18
Polybius, 23, 40, 41, 44
Pompeius Lenaeus, 247
Pomponius Mela, 196
pornography, 6, 9, 142, 150–3
Posidonius the Macedonian, 41
practical applicability, 1–14, 19–24, 38, 59–61, 186, 193, 195–8, 217, 220, 222, 228, 233
prokopē, 10
Prometheus, 224
Proserpina, 227
providential Nature, 11, 243
Psammetichus, 231
Ptolemy II Philadelphus, 43
Ptolemy, Claudius, 4, 6, 10, 186–214
Pythagoras, 166

Quintilian, 116, 126, 127, 130

Raphael, 75
ratiocinatio, 8, 54, 55, 57, 66, 116
recipe, 6, 9, 30–3, 94, 111, 139–61, 248
remedy, 18, 94, 113, 134, 138–58, 239–46
Rhine, 8, 68–87

Salpe, 149, 150, 157, 158
sambucas, 40
Saramago, Jose, 266
Scarpagnino, 75
scientia, 8, 66, 126
scopic paradigm, 24
Scribonius Largus, 139, 149
self-fashioning, 2, 15, 17, 18, 69, 86, 87, 220
sermo cottidianus, 222
sexuality, 9, 114, 123, 138, 142, 146–53, 224, 226
Silvinus, 218, 221
Simplicius, 167, 168, 170, 178, 182, 183

Sitz im Leben, 7
Socrates, 165, 173
Soranus of Ephesus, 150, 152, 158
speech genre, 13, 19, 24
Statilius, 143, 146, 161
Stoic philosophy, 9, 164–8, 181–3, 233, 236, 239
surgical, 5, 8, 94, 95, 97, 100–8
sympathy and antipathy, 9, 234–44

teacher and student, 3–5, 116–21, 131, 135, 166, 169, 196, 232, 233
Theodorus Priscianus, 139, 142, 248
Theophrastus, 236, 237
theōria, 14
therapeutics, 4, 8, 93–6, 99, 100, 106, 238–42, 246
Tibullus, 120
Topsell, 264
tortoise, 28
translatability, 8, 11, 54, 61, 70, 72, 81, 104, 186, 252, 260, 261
trial and error, 4, 50, 97, 220
tuchē/τύχη, 23, 97

unicorn, *monoceros* 11, 252, 257–67
urbanitas, 220
utility, 9–11, 18, 22, 76, 112–35, 142, 233–5, 245–8

Valgius, Gaius, 247
Varro, 75, 145, 221
Vergil, 10, 15, 31, 125, 219–28, 238
virtus, 23, 118
visualization, 32, 34, 78, 87
Vitruvius, 7, 8, 14, 15, 17, 27–52, 53–67, 75–88, 116, 121, 125, 131, 247, 252
viva vox, 5
volumen, 60

washer, 35
weather prediction, 3, 4, 6, 10, 200–14
Wissenschaftstexte, 16, 50
Wissensvermittlung, 16, 18, 19, 67
women, 9, 64, 84, 112, 136, 138, 154, 243

Xenophon, 23, 138, 141, 173

zoology, 1, 10, 11, 252, 256, 259, 260, 262, 265
Zopyrus of Tarentum, 32, 33, 35, 50